Lecture Notes
in Computational Science
and Engineering

6

Editors
M. Griebel, Bonn
D. E. Keyes, Norfolk
R. M. Nieminen, Espoo
D. Roose, Leuven
T. Schlick, New York

Springer-Verlag Berlin Heidelberg GmbH

Stefan Turek

Efficient Solvers for Incompressible Flow Problems

An Algorithmic and Computational Approach

 Springer

Author

Stefan Turek
Institut für Angewandte Mathematik
Universität Heidelberg
Im Neuenheimer Feld 294
D-69120 Heidelberg, Germany
e-mail: ture@gaia.iwr.uni-heidelberg.de
http://gaia.iwr.uni-heidelberg.de/~ture

```
Library of Congress Cataloging-in-Publication Data
Turek, Stefan.
   Efficient solvers for incompressible flow problems / an
algorithmic and computational approach / Stefan Turek.
      p.   cm. -- (Lecture notes in computational science and
engineering ; 6)
   ISBN 978-3-642-63573-1        ISBN 978-3-642-58393-3 (eBook)
   DOI 10.1007/978-3-642-58393-3
   1. Fluid dynamics.   I. Title.   II. Series.
TA357.T895   1999
620.1'064--dc21                                          99-21146
                                                            CIP
```

Mathematics Subject Classification (1991): 76Mxx, 65Mxx, 65Nxx

Additional material to this book can be downloaded from http://extras.springer.com.

© Springer-Verlag Berlin Heidelberg 1999
Originally published by Springer-Verlag Berlin Heidelberg New York in 1999
Softcover reprint of the hardcover 1st edition 1999

Cover Design: Friedhelm Steinen-Broo, Estudio Calamar, Spain
Cover production: *design & production* GmbH, Heidelberg
Typeset by the editors using a Springer TeX macro package
SPIN 10555489   46/3143 – 5 4 3 2 1 0 – Printed on acid-free paper

Preface

The scope of this book is to discuss recent numerical and algorithmic tools for the solution of certain flow problems arising in *Computational Fluid Dynamics* (CFD). Here, we mainly restrict ourselves to the case of the incompressible Navier–Stokes equations,

$$\mathbf{u}_t - \nu\Delta\mathbf{u} + \mathbf{u}\cdot\nabla\mathbf{u} + \nabla p = \mathbf{f} \quad , \quad \nabla\cdot\mathbf{u} = 0 . \tag{1}$$

These *basic* equations already play an important role in CFD, both for mathematicians as well as for more practical scientists: Physically important facts with "real life" character can be described by them, including also economical aspects in industrial applications. On the other hand, the equations in (1) provide the complete spectrum of numerical problems nowadays concerning the mathematical treatment of partial differential equations.

Although this field of research may appear to be a small part only inside of CFD, it was and still is of great interest for mathematicians as well as engineers, physicists, computer scientists and many more: a fact which can be easily checked by counting the numerous publications. Nevertheless, our contribution has some unique characteristics since it contains a few of the latest results for the numerical solution of (complex) flow problems on modern computer platforms. In this book, our particular emphasis lies on the **solution process** of the resulting high dimensional discrete systems of equations which is often neglected in other works. Together with the included CDROM, which contains the 'FEATFLOW 1.1' software and parts of the 'Virtual Album of Fluid Motion', the interested reader may find a lot of suggestions for improving his own computational simulations.

Organisation:

Chapter 1 contains the motivation for our work. We provide the reader with detailed results from several recent Benchmark configurations for incompressible flow solvers. These are discussed in view of the numerical and

also computational problems of the existing mathematical methodology and
CFD software.

The mathematical description of a large variety of Navier–Stokes solvers is
the subject of Chapter 2. The essential point in our approach is a strict splitting of tasks, namely the **outer control part** which is responsible for the
global convergence and accuracy of the overall problem, and the **inner solver
engine** which has to provide approximate solutions with respect to a given
(discrete) framework. The aim is to demonstrate how classical schemes, as
for instance introduced by Chorin, Van Kan or by Vanka, can be generalized
and essentially improved, as "pure solvers" and also as powerful ingredients
in the modern mathematical discretization context. We concentrate on the
algorithmic aspects concerning the solution process while the other part related to discretization techniques is discussed more in detail in Chapter 3.

The tool for a better understanding of the existing solver methodology is a
Navier–Stokes tree which contains most of the employed CFD techniques as
subtrees. The basic assumption is that we can reduce the various solution
schemes for the incompressible Navier–Stokes equations to - among others -
the treatment of discrete nonlinear saddle point problems,

$$Su + kBp = \mathbf{g} \quad , \quad B^T \mathbf{u} = 0, \tag{2}$$

with matrices B and B^T as discrete analogues of the operators ∇ and $\nabla\cdot$,
time step k and the velocity matrix S coming from the discretized momentum equations. Then, the various approaches can be characterized through
differences in the

- treatment of the nonlinearity,

- treatment of the incompressibility,

- complete outer control.

Having reached well-known tested ground for numerical analysts, namely
the **solution of discrete systems of equations,** we can continue with
standard techniques derived from Numerical Linear Algebra. We may treat
the nonlinearity using some *fixed point defect correction* techniques or other
quasi–Newton variants, and apply the general *pressure Schur complement*
(PSC) approach which formally transforms the original coupled system of
equations into an equivalent scalar equation for the pressure only,

$$B^T S^{-1} B p = \frac{1}{k} B^T S^{-1} \mathbf{g}. \tag{3}$$

Then, the velocity **u** can be derived from p once calculated. As it is well-known for scalar linear systems, arising for instance from Poisson or transport–diffusion problems, we apply the simple *preconditioned Richardson* iteration with certain preconditioners C^{-1},

$$p^l = p^{l-1} - C^{-1}(B^T S^{-1} B p^{l-1} - \frac{1}{k} B^T S^{-1} \mathbf{g}). \tag{4}$$

This general *defect correction* approach is our basic iteration for the following, and all derived techniques concentrate on the "numerical linear algebraic" task of accelerating this simple iteration scheme. As usual, the first step to increase efficiency is to derive better preconditioners C^{-1}. Two different approaches are proposed:

1. We construct – on discrete and/or continuous level – globally defined operators of the type $C_i := B^T \tilde{S}_i^{-1} B$ and use them in an additive way. These are the **global pressure Schur complement** methods (*global PSC*) which contain *projection-like schemes* (or *fractional step, pressure correction*) as proposed by Chorin [22] or Van Kan [115].

2. We construct local preconditioners $C_i^{-1} := B_{|\Omega_i}^T S_{|\Omega_i}^{-1} B_{|\Omega_i}$ on certain patches Ω_i and combine them in the typical way related to block Jabobi- or Gauß–Seidel schemes. These are **local pressure Schur complement** methods (*local PSC*) and include schemes as for instance the *Vanka* smoother [114].

The next step is to accelerate these simple schemes as preconditioners in Krylov space methods or, often significantly better, as *smoothers* in the standard multigrid context. We explain this multilevel approach more in detail since this technique is the crucial step towards very efficient and robust CFD solvers.

In Chapter 3, we discuss other important tools which are necessary in our framework of numerical solution techniques for incompressible flow problems. While we have mainly concentrated so far on the solution process of discretized Navier–Stokes problems, we examine supplementary issues as discretization techniques, error control mechanisms, adaptivity and other necessary numerical ingredients:

1. **Finite element** spaces (including approximation and stability properties of Stokes elements, the nonconforming $\tilde{Q}1/Q0$ finite elements, stabilization techniques for convective terms via upwind or streamline–diffusion techniques, explicit construction of discretely divergence–free subspaces, a posteriori error control mechanisms).

2. **Time discretization** techniques (including the Fractional–step–θ–scheme and other One–step θ–schemes, adaptive time step control).

3. **Nonlinear iteration** schemes (including adaptive fixed point defect correction techniques, quasi–Newton schemes, stopping criterions, linearization techniques for nonstationary problems, least square CG methods).

4. **Multigrid** tools (including properties of simple basic iterations as smoothers or as preconditioners in Krylov–space methods, construction of grid transfer operators and coarse grid matrices, adaptive step-length control of the correction).

5. **Boundary conditions** (including natural *do nothing* conditions, pressure drop and flux settings, iterative implementation techniques, treatment of moving boundaries).

Having derived all necessary conmponents, one can arrange the 'Navier-Stokes solvers' in a scale which is mainly oriented at their stability and robustness. However, the probably more important question arising in the numerical treatment of the Navier–Stokes equations is:

> *'What are the total numerical cost to obtain a certain accuracy?*
> *The answer involves the measurement of number of time steps,*
> *nonlinear iteration steps and linear multigrid sweeps, but the final*
> *measure is the elapsed CPU time to achieve a desired accuracy!'*

We have to remark explicitly that all tests and resulting conclusions are for **fixed spatial meshes**. These are systematically varied in order to simulate the most usual instances which may appear in fully adaptive approaches. However, up to now, no adaptive configuration has been included. It is obvious that our "optimality" statements are not complete since they neglect the "optimal" spatial mesh. Nevertheless, all tests in this book can be viewed as special exercises with the aim to derive the optimal solver for "any" fixed discretization. Since we examine explicitly the case of complex triangulations including also highly-stretched meshes, all conclusions are relevant for the future fully adaptive framework, too.

We present most of these numerical tests in Chapter 4, for several characteristic flow situations. The quality of the applied solution schemes is examined with respect to:

- the complexity of the domain, resp., the shape of the mesh (large aspect ratios!),

- the size of the viscosity parameter ν,

- the size of the performed time step k.

Based on this knowledge we can finally show that there are indeed "discrete Black Box" solution approaches - at least for fixed but arbitrary discrete frameworks - which work likewise robust and efficient for all examined flow configurations.

Acknowledgments:

Let me start with mentioning my fruitful work together with Rolf Rannacher in Heidelberg who mainly formed my mathematical background over the last 10 years. Let me also thank Heribert Blum in Dortmund who was my mentor in mathematical software engineering. Both, Blum and Rannacher, are mainly responsible for my understanding of today's numerical work.

Let me also emphasize Owen Walsh and John Heywood from UBC Vancouver, Friedhelm Schieweck and Lutz Tobiska from Magdeburg and Phil Gresho from the LLNL who are interlocutors since years, giving me a lot of inspiration and new ideas. Especially the work of Phil has to be pointed out, and this not only because of his important "link" function – as one of the editors of the 'International Journal for Numerical Methods in Fluids' – between the mathematical and the engineering CFD community.

Furthermore, let me mention Michael Schäfer from Darmstadt for the common work over the last years concerning the development and evaluation of the *1995 DFG Benchmark*, and also the IWR at Heidelberg, especially Bernhard Przywara and Stefan Schnadt, for providing the "infinite" compute power which is necessary for such numerical studies. And not to forget, my thanks go to all members of our *FEAST group*, particularly Peter Schreiber, Susanne Kilian, Hubertus Oswald, Rainer Schmachtel, Guohui Zhou, Andreas Prohl, Christian Becker, John Wallis, Ludmilla Rivkind and the many others at our institute.

Finally, this book is dedicated to Monika, my children and my parents who have to live with a *numerical computing scientist* for many years.

Heidelberg, October 1998   *Stefan Turek*

Contents

Notation

This book is sometimes written in a very technical style and it is mainly directed to the professional CFD specialist who is familiar with key words as *finite elements*, *multigrid* or *projection schemes* and some other special issues from mathematical and computational sciences. Nevertheless, a short explanation for many of these main topics in Computational Fluid Dynamics will be given in this book, in particular in the Chapter 'Other mathematical components'.

Furthermore, it is very natural that many abbreviations will be utilized. To give the reader a better chance that one can easier find the meaning of such technical terms, we list some of the most important items in the following list.

Notations for PDE's (Partial Differential Equations):

Ω	domain $\Omega \subset \mathbf{R}^d$ with space dimension $d = 2$ or $d = 3$
$\partial\Omega$	boundary of Ω
p	scalar pressure $p(x, y, t)$ in 2D, resp., $p(x, y, z, t)$ in 3D
$\mathbf{u}$	velocity field $\mathbf{u}(x, y, z, t)$ with components $(u_1, \dots, u_d)$
$\mathbf{u}_t$	time derivative operator $\dfrac{\partial \mathbf{u}}{\partial t}$
Δu_i	Laplacian operator $\dfrac{\partial^2 u_i}{\partial^2 x} + \dfrac{\partial^2 u_i}{\partial^2 y} + \dfrac{\partial^2 u_i}{\partial^2 z}$
∇p	gradient operator $(\dfrac{\partial p}{\partial x}, \dfrac{\partial p}{\partial y}, \dfrac{\partial p}{\partial z})^T$
$\nabla \cdot \mathbf{u}$	divergence operator $\dfrac{\partial u_1}{\partial x} + \dfrac{\partial u_2}{\partial y} + \dfrac{\partial u_3}{\partial z}$
$\mathbf{v} \cdot \nabla u_i$	transport operator $v_1 \cdot \dfrac{\partial u_i}{\partial x} + v_2 \cdot \dfrac{\partial u_i}{\partial y} + v_3 \cdot \dfrac{\partial u_i}{\partial z}$
ν	viscosity parameter
Re	Reynolds number, $Re = \dfrac{U \cdot L}{\nu}$, with U, L *characteristic* velocity and length

Notations for discretizations and finite element constructs:

h	mesh width parameter
$k, \Delta t$	time step
$L^2(\Omega), \mathbf{H}_0^1(\Omega)$	usual Lebesgue and Sobolev spaces
$a(\cdot, \cdot)$	bilinear form, mostly $a(\mathbf{u}, \mathbf{v}) := (\nabla\mathbf{u}, \nabla\mathbf{v})$
$b(\cdot, \cdot)$	bilinear form, mostly $b(p, \mathbf{v}) := -(p, \nabla\cdot\mathbf{v})$
$a_h(\cdot, \cdot), b_h(\cdot, \cdot)$	discrete counterparts
$(\cdot, \cdot), \|\cdot\|$	inner product, resp., norm of $L^2(\Omega)$
$\langle\cdot, \cdot\rangle_E$	euclidian scalar product
$\|\|\|\cdot\|\|\|$	discrete norms
$\mathbf{T}_h$	(regular) decomposition $\mathbf{T}_h = \bigcup\{T\}$ with simple elements T
H_h, L_h	discrete spaces for velocity and pressure ansatz functions
I_{2h}^h, I_{k-1}^k	prolongation operator
I_h^{2h}, I_k^{k-1}	restriction operator
NEL,NMT,NVT	number of elements, midpoints and vertices
FE,FV,FD	finite element, finite volume, finite difference
UPW	Upwind
SD	Streamline–diffusion
$\tilde{Q}1/Q0$	nonconforming velocity/piecewise constant pressure ansatz
$Q1/Q0$	conforming bilinear velocity/piecewise constant pressure ansatz
$Q1/Q1$	conforming bilinear velocity/conforming bilinear pressure ansatz
CN	Crank–Nicolson scheme
FS	Fractional–step–θ scheme
IE,BE	Implicit Euler/Backward Euler scheme

Notations for matrices and Numerical Linear Algebra:

M	velocity mass matrix, in the finite element context arising from (φ_i, φ_j)
M_l	*lumped* – that means diagonalized – velocity mass matrix
M_p	pressure mass matrix, analogous to M, with pressure ansatz functions
L, Δ_h	Laplacian matrix according to the Δ–operator
K	transport matrix according to the $(\mathbf{v}\cdot\nabla)$–operator
S	velocity matrix, typically $S := \alpha M + \theta_1\nu kL + \theta_2 kK$
B	gradient matrix according to the ∇–operator, $B = (B_1, \ldots, B_d)^T$

B^T	divergence matrix, equals transposed gradient matrix B
$B^T S^{-1} B$	discrete *pressure Schur complement* operator
P	reactive preconditioner $P := B^T M_l^{-1} B$
D,L,U	diagonal, lower, upper part of a given matrix
JAC	Jacobi scheme
GS	Gauß–Seidel scheme
SOR	SOR scheme
SSOR	SSOR scheme
ILU	ILU scheme
CG	conjugate gradient scheme
PCG	preconditioned conjugate gradient scheme
BiCGSTAB	BiCGSTAB scheme
GMRES	GMRES scheme
MG	multigrid/multilevel scheme

Other notations:

AR	aspect ratio
VR	volume ratio
PSC	*pressure Schur complement*
MPSC	*multilevel pressure Schur complement*
SPSC	*single (grid) pressure Schur complement*
c_a	lift coefficient ('*Auftrieb*')
c_w	drag coefficient ('*Widerstand*')
PP2D	*Discrete projection scheme* in the FEATFLOW package ($\sim$ global MPSC)
CC2D	*coupled Galerkin scheme* in the FEATFLOW package ($\sim$ local MPSC)

Chapter 1

Motivation for current research

The subject of the following chapters is the numerical solution of the incompressible Navier–Stokes equations. This system of equations describes – in a domain $\Omega \subset \mathbf{R}^d$, $d = 2$ or 3, and for a time intervall $(t_0, T]$ – the velocity $\mathbf{u}(x_1, \ldots, x_d, t)$, with components $(u_1, \ldots, u_d)$, and the pressure $P(x_1, \ldots, x_d, t)$ of an incompressible fluid, for given physical properties (kinematic viscosity ν, density ρ) and prescribed initial and boundary values:

$$\rho \frac{\partial u_i}{\partial t} + \rho \frac{\partial}{\partial x_j} (u_j u_i) = \rho \nu \frac{\partial}{\partial x_j} \left(\frac{\partial u_i}{\partial x_j} + \frac{\partial u_j}{\partial x_i} \right) - \frac{\partial P}{\partial x_i} \quad , \quad \frac{\partial u_i}{\partial x_i} = 0 \quad (1.1)$$

Introducing the normalized pressure $p = P/\rho$ we prefer the following form of the incompressible Navier–Stokes equations which reads in the stationary case

$$-\nu \Delta \mathbf{u} + \mathbf{u} \cdot \nabla \mathbf{u} + \nabla p = \mathbf{f} \quad , \quad \nabla \cdot \mathbf{u} = 0 , \qquad (1.2)$$

and for nonstationary flows

$$\mathbf{u}_t - \nu \Delta \mathbf{u} + \mathbf{u} \cdot \nabla \mathbf{u} + \nabla p = \mathbf{f} \quad , \quad \nabla \cdot \mathbf{u} = 0 . \qquad (1.3)$$

These equations seem to have a very simple structure, nevertheless they provide "grand challenge" problems for mathematicians and physicists as well as engineers, and they are (among others) object of very intensive research activities. However, beside people coming from research areas also software specialists and providers of commercial codes work hard in this field, and all together meet at one common point, the *Computational Fluid Dynamics* (CFD): *The Navier–Stokes equations describe "real life" processes and help to understand nature, and the necessary key tool is a practicable CFD software which can perform numerical simulations accurately and efficiently on computers.*

Particularly for mathematicians, incompressible flow problems provide a great potential for research activities since they include a wide variety of difficulties which typically arise in the numerical treatment of partial differential equations and which consequently determine the quality of the resulting CFD tools. Thus, they are the perfect playground to test the efficiency of today's numerics. Going into detail, the problems which scientists and practitioners are confronted with concern the following mathematical aspects:

- time dependent partial differential equations in complex domains

- strongly nonlinear systems of equations

- saddle–point problems due to the incompressibility constraint

- local changes of the problem character in space and time

- temporarily stiff systems of differential equations

These characteristics impose great challenges on almost all fields of numerical and computational approaches. Among others, the following examplary issues have to be taken into consideration if efficient CFD tools shall be designed:

appropriate choices for the discrete velocity and pressure spaces (**LBB– condition**)

very high–dimensional nonlinear systems of equations (**several millions of unknowns**)

locally varying time steps (**implicit schemes**)

locally anisotropic spatial meshes (**boundary layers, complex geometries**)

efficient iterative solvers (*adapted to* **workstations** *and/or* **supercomputers**)

Active research in numerical and computational methods is going on since more than 30 years and the number of publications and software packages is enormous. Current trends in CFD are to combine these basic equations with even more complex components to simulate "harder" applications. Only to call some of them: models from physics and chemistry are added for simulating turbulence, multiphase flow, nonlinear fluids, combustion/detonation, free boundaries, weakly compressible effects, etc. Additionally, the mathematical community begins to incorporate error control mechanisms of a posteriori type which require the handling of fully adaptive concepts in space and time.

These physical and mathematical extensions have one common aspect: They all require very efficient <u>and</u> robust solution schemes for nonlinear or linearized Navier–Stokes–like systems. Someone not being a CFD insider might get the impression that the numerical solution of the basic (laminar) incompressible Navier–Stokes equations seems to be completely under control, including all the well-known difficulties described above. Therefore, before the embedding into more complex frameworks is intensified, let us examine the question:

> *'Are the existing solution algorithms for incompressible flow problems already optimal or is further, maybe even tremendous improvement necessary?'*

How to answer such a fundamental question? There are the following two possibilities:

1. Ask the practitioners in industry whether the existing software satisfies actually their needs (if one neglects the high–polished advertising materials).

This attempt will not give the final conclusive answer since in general the distributed commercial codes have not realized all the new improved numerical, algorithmic and computational ideas yet. Thus, they might not represent the current state of the art.

2. Seek or develop specific benchmark configurations which are near to "real life" problems and which can be performed by many groups/codes to obtain valid comparisons.

We concentrate on the second approach and introduce the *1995 DFG benchmark* which has been performed by many groups during the last two years. Since the participating groups and codes represent a large part of most recent research activities in this field, the following results provide still today a quite good overview concerning the state of the art in *Computational Fluid Dynamics* for the laminar incompressible Navier-Stokes equations.

1.1   Results and conclusions from Benchmark calculations

Under the DFG Priority Research Programme 'Flow Simulation on High Performance Computers', solution methods for various flow problems have been developed with considerable success. Several new techniques such as *unstructured grids, multigrid, operator splitting, domain decomposition* and *mesh adaptation* have been used in order to improve the performance of numerical methods. To facilitate the comparison of these solution approaches, a set of benchmark problems has been defined and all participants of the Priority Research Program working on incompressible flows have been invited to submit their solutions. We present the results of these computations contributed by altogether 17 research groups, 10 from the Priority Research Program and 7 from outside. The major purpose of the benchmark was to establish whether constructive conclusions can be drawn from a comparison of these results so that the solutions can be improved. It was not the aim to come to the conclusion that a particular solution A is better than a different solution B; the intention was rather to determine whether and why certain approaches are superior to others. The benchmark was particularly meant to stimulate future work.

In the first step, only incompressible laminar test cases in two and three dimensions have been selected which are not too complicated, but still contain most difficulties representative of industrial flows in this regime. In particular, characteristic quantities such as drag and lift coefficients have to be computed in order to measure the ability to produce quantitatively accurate results. This benchmark aims to develop objective criteria for the evaluation of the different algorithmic approaches. For this purpose, the participants have been asked to submit a fairly complete account of their computational results together with detailed information about the discretization and solution methods used.

As a result it should be possible, at least for this particular class of flows, to distinguish between *efficient* and *less efficient* solution approaches. Since this benchmark has proven to be successful it is going to be extended to include also certain turbulent flows. The recent version of the "turbulent" benchmark can be found in [88] and can be obtained by the author on request.

It was particularly hoped that this benchmark will provide the basis for reaching decisive answers to the following questions which are currently the subject of controversial discussion:

1. *Is it possible to calculate incompressible (laminar) flows accurately and efficiently by methods based on explicitly advancing momentum?*

2. *Can one construct an efficient solver for incompressible flows without employing multigrid components, at least for the Pressure–Poisson equation?*

3. *Do conventional finite difference methods have advantages over new finite element or finite volume techniques?*

4. *Can steady-state solutions be efficiently computed by pseudo-time-stepping techniques?*

5. *Is low-order treatment of the convection competitive, at least for low Reynolds numbers?*

6. *What is better for time stepping: fully coupled iteration or operator splitting?*

7. *Does it pay to use higher order discretizations in space or time?*

8. *What is the potential of using unstructured grids?*

9. *What is the potential of adaptive grid adaptation and adaptive time step selection?*

10. *What is the best for the nonlinearity: quasi-Newton iteration or nonlinear multigrid?*

These questions appear to be of vital importance in the construction of efficient and reliable solvers, particularly in three space dimensions. Everybody who is extensively consuming computer resources for numerical flow simulation should be interested.

We give a brief summary of the test configurations only. The complete information containing all definitions and results can be found in [87]. The fluid properties for an incompressible Newtonian fluid are identical for all test cases. For the 2D test cases the flow around a cylinder with circular cross–section is considered (see Fig. 1.1).

Some definitions are introduced to specify the values which have to be computed. $H = 0.41\,\mathrm{m}$ is the channel height and $D = 0.1\,\mathrm{m}$ is the cylinder

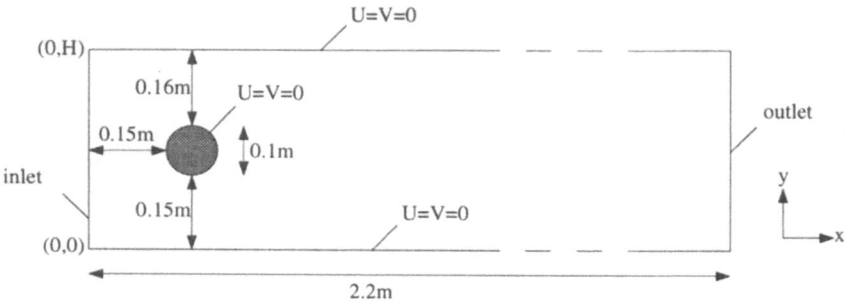

Figure 1.1: Geometry of 2D test cases with boundary conditions

diameter. The Reynolds number is defined by $Re = \overline{U}D/\nu$ with the mean velocity $\overline{U}(t) = 2U(0, H/2, t)/3$. The drag and lift forces are

$$F_D = \int_S (\rho\nu\frac{\partial v_t}{\partial n}n_y - Pn_x)\,dS \quad , \quad F_L = -\int_S (\rho\nu\frac{\partial v_t}{\partial n}n_x + Pn_y)\,dS\,, \quad (1.4)$$

while the drag and lift coefficients are calculated via

$$c_D = \frac{2F_w}{\rho\overline{U}^2 D} \quad , \quad c_L = \frac{2F_a}{\rho\overline{U}^2 D}\,. \quad (1.5)$$

The Strouhal number is defined as $St = Df/\overline{U}$, where f is the frequency of separation. The length of recirculation is $L_a = x_r - x_e$, where $x_e = 0.25$ is the x-coordinate of the end of the cylinder and x_r is the x-coordinate of the end of the recirculation area. As a further reference value the pressure difference $\Delta P = \Delta P(t) = P(x_a, y_a, t) - P(x_e, y_e, t)$ is defined, with the front and end point of the cylinder $(x_a, y_a) = (0.15, 0.2)$ and $(x_e, y_e) = (0.25, 0.2)$, respectively. The inflow profiles are parabolic with different scalings such that the resulting Reynolds numbers are $Re = 20$ (steady case), resp., $Re = 100$ (nonsteady cases). See [87] for a more detailed description.

For the 3D test cases the flow around a cylinder with circular (and squared [87]) cross–sections is considered. The problem configurations and boundary conditions are illustrated in Fig. 1.2.

Analogously to the 2D cases, some definitions are introduced to specify the values which have to be computed. The height and width of the channel is $H = 0.41\,\text{m}$, and the diameter of the cylinder is $D = 0.1\,\text{m}$. The characteristic

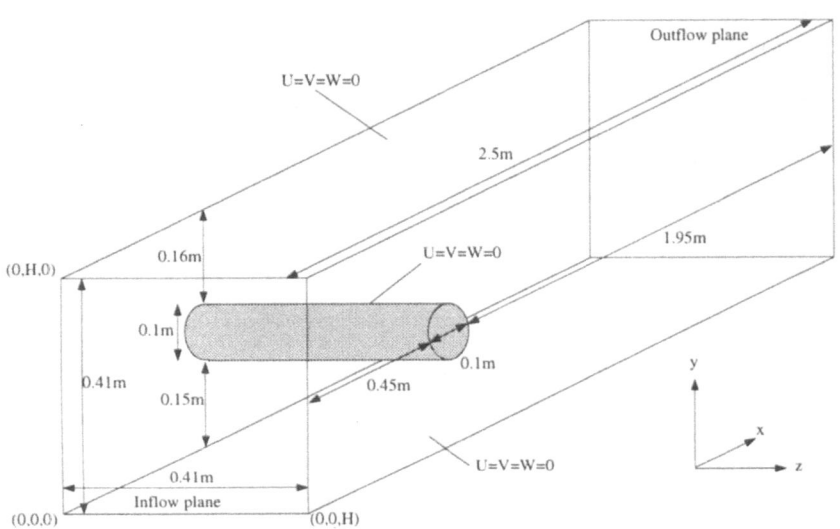

Figure 1.2: Geometry of 3D test cases ('Flow around a cylinder') with boundary conditions

velocity is $\overline{U}(t) = 4U(0, H/2, H/2, t)/9$, and the Reynolds number is defined by $Re = \overline{U}D/\nu$. The drag and lift forces are

$$
F_D = \int_S \left(\rho\nu \frac{\partial v_t}{\partial n} n_y - P n_x \right) dS \quad , \quad F_L = -\int_S \left(\rho\nu \frac{\partial v_t}{\partial n} n_x + P n_y \right) dS, \quad (1.6)
$$

leading to drag and lift coefficients via

$$
c_D = \frac{2F_w}{\rho \overline{U}^2 DH} \quad , \quad c_L = \frac{2F_a}{\rho \overline{U}^2 DH}. \quad (1.7)
$$

The Strouhal number is $St = Df/\overline{U}$ with the frequency of separation f, and a pressure difference is defined by $\Delta P = \Delta P(t) = P(x_a, y_a, z_a, t) - P(x_e, y_e, z_e, t)$ with coordinates $(x_a, y_a, z_a) = (0.45, 0.20, 0.205)$ and $(x_e, y_e, z_e) = (0.55, 0.20, 0.205)$. Again, the inflow profiles are parabolic with different scalings such that the resulting Reynolds numbers are $Re = 20$ (steady case), resp., $Re = 100$ (nonsteady cases).

Following instructions (see [87]) concerning the computations were given to the participants:

- In the case of the steady calculations, the results have to be presented for 3 successively coarsened spatial meshes.

- The unsteady results have to be presented for 3 successively coarsened spatial meshes with a finest time discretization and also for 2 successively coarsened time discretizations together with the finest spatial mesh.

- The outflow conditions and stopping criterions can be chosen by the user.

In Table 1.1 the numerical methods and details of the implementations of the participating groups are summarized (we are group 10). Many mathematical and engineering research codes of "prominent" groups have participated, and even with some commercial codes, for instance STAR–CD and TASCFLOW, comparisons could be performed. The following abbreviations are used in the tables: Finite difference method (FD), Finite volume method (FV), Finite element method (FE), Navier-Stokes equations (NS) and Multigrid method (MG). PEAK means the peak performance in MFlops and LINP the LIN-PACK 1000 MFlop-rate (LINPACK 1000 source code can be obtained by the authors).

The complete results of the benchmark computations are summarized in [87]: Exemplarily we show the results for three test configurations in the following tables. The number in the first column refers to the methods given in Table 1.1. The last column contains the performance of the computer used (as given by the contributors), either the LINPACK 1000 Mflop rate (LINP) or the peak performance (PEAK), which of course should be taken into account when comparing the different computing times. The column "unknowns" refers to the total number, i.e., the sum of degrees of freedom for all velocity and pressure components. The CPU timings are all given in seconds. In the last row of each table, estimated intervals for the "exact" results are indicated (as suggested by the authors in [87] on the basis of the obtained solutions).

On the basis of the results obtained by these benchmark computations some specific conclusions can be drawn. These have to be considered with care, as the provided results depend on parameters which are not available for the author, e.g., design of the grids, setting of stopping criteria, quality of implementation, etc. For five of the ten questions above the answers seem to be clear and were accepted by all groups participating:

1. *In order to compute incompressible flows of the present (laminar) type accurately and efficiently, one should use implicit methods. The step size restriction enforced by explicit time stepping can render this approach highly inefficient, as the physical time scale may be much larger*

	Space discretization	Time discretization	Solver	Implementation
1	FD, blockstructured non-staggered QUICK upwinding	fully implicit 2nd ord. equidistant	artificial compressibility expl. 5-step Runge-Kutta FAS-MG (steady) line-Jacobi (unsteady)	serial Fujitsu VPP500 1600 PEAK
2	FV, blockstructured 2nd ord. upwindig	implicit Euler equidistant	ILU with algebraic MG for linear problems	serial IBM RS6000/370 37 LINP
3	FV, blockstructured non-staggered QUICK upwinding	fully implicit 2nd ord. equidistant	stream function form fixed-point iteration ILU for lin. subproblems	serial SGI-Indigo2 75 PEAK parallel Cray T3D/16 16x88 LINP
4	FE, blockstructured 4Q1-Q1 BTD stabilisation	Projection 2 (Gresho) Crank-Nicolson (diff.) explicit Euler (conv.) adaptive	pseudo time step (steady) PCG for lin. subproblems hierarch. preconditioning	parallel GC/PP32 32x13.9 LINP
5a	FV, unstructured 1st ord. upwind	STARCD software	pressure correction	serial HP 735 13 LINP
5b	FV, unstructured CDS	STARCD software	pressure correction	serial HP 735 13 LINP
6	Lattice BGK equidistant orth. grid	explicit equidistant	gaskinetic solution of BGK-Boltzmann equation evol. of distribution funct.	serial HP735 13 LINP
7a	FV, blockstructured non-staggered CDS with def. corr.	Crank-Nicolson equidistant	nonlinear MG SIMPLE smoothing ILU for lin. subproblems	serial HP735 13 LINP
7b	FV, blockstructured non-staggered CDS with def. corr.	fully implicit 2nd ord. equidistant	nonlinear MG SIMPLE smoothing ILU for lin. subproblems	parallel GC/PP128,32,8 128x13.9 LINP
8a	FE, unstructured P2-P1 (Taylor-Hood) CDS	2nd order fract. step operator splitting equidistant	nonlinear GMRES PCG for lin. subproblems	serial SGI R4000 8.3 LINP
8b	FE, unstructured P2-P1 (Taylor-Hood) CDS adaptive refinement	2nd order fract. step operator splitting equidistant	nonlinear GMRES PCG for lin. subproblems	serial SGI R4400 13.2 LINP IBM RS6000/590 58 LINP
9a	FV, blockstructured CDS	fully implicit 2nd ord. equidistant	SIMPLE ILU for lin. subproblems	serial IBM RS6000/250 34 LINP
9b	FV, unstructured CDS	fully implicit 2nd ord. equidistant	SIMPLE ILU-CGSTAB for lin. subproblems	serial IBM RS6000/590 90 LINP

Table 1.1: Numerical methods and implementation of participating groups

	Space discretization	Time discretization	Solver	Implementation
10	FE, blockstructured Q1(rot)-Q0 adaptive upwind	2nd order fract. step projection method adaptive	fixed-point iteration MG for lin. NS with Vanka smoother (steady) MG for scalar lin. subproblems (unsteady)	serial IBM RS6000/590 90 LINP
11	FV, structured CDS with momentum interpolation	explicit 3rd ord. Runge-Kutta equidistant	SIMPLE ILU for lin. subprobl.	serial SNI S600/20 5000 PEAK
12	FV, unstructured non-staggered adapt. 2nd ord. DISC deferred correction	–	SIMPLEC ILU-BICGSTAB for lin. subproblems	serial SUN SS10 5.5 LINP
13a	FD, structured	explicit Euler equidistant	stream function form pseudo time step SOR for lin. subprobl.	serial IBM RS6000/590 90 LINP
13b	FD, structured	explicit 4th order Runge-Kutta-Gill equidistant	stream function form SOR for lin. subprobl.	serial IBM RS6000/590 90 LINP
14a	FE, blockstructured P1-P0 (Crouzeix-Raviart) 1st order upwind	–	nonlinear MG Vanka smoother	serial HP737/125 6.6 LINP
14b	FE, blockstructured Q1(rot)-Q0 1st order upwind		fixed-point iteration MG for lin. NS with Vanka smoother	parallel GC/PP96 96x13.9 LINP
14c	FE, unstructured P1-P0 (Cr.-Rav.) Samarskij upwind adaptive refinement	BDF(2), equidistant	fixed-point iteration GMRES for pressure Schur-complement lin. MG for velocity	parallel GC/PP24 24x13.9 LINP
15a	FD, structured staggered, orthogonal CDS/UDS flux-blend.	explicit Euler adaptive	SOR for pressure	serial HP720 7.4 LINP
15b	FD, structured staggered, orthogonal CDS/UDS flux blend.	explicit Euler adaptive	SOR for pressure	parallel HP720 cluster 8x7.4 LINP
16	FV, blockstructured CDS	explicit 2nd ord. leap-frog time-lagged diff.	pressure correction Gauss-Seidel for lin. subproblems	serial SGI Indigo 9.6 LINP Convex C3820 19.2 LINP
17	FV, unstructured adaptive upwind	–	fixed-point iteration MG for lin. NS $BILU_\beta$ smoother	serial SGI R4400 8.3 LINP

Table 1.1: (continued)

	Unknowns	c_D	c_L	L_a	ΔP	Mem.	CPU time	MFlop rate
1	200607	5.5567	0.0106	0.0845	0.1172	15	788	1600 PEAK
	51159	5.5567	0.0106	0.0843	0.1172	4	273	
	13299	5.5661	0.0105	0.0835	0.1169	1	144	
3a	10800	5.6000	0.0120	0.0720	0.1180	2.5	121	75 PEAK
4	297472	5.5678	0.0105	0.0847	0.1179	137	31000	445 LINP
	75008	5.5606	0.0107	0.0849	0.1184	73	8000	
	19008	5.5528	0.0118	0.0857	0.1199	57	2000	
6	1314720	5.8190	0.0110	0.0870	0.1230	40	80374	13 LINP
	332640	5.7740	0.0030	0.0830	0.1230	10	10461	
	85140	5.7890	-0.0060	0.0870	0.1230	2.6	1262	
7a	294912	5.5846	0.0106	0.0846	0.1176	75	192	13 LINP
	73728	5.5852	0.0105	0.0845	0.1176	19	47	
	18432	5.5755	0.0102	0.0842	0.1175	5	13	
8a	20487	5.5760	0.0110	0.0848	0.1170	9.0	2574	8.3 LINP
	6297	5.5710	0.0130	0.0846	0.1160	2.9	362	
	2298	5.4450	0.0200	0.0810	0.1110	1.3	109	
9a	240000	5.5803	0.0106	0.0847	0.1175	53	9200	34 LINP
	60000	5.5786	0.0106	0.0847	0.1173	10	1400	
	15000	5.5612	0.0109	0.0848	0.1166	2.5	200	
10	2665728	5.5755	0.0106	0.0780	0.1173	350	677	90 LINP
	667264	5.5718	0.0105	0.0770	0.1169	89	169	
	167232	5.5657	0.0102	0.0730	0.1161	22	52	
	42016	5.5608	0.0091	0.0660	0.1139	5	18	
12	32592	5.5069	0.0132	0.0830	0.1155	18	1796	5.5 LINP
	26970	5.5125	0.0056	0.0827	0.1154	15	1099	
	22212	5.6026	-0.0031	0.0815	0.1167	13	3437	
13a	25410	5.6145	0.0159	0.8315	3.0002	4	14203	90 LINP
	12738	5.6114	0.0169	0.8224	2.9943	2	3018	
	6562	5.7377	0.0514	0.8107	3.2277	1		
14a	3077504	5.6323	0.0137	0.0782	0.1159	214	15300	6.6 LINP
	768704	5.6382	0.0102	0.0775	0.1156	53	5490	
	191840	5.5919	-0.0009	0.0750	0.1143	13	2800	
14b	30775296	5.5902	0.0108	0.0853	0.1174	5340	1534	1334 LINP
	7695104	5.6010	0.0110	0.0844	0.1174	1341	400	
	1922432	5.6227	0.0113	0.0833	0.1172	338	119	
14c	797010	5.5708	0.0167	0.0837	0.1168	460	8000	334 LINP
	363457	5.5598	0.0142	0.0835	0.1166	230	3290	
	176396	5.5106	0.0046	0.0835	0.1150	110	2560	
15a	432960	5.5602	0.0329	0.0730	0.1054	4.4	179986	7.4 LINP
	108240	5.6300	0.0751	0.0720	0.1037	1.1	13593	
	27060	5.7769	0.2085	0.0680	0.0998	0.3	688	
17	111342	5.5610	0.0107		0.1170	87	2568	8.3 LINP
	60804	5.5520	0.0102		0.1168	47	1092	
	19416	5.5160	0.0099		0.1158	15	373	
	lower bound	5.5700	0.0104	0.0842	0.1172			
	upper bound	5.5900	0.0110	0.0852	0.1176			

Table 1.2: Results for steady test case 2D-1

	Unknowns		c_{Dmax}	c_{Lmax}	St	ΔP	Mem.	CPU time	MFlop rate
	Space	Time							
1	267476	67	3.2224	0.9672	0.2995	2.4814	—	—	1600 PEAK
	267476	34	3.2030	0.9223	0.2941	2.4664	—	—	
	267476	18	3.1605	0.8026	0.2901	2.4466	—	—	
	68212	67	3.2171	0.9591	0.2995	2.5009	—	—	
	17732	68	3.2168	0.9295	0.2979	2.5573	—	—	
3	12800	34	3.2200	0.9720	0.2960	2.4700	2.5	789	75 PEAK
4	297472	670	3.2460	0.9840	0.2985	2.4900	137	6600	445 LINP
	297472	338	3.2710	0.9800	0.2959	2.4870	137	3400	
	297472	172	3.3200	0.9720	0.2907	2.4810	137	1700	
	75008	670	3.2410	0.9910	0.2985	2.5020	73	2350	
	19008	674	3.2320	1.0260	0.2967	2.5320	57	1350	
6	332640	12000	4.1210	1.6120	0.3330	3.1420	10	10086	13 LINP
	85140	6000	4.7330	2.0600	0.3380	3.4300	2.6	1259	
7a	294912	36	3.2358	1.0069	0.3003	2.4892	75	6167	13 LINP
	294912	19	3.2356	1.0000	0.2973	2.4871	75	6391	
	294912	10	3.2152	0.9028	0.2881	2.4715	75	4994	
	73728	36	3.2443	1.0261	0.2994	2.4929	19	1946	
	18432	36	3.2706	1.0695	0.2968	2.5035	5	445	
8a	29084	66	3.2240	1.0060	0.3020	2.4860	11	4992	8.3 LINP
	29084	33	3.2470	1.0740	0.3030	2.5010	11	3777	
	29084	16	3.2900	1.2500	0.3130	2.5700	11	3217	
	8764	66	3.1740	0.9640	0.3000	2.4630	3.6	1000	
	2978	70	2.8920	0.5540	0.2890	2.2870	1.5	339	
9a	240000	5000	3.2267	0.9862	0.3017	2.4833	53	32500	34 LINP
	60000	10000	3.2232	0.9830	0.3012	2.4773	10	8550	
	60000	5000	3.2232	0.9832	0.3012	2.4773	10	4500	
	60000	2500	3.2232	0.9836	0.3012	2.4773	10	3400	
	15000	5000	3.2058	0.9651	0.2994	2.4587	2.5	3240	
10	667264	612	3.2314	0.9999	0.2973	2.4707	128	8545	90 LINP
	667264	204	3.2351	1.0123	0.2957	2.4734	128	2850	
	667264	68	3.2771	1.1205	0.2997	2.4961	128	1065	
	167232	188	3.2498	1.0081	0.2927	2.4410	32	655	
	42016	164	3.2970	0.8492	0.2713	2.3423	8	147	
13b	25410	6755	3.1822	1.0692	0.2960	2.6066	5.1	44710	90 LINP
	25410	3877	3.1895	1.0883	0.2968	2.6057	4.8	27175	
	25410	1678	3.2043	1.1268	0.2979	2.5307	4.7		
	12738	6799	3.1945	1.1233	0.2941	2.6140	2.9	13045	
	6562	7223	3.1317	1.2961	0.2768	3.0253	1.8		
15a	432960	7790	3.0804	0.7256	0.2778	2.1330	4.4	108844	7.4 LINP
	108240	4003	3.1677	0.6880	0.2646	2.0954	1.1	34876	
	108240	3859	3.1096	0.8249	0.2841	2.1105	1.1	58003	
	27060	1985	3.2544	0.5658	0.2336	1.9727	0.3	3796	
	27060	1670	3.1759	0.7656	0.2740	1.9961	0.3	4188	
	lower bound		3.2200	0.9900	0.2950	2.4600			
	upper bound		3.2400	1.0100	0.3050	2.5000			

Table 1.3: Results for time–periodic test case 2D-2

	Unknowns		c_{Dmax}	c_{Lmax}	ΔP	Mem.	CPU time	MFlop rate
	Space	Time						
1	630564	800	3.2826	0.0027	-0.1117	79	156460	1600 PEAK
3	608496	1600	3.2590	0.0026	-0.1072	74	76142	1408 LINP
	608496	800	3.2590	0.0026	-0.1157	74	50764	
6	6303750	18000	4.1600	0.0200		43	142646	13 LINP
7b	1572864	1600	3.3011	0.0026	-0.1102	518	149923	445 LINP
	1572864	800	3.3008	0.0026	-0.1105	518	93055	445 LINP
	1572864	400	3.3006	0.0026	-0.1107	518	62026	445 LINP
	196608	1600	3.3053	0.0028	-0.1066	71	63057	111 LINP
8a	362613	1000	3.2340	0.0028	-0.1114	126	347000	58 LINP
8b	199802	1000	3.2120	0.0122	-0.1112	105	846000	13.2 LINP
	98637	1000	3.2350	0.0123	-0.1114	39	243000	
10	6116608	668	3.2802	0.0034	-0.0959	840	164837	90 LINP
	6116608	272	3.3748	0.0360	-0.0603	840	77538	
	6116608	60	2.7312	0.0069	-0.0682	840	29742	
	771392	724	3.3323	0.0033	-0.0766	105	24745	
	98128	660	3.4200	0.0040	-0.0407	13	5687	
	lower bound		3.2000	0.0020	-0.0900			
	upper bound		3.3000	0.0040	-0.1100			

Table 1.4: Results for unsteady test case 3D-3

than the maximum possible time step in the explicit algorithm. This is obvious from the results for the stationary cases in 2D and 3D, and also for the nonstationary cases in 2D. For the nonstationary cases in 3D only too few results on apparently too coarse meshes have been provided, in order to draw clear conclusions.

2. *Flow solvers based on conventional iterative methods for the linear sub-problems have on sufficiently fine grids no chance against those employing suitable multigrid techniques. Multigrid can allow computations on workstations (if the problem fits into the RAM) for which otherwise supercomputers would have to be used. In the submitted solutions the supercomputers have mainly been used for their high CPU power but not for their large storage capacities. For example, in test case 3D-3 (Table 1.4) the solutions 1 and 3 require with about 600,000 unknowns on supercomputers significantly more CPU time than the solution 10 with the same number of unknowns on a workstation.*

3. *The most efficient and accurate solutions are based either on finite element or finite volume discretizations on contour adapted grids.*

4. *The computation of steady solutions by pseudo time-stepping techniques is inefficient compared with using directly a quasi-Newton iteration as stationary solver.*

5. *For computing sensitive quantities such as drag and lift coefficients, higher order treatment of the convective term is indispensable. The*

> *use of only first order upwinding (or crude approximation of curved boundaries) does not lead to satisfactory accuracy even on very fine meshes (several million unknowns in 2D).*

For the remaining five questions the answers are not so clear. More test calculations will be necessary to reach more decisive conclusions. The following preliminary interpretations of the results obtained so far may become the subject of further discussion:

6. *In computing nonstationary solutions, the use of operator splitting (pressure correction) schemes tends to be superior to the more expensive fully coupled approach, but this may depend on the problem as well as the quantity to be calculated. Further, as fully coupled methods also use iterative correction within each time step (possibly adaptively controlled), the distinction between fully coupled and operator splitting approaches is not so clear.*

7. *The use of higher than second-order discretizations in space appears promising with respect to accuracy, but there remains the question of how to solve efficiently the resulting algebraic problems (see the results of 8 for all test cases). The results provided for this benchmark are too sparse to allow a definite answer.*

8. *The most efficient solutions in this benchmark have been obtained on blockwise structured grids which are particularly suited for multigrid algorithms. There is no indication that fully unstructured grids might be superior for this type of problem, particularly with respect to solution efficiency (compare the CPU times reported for the solutions 7 and 9 in 2D). The winners in the future may be hierarchically structured grids which allow locally adaptive mesh refinement together with optimal multigrid solvers.*

9. *From the contributed solutions to this benchmark there is no indication that a posteriori grid adaptation in space is superior to good hand-made grids (see the results of 14c). This, however, may dramatically change in future, particularly in 3D. Intensive development in this direction is currently in progress. For nonstationary calculations, adaptive time step selection is advisable in order to achieve reliability and efficiency.*

10. *The treatment of the nonlinearity by nonlinear multigrid has no clear advantage over the quasi-Newton iteration with multigrid for the linear subproblems (compare the results of 7 with those of 10). Again, it is the extensive use of well-tuned multigrid (wheresoever in the algorithm) which is decisive for the overall efficiency of the method.*

Let us give some additional remarks to the presented benchmark results. Although this benchmark has been fairly successful as it has made possible some solidly based comparisons between various solution approaches, it still needs further development. Particularly in the nonstationary test cases, further characteristic quantities (e.g., time averaged values) should be computed as in some cases, by chance, maximum values may be obtained with good accuracy even without capturing the general pattern of the flow at all. This is an open problem in the design of nonstationary CFD benchmarks!

We can state as a main result from these benchmark tests that even in the laminar case the chosen nonstationary 3D problems have proven to be much harder than expected. In particular, it was apparently not possible to achieve reliable reference solutions for all test cases in 3D. Almost all codes were able to provide qualitatively good results, looking only at velocity or pressure values and general flow patterns, but many failed for the precise prediction of the lift coefficient. "Unfortunately", this quantity is not a mathematical invention, but a physical value with practical importance. It can be seen as a prototype for other important quantities in CFD simulations as the drag coefficient of wings which is of major importance for the economical aircraft design. This lift value could be approximated in the steady 3D and the nonsteady 2D cases, but utilizing heavy computer power only. In the nonstationary 3D cases we could not obtain reference results – the given results differ by 20% and more – and no experimental values are available at this time.

The consequence is that the benchmark is still "open"! This is a somewhat surprising result. It should not be forgotten that we talk about flow calculations for Reynolds numbers 20 (!) and 100 (!) which are still far away from many realistic situations. The required improvements must directly concern the algorithmic components since all tests show that the use of GigaFlop supercomputers is not sufficient at all. At the same time, also recent numerical improvements such as *optimal multigrid* or *solution adapted meshes* must be applied.

These conclusions seem to be valid not only for this special benchmark, but also for other complex flow problems with "real life" character. Asking practitioners from automotive industry who might have a more complete overlook over CFD tools, they (inofficially) say:

> *'There is no software available which can provide guaranteed lift and drag coefficients on a car–body with a error tolerance of less than 20%; often the sign of the lift cannot even be predicted. Hence, we stopped flow problems around objects and use simulation tools for interior flow problems only, for instance for modelling heating devices or acoustic behaviour in car cabins. Here, we are content with a qualitatively good prediction!'*

This indicates that *Computational Fluid Dynamics*, at least restricted to the case of laminar incompressible flow, is far away from replacing expensive (wind tunnel) experiments as predicted some years before. The high-lighted advertisements must be neglected since they give a wrong impression of modern CFD usage. First of all, we have to return to simpler problems as the (basic) Navier–Stokes equations, to learn to control them numerically before harder problems containing turbulence and complex physical and chemical components should be tried. Without tremendous improvements, CFD may be useless in many situations.

Let us finish this Section with the following recommendations for the development of future benchmarks in this research field:

- The reference values of a benchmark calculation have to be unknown as long as possible. Otherwise, participants know what to look for.

- Benchmark calculations have to be performed for a sequence of spatial and temporal meshes. It often happened that the results on the coarsest mesh were the best, with correspondingly the smallest computational work. While the number of unknowns and hence the computational effort is increased, the resulting quality may decrease. Consequently, as long as the reference results are unknown (and no quantitative error control is available), everybody continues as usual with further refinements.

- The commercial codes have to be forced to compare with research codes. Nobody expects the same efficiency as can be reached by special research codes. However, it cannot be accepted that some codes claim to "solve everything" but then they fail for a laminar flow calculation with Reynolds number $Re = 20$. Furthermore, the same is true for research codes including new mathematical approaches.

1.2   Numerical and algorithmic bottlenecks of CFD tools

What are the reasons that many participants failed to produce the "right" results for the described benchmark, especially for the 3D cases? Why cannot nonstationary flows be simulated precisely, even being laminar? The conclusions from above seem to indicate the following three main reasons:

- Bad efficiency of the performed discretization and solver techniques.

- Unnecessary large number of unknowns and time steps.

- Inefficient implementations.

Efficiency of discretization and solver techniques:

The combination of chosen discretization and solution schemes is often not "optimal" and the resulting CPU requirements are too large. While in two dimensions stationary and (sometimes) nonstationary problems can be solved by brute (computer) force, the same "technique" is often impossible in 3D. In 2D, it makes (almost) no difference if a calculation takes 10 seconds or 5 minutes, but in 3D the analogous relation results in differences of 1 day or 1 month! Further, the benchmark calculations indicate clearly that *supercomputer power* without *algorithmic efficiency* is not enough. The next generation of processores may provide a speedup factor of 10, while by numerical and algorithmic improvements a speedup of factor 100 and more should be expected.

Therefore, improved solvers on multilevel basis with problem adapted smoothers and intergrid transfer operators have to be derived, hereby taking into account their compatibility with appropriate discretization methods in space and time. We lay our particular emphasis on this strong connection between solution and discretization methods since optimal approaches must be a compromise between these two different topics. This is in contrast to many other works in which case the authors stop after performing the discretization process and let the reader alone with the advise of applying "some standard solution tools".

Accuracy and Robustness:

The second reason for the bad run-time behaviour is that in many calculations based on *grid discretization* methods a large number of mesh points is employed. Especially for methods as finite elements, finite volumes or finite differences, often too many spatial unknowns as sum of velocity and pressure degrees of freedom are addressed. Since for most standard solution schemes the CPU time grows at least linearly with the degrees of freedom, it is obvious that this huge number of unknowns automatically requires large CPU times. Therefore, a simple remedy is to decrease the number of discretization points, but how to do this in a safe way? And let us even assume that an error control mechanism is available which controls adaptively – during the calculation – the number and the position of the grid points, and that the calculation can be performed with the minimal number of unknowns. However, then we still have to ask for the numerical robustness:

- Since we have no a priori knowledge about the grid, the error control may lead to very anisotropic meshes containing elements with large deformations. These special elements are often used for a better approximation of boundary layers near walls or of small-scale structures. However, can the available solvers work efficiently on such meshes?

- Suppose that a temporal error control allows large time steps and that we work with a fully implicit time stepping scheme, without restrictions onto the size of the possible time steps due to stability reasons. However, do we get restrictions for the possible time steps because of insufficient robustness of the available solvers? And what is the resulting temporal behaviour in semi-implicit approaches, as for instance *pressure correction*, *projection* or *fractional step* methods, especially if highly stretched elements are used?

Both topics, solver and discretization, address the typical "mathematical legs" of CFD tools: **accuracy, robustness** and **efficiency**. However, there is another important component for successful software: the implemented code.

Implementation and Hardware:

The third reason for the extensive CPU requirements of many software packages is the quality of the actual implementation, including the employed data structure. Many codes use fully unstructured grids to approximate complex domains with locally varying structures. Particularly in mathematical research codes the trend is definitely to apply adaptivity concepts for all components. These "unstructured" features are mostly based on very structured data types (spanning trees with recursive father/son relationship) which are easy to handle by the code developer. However, the idea of "unstructured behaviour" and adaptivity induces a significant performance overhead on modern computer architectures: presently we have reached a state where on most platforms a memory access is significantly more expensive than many floating point operations. All numerical performance depends heavily on exploiting internal parallelism (pipelining and superscalar execution) and hierarchical memory architectures and caches. Only then, the maximum performance of several hundreds MFlop Peak performance can be approximately reached (see [26] and especially [1], [2], [63], [112]). This applies to PC's and workstations as well as high performance vector- or parallel computers.

However, this development is not taken into account by most of the recent codes for general domains and configurations, and in fact they gain no advantage of the new processor technologies. Furthermore, this loss is even unnecessary since unstructured data organization is needed only in some small parts of the domain, for instance near walls to resolve complex geometrical details or to gain high accuracy in boundary layers. But far away from them, the flow is smooth in general and structured grids lead to highly accurate results. For the numerical software developer, this has as consequence that even on low and midrange computers it is becoming important to use as far as possible uniform data structures to reach more accuracy and optimally fast execution times. An indirect memory access through a pointer or an index array may become many times more expensive than a floating point operation. Hence, we will introduce some *subdomain/patch* oriented solution techniques (see [1] and also [65]) which are a compromise between rigid uniform mesh strucures to gain processor efficiency and solution robustness and unstructured "adaptive" macro–grids to obtain accuracy with quasi–optimal number of unknowns.

Summarizing these last considerations, three major aspects have to be taken into account if improved CFD software shall be developed (in brackets we address the research fields which are challenged):

1. **algorithmic design** ($\rightarrow$ numerical solution schemes)

2. **error control** ($\rightarrow$ discretization techniques)

3. **software engineering** ($\rightarrow$ implementation strategies)

The main part of this book treats the aspect of examining *algorithmic tools* for the pure solution process and special numerical solution techniques for Navier–Stokes-type problems. The field of *error control* is discussed, too, however we provide only the major background and the general framework which is based on current research activities: This issue will be one of the most important topics for future numerical methods, not only restricted to CFD and the Navier–Stokes equations. The aspect of an improved *software engineering* is partially realized in our own software project FEATFLOW (see [111]) which is the basis of our calculations. However, another software project (FEAST [112]) is just under development which will contain the practical realization of all ideas presented in this book.

Let us discuss more in detail some important aspects of *algorithmic design*. We will provide results which demonstrate that insufficient numerical and algorithmic considerations can cause complete failures of codes under special circumstances. These problems are highlighted for two solution schemes which are widely used by the CFD community.

The first one is using the so–called *Vanka* smoother (see [114]). This scheme is a simple iterative relaxation method which acts directly on element–level (that means, working on each separate mesh cell or clustering certain velocity and pressure components in a similar way), and which updates subsequently some specific velocity and pressure components. This local procedure is embedded into an outer Gauß-Seidel relaxation (if performed on workstations) or an outer Jacobi iteration (on parallel systems). This approach behaves like a simple block iteration procedure for systems of equations and has been introduced by Vanka [114]. It is often applied by the mathematical community in combination with Galerkin–schemes if coupled approaches for Stokes or Oseen equations are performed. Particularly in combination with an outer *quasi-Newton* iteration for the treatment of the nonlinearity, a very efficient direct solver for stationary incompressible Navier–Stokes equations in 2D and 3D can be constructed (see the results of the previous benchmark). Since this classical iteration scheme can be interpreted as belonging to our general class of *Pressure Schur Complement* solvers (local MPSC) we skip a more detailed explanation and refer to the next chapter. At this point, we discuss only briefly the advantages and disadvantages of this classical scheme.

Advantages of coupled multigrid with the *Vanka* smoother:

- In connection with an outer *quasi-Newton* iteration, very efficient direct solvers for stationary flow problems are possible. Hence, this solver is mainly suited for low Reynolds number flow or for coupled (Galerkin) approaches for nonsteady problems which force to solve generalized stationary Navier–Stokes problems in every time step.

- The convergence behaviour of multigrid solvers with the *Vanka* smoother is (almost) independent of the Reynolds number. That means that the convergence rates and hence the numerical cost are (almost) the same for Stokes as well as for moderate Reynolds number problems (as long as the flow is steady!). In fact, for moderate Reynolds numbers the convergence rates even improve in many cases.

Disadvantages of coupled multigrid with the *Vanka* smoother:

- The last point, the independence of the convergence rates from the viscosity parameter, is a large disadvantage for fully nonstationary problems with small time steps. In contrast to certain other methods, for instance the projection schemes, the convergence rates do not improve for time step sizes tending to zero. As a result, these direct multigrid approaches consume much more CPU times than the following pressure correction/projection type methods and are not competitive.

- Due to the inherent Gauß-Seidel/Jacobi character there is a strong relation to the underlying mesh. As well known from scalar model equations of diffusion or convection type, the convergence rates for these schemes degenerate, especially if large *aspect ratios* or large jumps in element sizes occur. For scalar problems the remedy may be an incomplete ILU decomposition as smoother but how to perform efficiently this technique for indefinite coupled systems of equations?

- Due to the very local (elementwise) character of this iteration scheme, it is difficult to link it together with machine–optimized libraries as the BLAS or the LAPACK package. Most of the CPU time is consumed first by indirect memory access and secondly by solving small subproblems in contrast to the available large cache. Hence, it is very difficult to obtain the full power of modern superscalar architectures.

Another typical approach, particularly for nonsteady flows but also working for steady equations, are the so called *pressure correction, projection* and *fractional step* schemes. Instead of relaxing simultanously the pressure and velocity components, the overall coupled problem is divided into simpler scalar equations of Poisson or convection–diffusion type. These can be solved in a very robust and efficient manner with standard tools which are well known from the solution process for these scalar equations. These operator splitting techniques are favourized in engineering and commercial codes since in many applications, especially for high Reynolds numbers, they seem to provide results in a very fast and robust fashion.

Moreover, the implementation is straightforward and can be adapted to modern workstations and vector- or parallel supercomputers, since standard tools as conjugate gradient or multigrid in combination with BLAS routines can be taken over from numerical specialists.

Our experience with these projection schemes, in particular for highly nonsteady flows, is very posititive, and in fact we successfully performed the previous benchmark calculations with a similar approach, the *discrete projection methods* (see [102]). However, these methods can cause some heavy trouble, too.

Properties of operator splitting techniques:

- These schemes work perfectly for large Reynolds numbers, resp., small viscosity parameters which require small time steps by physical reasons. But they often fail for steady or nonsteady flows with large viscosity parameter.

- It is not clear how to apply an error control in time. Obviously, a time step control which is based on the typical local extrapolation principle – by comparing calculations for time steps Δt and $\Delta t/2$ – may work sufficiently. However, the same approach can also fail completely as our numerical examples will show. Even in the case that these mechanisms work they do not allow a quantitatively precise prediction of flow parameters depending on time since they are acting only as *error indicators*.

- The behaviour as time stepping scheme is often dependent on the spatial mesh. The time step size may have to be adapted if locally the spatial mesh width has been changed. Physical reasons do not enforce this restriction, but the numerical behaviour is significantly influenced!

- The balance of the resulting CPU times for the solution process (solving the Pressure–Poisson problem and Helmholtz/convection-diffusion/Burgers equations for the intermediate velocity) and the other algorithmic components (generating defect vectors and matrices in every time step) may fail. This problem arises if the solvers are highly optimized, but the flow requires such small time steps that the "non-solver" components dominate the CPU cost. In this case, no optimal performance can be obtained, especially if we take into account that these "memory access" intensive components usually cannot exploit the fast processor facilities.

As mentioned before, the later numerical examples will prove the assertions above. Here, we want to shed more light upon the last statement, concerning the balance of separate components inside of a CFD code. The examples from the following *FEATFLOW Benchmark* will provide another insight into the process of algorithmic design and shall demonstrate which further computational problems have to be taken into account if optimal schemes and codes have to be designed.

The *FEATFLOW Benchmark* is a set of test calculations which is very similar to the proposed *1995 DFG Benchmark* 'Flow around a cylinder' in the previous section. However in contrast to the original one, all input parameters and even the triangulation are fixed now. The remaining interesting point are the total CPU cost and how they are distributed to the separate tasks in a CFD code, i.e., mesh generation, assembling of matrices and right hand sides, solving linear subproblems or postprocessing steps.

The original idea behind this benchmark was to have a tool at our institute which helps us in selecting the "optimal" workstations for our classes of problems:

> *'Which workstation, that means in practice, which processor and which compiler options, give us the "best possible performance" without changing explicitly our code?'*

Beside the fact that we obtained some valuable knowledge about the practical behaviour of hardware and software components, we additionally figured out that this performance analysis influences also the design of the numerical and algorithmic components. One of the most surprising results was that our preferred nonstationary solution method, the *discrete projection scheme* (see [102]), does not always run optimally on modern workstation platforms. This was a further reason to search for improved methods which are developed and described in the next chapter. It is a very important aspect of *scientific computing* which stands outside of mathematical experience and which should be explicitly mentioned in this context.

The precise configurations are described at our FEATFLOW Homepage (see below); here we only provide the reader with the most important details. We apply a coupled multigrid approach with the mentioned *Vanka* smoother in a direct steady solver (CC2D or CC3D), and the *discrete projection scheme* as nonstationary scheme to iterate the stationary limit (PP2D/PP3D). The Reynolds number for these tests is about $Re = 20$, and all calculations are performed with the nonconforming *rotated multilinear* finite elements. Additionally, we apply stabilization techniques of *upwind* (UPW) or *streamline diffusion* (SD) type for the convective term, and we use SOR or ILU–schemes as smoothers in the multigrid solver for the scalar subproblems if applying the *projection* solver. The following abbreviations for the elapsed time in seconds are used:

total	$\sim$ complete CPU time
PRO	$\sim$ mesh generation, initialization phase and postprocessing
LC	$\sim$ modifications of matrices (in sparse storage technique) or right hand sides
MAT	$\sim$ CPU time for matrix generations
ILU	$\sim$ building up ILU decompositions
C-MG	$\sim$ elapsed time in the CC2D/CC3D tests with the *Vanka* smoother
U-MG	$\sim$ CPU times for solving velocity, resp., pressure subproblems via
P-MG	multigrid if the *discrete projection schemes* PP2D/PP3D are applied

Exemplarily, we show some results for the direct stationary approaches CC2D and for the nonstationary schemes PP3D. The complete results can be found at:

---
http://www.iwr.uni-heidelberg.de/~featflow

We invite all 'FEATFLOW users' to participate at this special computer benchmark. This benchmark is part of the complete FEATFLOW software such that the run of this set of tests can be easily performed by everybody. If anyone gets new results we are very interested in obtaining the measured

Type	total	PRO	LC	MAT	C-MG
IBM RS6000/597 (170 MHz)	149	9	11	10	118
DEC WS (500 MHz)	202	4	19	8	171
SUN ULTRA 30 (300 MHz)	263	13	22	9	219
PC DEC Alpha (533 MHz)	303	7	24	13	259
SUN ULTRA 450 (250 MHz)	313	15	26	10	262
IBM RS6000/590 (66 MHz)	319	23	23	24	249
SGI PowInd2/10K (195 MHz)	409	12	38	15	343
PC PENTIUM II (266 MHz)	413	13	38	18	344
SUN ULTRA1 (170 MHz)	430	22	32	15	361
HP C110	677	10	60	23	584
SGI PowInd2/8K (75 MHz)	698	34	74	28	571
SGI r4400 (250 MHz)	750	30	60	24	635
IBM 250/601 PPC	832	43	67	41	681
SUN SS10/51	1676	59	130	55	1432

Table 1.5: (Selected) results for test case CC2D–UPW

Type	total	PRO	LC	ILU	MAT	U-MG	P-MG
IBM RS6000/597 (170 MHz)	2748	10	94	271	1212	812	347
DEC WS (500 MHz)	3604	9	138	240	988	1563	619
SUN ULTRA 30 (300 MHz)	4642	12	165	325	1425	1971	740
PC DEC Alpha (533 MHz)	4684	12	206	276	1640	1856	696
SUN ULTRA 450 (250 MHz)	5457	14	183	383	1777	2277	821
IBM RS6000/590 (66 MHz)	5970	23	202	626	2973	1511	629
SGI PowInd2/10K (195 MHz)	6947	19	277	614	1578	3240	1209
SUN ULTRA1 (170 MHz)	6953	21	230	518	2320	2804	1056
PC PENTIUM II (266 MHz)	7211	17	294	472	3127	2324	973
HP C110	11458	34	411	820	3039	5120	2035
SGI PowInd2/8K (75 MHz)	13113	40	439	1184	4764	4907	1768
SGI r4400 (250 MHz)	13282	35	509	738	4864	5203	1923
SUN SS10/51	28213	77	1014	1647	10581	10823	4047

Table 1.6: (Selected) results for test case PP3D–SD

CPU times as well as the used compiler options to update the online database. So, new contributions are always welcome!

On the basis of (all) these results, the following conclusions can be drawn. However, it must be taken into consideration that these results are from the years 1997/1998 and the next generation of workstations will have come. Nevertheless, the main conclusions will remain the same, especially with respect to the "algorithmic re–design".

Conclusions for hardware/software components:

- In contrast to the announced processor performance of the brochures, there are significant differences in the "real life" performance for our CFD applications. The CPU cost differ significantly by a factor 2 to

10! The results are in good agreement with other numerical tests which show even larger discrepancies. Check your configuration!

- There are significant differences in the performance of the *floating point* intensive parts (solving linear equations via multigrid) and the *memory access* intensive part for generating stiffness matrices and defect vectors. Test it for your own problems!

Conclusions for numerical/algorithmic components:

- On "fast" processors, as the IBM POWER2 architecture for instance, there may arise a bad balance between the "solver" and the "matrix generation" part. In combination with the streamline–diffusion technique which is very expensive in 3D (involving a trilinear form with 13 terms) it may happen that the "matrix/defect generation" part is more costly than our highly optimized solver modules. Hence, it makes no sense to search for even faster solution tools. This deficiency is due to the fact that the cost for matrix/defect generation are independent of the actual time step size, while the multigrid expenses decrease with time steps getting smaller. For a better balance of these different parts we might increase the size of the time steps. However, these are chosen by an adaptive time stepping criterion and are strongly related to the employed *projection scheme*. The remedy is to apply a method which solves several times subproblems with the same matrices, while the matrix/defect generation is performed only once during this period. This approach leads to the subsequent global *MPSC* schemes.

- If a coupled multigrid approach with the *Vanka* smoother is performed for solving stationary Navier–Stokes equations, the multigrid and in particular the smoothing part consumes most of the time (about 90%), and in comparison much more CPU time than corresponding solution steps in *projection* methods. This is due to the fact that these special smoothing steps are very recursive operations which require cost–intensive memory access and small local subproblems only. Hence, they often cannot exploit the fast processor performance. For steady problems with their typical short run-time characteristics, this discrepancy is not so important, but these schemes are much too expensive for nonstationary problems, in comparison to *operator splitting* approaches (see also [102]).

These benchmark calculations show the importance of numerical experience if optimal codes are asked for. Both classes of methods, the *operator splitting* and the *direct coupled* approach, work satisfactorily for many classes of problems, for fully nonsteady and for steady cases. However, their "optimal range" is different and both approaches have their own special advantages and disadvantages. Nevertheless, they can be first generalized and then optimized also for flow problems which are inbetween or which change their character during the calculation. This generalization and subsequent optimization with respect to efficiency, robustness and accuracy on the one hand and computational details of the implementation on the other are subject of the following chapters.

Chapter 2

Derivation of Navier–Stokes solvers

We consider numerical solution techniques for the following (laminar) Navier–Stokes systems, all endowed with the incompressibility constraint:

nonstationary Navier–Stokes equations,

$$\mathbf{u}_t - \nu\Delta\mathbf{u} + \mathbf{u}\cdot\nabla\mathbf{u} + \nabla p = \mathbf{f} \quad , \quad \nabla\cdot\mathbf{u} = 0 , \tag{2.1}$$

generalized **stationary Navier–Stokes** equations,

$$\alpha\mathbf{u} - \nu\Delta\mathbf{u} + \mathbf{u}\cdot\nabla\mathbf{u} + \nabla p = \mathbf{f} \quad , \quad \nabla\cdot\mathbf{u} = 0 , \tag{2.2}$$

linearized Navier–Stokes/**Oseen** equations (with given $\mathbf{U}$),

$$\mathbf{u}_t - \nu\Delta\mathbf{u} + \mathbf{U}\cdot\nabla\mathbf{u} + \nabla p = \mathbf{f} \quad , \quad \nabla\cdot\mathbf{u} = 0 , \tag{2.3}$$

(stationary or nonstationary) **Stokes** equations,

$$\mathbf{u}_t - \nu\Delta\mathbf{u} + \nabla p = \mathbf{f} \quad , \quad \nabla\cdot\mathbf{u} = 0 , \tag{2.4}$$

divergence-free L^2**-projections**,

$$\mathbf{u} + \nabla p = \mathbf{f} \quad , \quad \nabla\cdot\mathbf{u} = 0 . \tag{2.5}$$

Furthermore, the numerical schemes which will be proposed later, work analogously in the case of inhomogeneous data g for the incompressibility constraint, for example

$$\mathbf{u}_t - \nu\Delta\mathbf{u} + \mathbf{u}\cdot\nabla\mathbf{u} + \nabla p = \mathbf{f} \quad , \quad \nabla\cdot\mathbf{u} = g. \qquad (2.6)$$

All techniques in the following sections can be applied to these described classes of incompressible flow problems. Additionally, the partial differential equations have to completed with several "standard" features which have to be precisely defined with regard to the current application. These are:

- viscosity parameter ν,

- domain $\Omega \subset \mathbf{R}^d, d = 2$ or 3, with boundary $\partial\Omega$,

- boundary values for $\mathbf{u}$ (Dirichlet boundary conditions) or for $\mathbf{u}$ and p (natural b.c.'s),

- time intervall $(t_0, T]$,

- initial condition $\mathbf{u}_0$ (and p_0 if necessary) for $t = t_0$,

- given convective direction $\mathbf{U}$ in the linearized case.

Moreover, the prescribed settings for the solution of these problems can include topics as:

- a "strategy" for choosing underlying spatial meshes with respect to Ω.

- a "strategy" for selecting the time steps with respect to $(t_0, T]$.

What is the meaning of these "parameter settings" and "strategies" in the case of the Navier–Stokes equations? The idea behind is that we split the complete numerical solution process of such flow problems into two topics:

1. *the* **outer control** *part ("$u - u_h$") which performs all discretization tasks and which defines the listed items. This may also include error control mechanisms or more complex frameworks, as for instance turbulence models, free surface or (weakly) compressible problems which are reduced – among others in an operator splitting – to the solution of laminar incompressible subproblems of the described type, for instance as preconditioners. To make this point clear: Only this* outer *part is responsible for the adaptivity with respect to the mesh and hence for the optimality of the difference between the discrete solution "u_h" (if solved sufficiently accurate in the following* inner *part) and the exact solution "u".*

2. *the* **inner solution** *part (* "$u_h - u_h^l$" *) which has the exclusive task to provide approximate solutions of such (discretized) Navier–Stokes-like systems for a given framework, including prescribed discretization strategies in space and time. This inner part can provide the optimal solution for a fixed mesh (by the* outer *part), but cannot give arbitrarily accurate solutions "*u_h^l*" with respect to the exact solution "*u*".*

Since the understanding of this "splitting of tasks" is essential for the following considerations, we try to explain these two different topics and the philosophy behind more in detail.

The outer "discretization demon":

This topic can be viewed as a "demon" which splits the complete flow problem into different subproblems. For instance, let us assume that we have designed a spatial mesh, that we have applied certain discretization techniques for pressure and velocity, and if necessary for other unknown quantities occuring in the continuous problem. Further, that we have already prescribed boundary values, initial conditions and all other input parameters which are necessary to treat the resulting discrete problem. Then the subsequent solution process will, if possible at all, provide approximate discrete solutions for this framework.

Several questions may occur: What about the *numerical* error (that means how accurate is our approximate solution in comparison to the exact discrete solution of the defined framework)? What about the *real* error (how accurate is the approximate solution compared with the unknown exact solution of the continuous problem)? And – if necessary – how can the numerical or real error be diminished? The result of this *outer control* may be that a new spatial mesh is generated which is locally or globally refined or coarsened, or that a new sequence of locally varying time steps is prescribed. Then, the *inner solution* process is initiated to do a restart. This is, for instance, the typical run of an a posteriori error control mechanism with adaptive mesh modification which is directed by mathematical or by "handwaving" issues. Another background may be the embedding of the laminar incompressible Navier–Stokes equations into more complex frameworks, as for instance the *low Reynolds number* $k - \epsilon$ *turbulence model* (see [88]).

In this case, additional equations for the *turbulence kinetic energy k* and *dissipation rate* ϵ and the corresponding relations with respect to **u**, p and the turbulent viscosity are given. A typical approach is to split the complete system into incompressible laminar subproblems for **u** and p, and transport–diffusion equations for k and ϵ, controlled by an overall outer iteration. This leads to a sequence of standard Navier–Stokes-like systems or linearized versions which have to be solved, with different right hand sides and viscosity parameters, and probably on different meshes. However, this solution process is the task of the "inner" part.

The inner "solution engine":

As indicated above, the task of this topic is only to provide an approximate solution of a discretized Navier–Stokes-like subproblem as introduced at the beginning. The problem to be solved is not a general one, but it is precisely fixed. It may include a given mesh, a prescribed sequence of time steps (in the nonstationary case), a stabilization technique for the convective term of *upwind* or *streamline–diffusion* type, boundary values and many more input parameters.

However, there is one important feature: the solver "responds" with an approximate solution due to the given parameters and discretization strategies only. The differences between the resulting approximation of the prescribed framework and the exact continuous solution are neglected. To estimate the quality of the approximate solution with respect to these differences, this is the task of the "outer" part.

The motivation for this splitting of tasks is obvious. While the first outer part is the responsible "demon" for the convergence and the accuracy with respect to the overall problem, the second inner part is nothing else than a "solver engine" for a fixed configuration. Consequently, the three major requirements for numerical algorithms – **accuracy**, **robustness** and **efficiency** – are differently distributed on these two tasks.

The question of accuracy with respect to the underlying continuous problem is mainly treated by the outer control part. Nevertheless, efficiency and robustness depend obviously on this part, too. In contrast, the speed and stability behaviour of solution schemes and their implemented counterparts are mainly directed by the "solver engine". We lay our particular emphasis on this "solver part" and neglect at the moment the question of error control and accuracy, mainly by the following reasons:

Before one starts with fully adaptive error control for partial differential equations or with defect correction/operator splitting techniques for more complex CFD problems, one should be able to guarantee that the related subproblems for arbitrary configurations can be solved at all. What is the use of adaptive grid refinement if the corresponding discrete linear or nonlinear systems cannot be successfully treated or only with huge numerical cost? Before these "new" complex tasks are tackled, the laminar homeworks have to be under control.

Following this motivation, this book mainly concentrates to provide appropriate discretization and solution tools for laminar incompressible flow problems as introduced at the beginning of this chapter. The derived techniques shall perform (almost) independently of:

- the complexity of the geometry and the shape of the underlying mesh
- the size of involved time stepping parameters
- initial and boundary values and right hand side vectors
- the viscosity parameter (?)

In particular, the last item requires some further comments. It is clear that we cannot provide numerical techniques which are independent of the Reynolds number, especially with respect to nonlinear effects ("turbulence"). However for a certain flow regime, we can provide numerical techniques which work efficiently for moderately large and also for small Reynolds numbers. Then, these techniques can be at least applied to higher Reynolds number flows, hereby requiring stabilization techniques as *upwinding* or *streamline diffusion*. What we want to guarantee is that for a prescribed framework the discrete equations can be solved! Then, the task of the "outer control part" is to provide estimates for the quality of this discrete approximation.

What we want to exclude is that for a given viscosity parameter a multigrid tool for solving the underlying convection–diffusion problems is diverging, or that the pressure correction scheme explodes unexpectedly for very small Reynolds numbers. Moreover, we cannot accept that a linear solver fails with respect to complex geometrical details of the domain Ω or on anisotropic meshes. That is what we claim robustness and efficiency in this "discrete Black Box" solver context, and this is the aim of our search for optimality in this book.

Although the following sections concentrate mainly on such solution strategies, we expect the *solution control part* to be a very important topic in the future, for mathematicians due to many new mathematical problems as well as for CFD users due to the expected reliability of computer simulations. However, these are still dreams for the future and first of all, the mathematical basic research and related numerical calculations for testing and improving the recent theoretical results have to be performed. A short summary of the recent methodology and the overall philosophy behind error control and adaptivity are subject of the Chapter 3, 'Other mathematical components'.

Coming back to the demand for robust and efficient solution techniques, first of all we try to find a generalization of existing algorithms. Having found this general description tool, we will be able to look for improvement strategies. The key for success will be to understand first the advantages and disadvantages of the individual schemes and the corresponding dependencies on parameters and certain frameworks, and then to combine and modify them with the aim to improve their quality. There will be two major results:

1. We give explicit advices for CFD users, concerning the question which of the available schemes should be employed, under which circumstances. These include aspects as low/high viscosities, constant/locally varying time scales and uniform/locally anisotropic spatial meshes.

2. We introduce very robust and efficient new schemes which represent the "best possible" compromise and which may serve as "discrete Black Box" solvers in the meaning above.

2.1   Mathematical description of Navier–Stokes solvers

We examine in the following numerical techniques for the Navier–Stokes equations,

$$\mathbf{u}_t - \nu \Delta \mathbf{u} + \mathbf{u} \cdot \nabla \mathbf{u} + \nabla p = \mathbf{f} \quad , \quad \nabla \cdot \mathbf{u} = 0 \quad , \quad \text{in } \Omega \times (0, T], \quad (2.7)$$

or similar problems which were introduced in (2.1) – (2.6) at the beginning of this chapter. As motivated in the previous section, we have to keep in mind that the main task of the subsequent approaches is to provide approximate solution due to certain frameworks which may already include discretization techniques. The question of accuracy of the provided approximate result with respect to the continuous solution is not subject of the following solution part. This control process is task of an "outer demon" as described above.

We concentrate on the 2D case which has proven to be representative for the 3D problems through our theoretical results, numerical and also computational experience. In particular, the structure of our 3D code is almost identical to the 2D version which both are part of the FEATFLOW package (see [111]) such that the restriction on 2D is justified in our case. Many important components as *time discretization schemes, spatial discretization, finite element spaces, Stokes elements, multigrid tools, nonlinear solvers*, etc., will be shortly introduced in this section to allow a better understanding by providing explicit examples. However, many more additional algorithmic tools and components are necessary for a fully developed Navier–Stokes solver, and some of them are explained more in detail in the following Chapter 'Other mathematical components'.

Apart from schemes like *discontinuous space–time Galerkin methods* (see Johnson [61]) and *characteristic methods* (see Pironneau [78]) which will be incorporated later, the common solution approach is a separate discretization in space and time: We first (semi-) discretize in time by one of the usual methods known from the treatment of ordinary differential equations, such as the Forward or Backward Euler-, the Crank–Nicolson- or Fractional–step–θ–scheme. Then, we obtain a sequence of generalized stationary Navier–Stokes equations with prescribed boundary values for every time step, which read in the case of the *One-step θ–schemes*:

Given $\mathbf{u}^n$ and the time step $k = t_{n+1} - t_n$, then solve for $\mathbf{u} = \mathbf{u}^{n+1}$ and $p = p^{n+1}$

$$\frac{\mathbf{u} - \mathbf{u}^n}{k} + \theta[-\nu\Delta\mathbf{u} + \mathbf{u} \cdot \nabla\mathbf{u}] + \nabla p = \mathbf{g}^{n+1}, \quad \nabla\cdot\mathbf{u} = 0 \quad , \quad \text{in } \Omega \qquad (2.8)$$

with right hand side

$$\mathbf{g}^{n+1} := \theta\mathbf{f}^{n+1} + (1 - \theta)\mathbf{f}^n - (1 - \theta)[-\nu\Delta\mathbf{u}^n + \mathbf{u}^n \cdot \nabla\mathbf{u}^n]. \qquad (2.9)$$

The parameter θ has to be chosen depending on the time–stepping scheme, e.g., $\theta = 1$ for the Backward Euler, or $\theta = 1/2$ for the Crank–Nicolson-scheme. The pressure term $\nabla p = \nabla p^{n+1}$ may be replaced by $\theta\nabla p^{n+1} + (1 - \theta)\nabla p^n$ but, with appropriate postprocessing, both strategies lead to solutions of the same accuracy. In all cases, we end up with the task of solving, in each time step, a nonlinear saddle point problem which has to be discretized in space.

Given $\mathbf{u}^n$, parameters $k = k(t_{n+1})$, $\theta = \theta(t_{n+1})$ and $\theta_i = \theta_i(t_{n+1}), i = 1, \ldots, 3$, then solve for $\mathbf{u} = \mathbf{u}^{n+1}$ and $p = p^{n+1}$

$$[I + \theta k N(\mathbf{u})]\mathbf{u} + k\nabla p = [I - \theta_1 k N(\mathbf{u}^n)]\mathbf{u}^n + \theta_2 k\mathbf{f}^{n+1} + \theta_3 k\mathbf{f}^n, \quad \nabla\cdot\mathbf{u} = 0.$$
$$\tag{2.10}$$

Here and in the following, we use the compact form for the diffusive and convective part

$$N(\mathbf{v})\mathbf{u} := -\nu\Delta\mathbf{u} + \mathbf{v}\cdot\nabla\mathbf{u}. \tag{2.11}$$

For the spatial discretization, we choose a finite element approach. However other approaches as finite volumes, finite differences or spectral methods are possible, too. In setting up a finite element model of the Navier–Stokes equations, one starts with a variational formulation. On the finite mesh T_h (triangles, quadrilaterals or their analogues in 3D) covering the domain Ω with local element width h, one defines polynomial trial functions for velocity and pressure.

These spaces H_h and L_h should lead to numerically stable approximations, as $h \to 0$, i.e., they should satisfy the *Babuška–Brezzi condition* [36] with a mesh–independent constant γ,

$$\min_{p_h \in L_h} \max_{\mathbf{v}_h \in H_h} \frac{(p_h, \nabla\cdot\mathbf{v}_h)}{\|p_h\|_0 \|\nabla\mathbf{v}_h\|_0} \geq \gamma > 0. \tag{2.12}$$

However, due to certain stabilization techniques (see [58]), many other approaches, including also unstable) equal order spaces, are allowed. Our favourized candidate is a quadrilateral element which (in 2D) uses piecewise *rotated bilinear* shape functions for the velocities, spanned by $\langle x^2 - y^2, x, y, 1\rangle$, and piecewise constant pressure approximations (see Figure 2.1).

The nodal values are the mean values of the velocity vector over the element edges, and the mean values of the pressure over the elements rendering this approach "nonconforming". This element is the natural quadrilateral analogue of the well–known triangular Stokes element of Crouzeix–Raviart [24] and can easily be defined in three space dimensions. A convergence analysis is given in [84] and computational results are reported in [93] and [94].

This element pair has several important features. It admits simple upwind strategies which may lead to matrices with certain M–matrix properties. Further, efficient multigrid solvers are available which work satisfactorily over the whole range of relevant Reynolds numbers, $1 \leq Re \leq 10^5$, and also on nonuniform meshes. In Section 2.5 we show by a complexity analysis that this

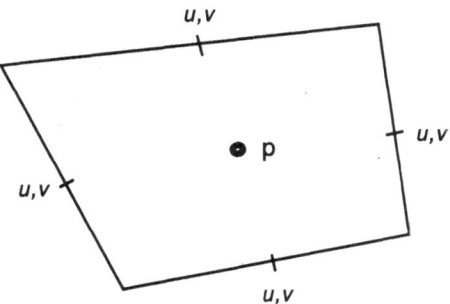

Figure 2.1: Nodal points of the nonconforming finite element pair $\tilde{Q}1/Q0$

pair of elements is most efficient in the case of highly nonstationary flows. In combination with the global *Multilevel Pressure Schur Complement* (MPSC) techniques, it works very robust and efficient in a multigrid code on highly stretched and anisotropic grids, too. These aspects and the relationship to other spatial discretization schemes will be discussed later. Using the same symbols $\mathbf{u}$ and p also for the coefficient vectors in the nodal representation for the functions $\mathbf{u}$ and p, the discrete version of problem (2.10) may be written as a coupled (nonlinear) algebraic system of the form:

Given $\mathbf{u}^n$, right hand side $\mathbf{g}$ and time step k, then solve for $\mathbf{u} = \mathbf{u}^{n+1}$ and $p = p^{n+1}$

$$Su + kBp = \mathbf{g}, \quad B^T \mathbf{u} = 0, \tag{2.13}$$

with matrix S and right hand side $\mathbf{g}$ such that

$$Su = [M + \theta k N(\mathbf{u})]\mathbf{u} \quad , \quad \mathbf{g} = [M - \theta_1 k N(\mathbf{u}^n)]\mathbf{u}^n + \theta_2 k \mathbf{f}^{n+1} + \theta_3 k \mathbf{f}^n . \tag{2.14}$$

Here, M is the *mass* matrix and $N(.)$ the matrix containing the diffusive and convective parts corresponding to the nonlinear form in (2.11). For dominant transport the convection part may include some stabilization, for instance, certain upwind mechanism (see [101]). B is the *gradient* matrix, and $-B^T$ the transposed *divergence* matrix. With M_l we denote the lumped (velocity) mass matrix which is a diagonal matrix. Two possible approaches for solving these discrete nonlinear and indefinite problems are:

1) We first treat the nonlinearity by an outer nonlinear iteration of fixed point- or quasi–Newton type or by a linearization technique through extrapolation in time, and we obtain linear subproblems (Oseen equations) which can be solved by a direct coupled or a splitting approach separately for velocity and pressure.

2) We first split the coupled problem and obtain definite problems in u (Burgers equations) as well as in p (Pressure–Poisson problems). Then we treat the nonlinear problems in u by an appropriate nonlinear iteration or a linearization technique.

The question of how often we have to perform these outer splitting steps leads to additional "free parameters" such that altogether the following topics have to be taken into account:

- **treatment of the nonlinearity:** fully nonlinear solution by Newton-like methods, linearization via extraploation in time, number of nonlinear steps, resp., nonlinear convergence criteria, type of linearization, etc.

- **treatment of the incompressibility:** direct coupled approach with simultaneous treatment of velocity and pressure, operator splitting approach via pressure correction/projection ansatz, choice of pressure correction subproblems, etc.

- **complete outer control:** convergence criteria for the overall outer iteration, number of splitting steps, convergence control, embedding into multigrid, etc.

These different choices lead to a large variety of schemes all of which are occuring in practice since years: projection schemes á la Chorin [22], Van Kan [115] or Gresho [41], pressure correction schemes with SIMPLE iteration ([29],[76]), fractional step methods, explicit advance of the momentum, semi–implicit treatment of the convective part, characteristic methods for the nonlinearity, fully implicit Galerkin approach, direct multigrid with SIMPLE or Vanka–smoother, Crank–Nicolson or multi-stage Runge–Kutta for the time discretization, spectral elements, finite elements/differences/volumes and so on. Theoretical considerations can provide some ideas concerning stability of these schemes, convergence rates for subproblems, necessary time steps or qualitative behaviour for large Reynolds numbers, but a complete analysis or quantitative prediction of real CPU cost is not possible today.

One possibility to make a judgement is to perform numerical tests, at least for some classes of problems which seem to be representative. This helps in figuring out numerical and computational problems of these schemes, being theoretically predictable or not; but it does not improve automatically these methods or helps in designing some better approaches. Our approach for a

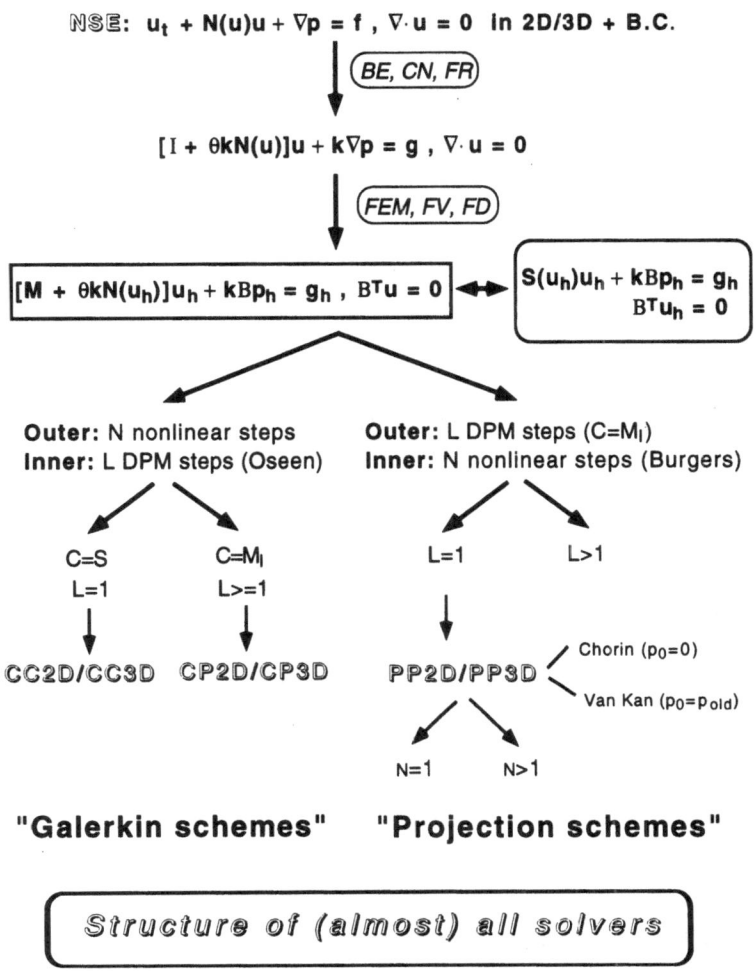

Figure 2.2: The Navier–Stokes tree

generalized ansatz which helps essentially in first understanding and then in improving existing Navier–Stokes solution schemes, is illustrated in the following **Navier-Stokes tree**. Many schemes follow the explained algorithmic approach of separately discretizing in space and time and can be represented by this flow chart, leading to many of the methods mentioned above. Indeed, most of the existing Navier-Stokes solvers can be interpreted by this tree structure. So, it is a very powerful tool since on its basis several *subtrees* can be combined in a new way with the aim to obtain better run-time characteristics.

The main idea from this flow chart is the described strict separation between the spatial (by finite elements/differences/volumes or spectral elements) and the temporal discretization (Crank–Nicolson, Explicit/Implicit Euler, Fractional Step, BDF, Runge-Kutta, etc.) from the task of solving steady generalized Navier–Stokes–like problems

$$S\mathbf{u} + kBp = \mathbf{g} \quad , \quad B^T\mathbf{u} = 0 \,. \tag{2.15}$$

These are thought to be discrete versions of the equations which have been introduced at the beginning of this chapter, (2.1) – (2.6), see also (2.13) and (2.14). Additionally, also the previously excluded schemes like *discontinuous space–time Galerkin methods* [61] and *characteristic methods* [78] are now included since they involve the solution of such coupled problems in each time step.

Having decoupled the discretization from the solution process we see immediately that the topics *treatment of the nonlinearity, treatment of the incompressibility* and *complete outer control* can be interpreted as components of a special solver for the discrete coupled nonlinear problems of type (2.15), neglecting the mixture with discretization techniques or operator splitting approaches. As a consequence, we have to analyse the behaviour of these different prescribed techniques as certain "solver engines" to figure out their pros and cons.

2.2 The general "Pressure Schur Complement" approach

From now on, we concentrate on numerical solution techniques for the following discrete coupled problems, both nonlinear or linear,

$$S\mathbf{u} + kBp = \mathbf{g} \quad , \quad B^T\mathbf{u} = 0 \,. \tag{2.16}$$

The (block-) matrix S is acting on the unknown velocity vector $\mathbf{u}$, and the matrices B and $-B^T$ are discrete analogues of the operators ∇ and $\nabla\cdot$. The time step parameter k in front of Bp is not of explicit importance and can be neglected by scaling the pressure p in each time step. However, the structure of the matrix S is significant, namely defined by

$$S\mathbf{u} := [\alpha M + \theta_1 \nu k L + \theta_2 k K(\mathbf{u})]\mathbf{u}, \tag{2.17}$$

in which case θ_i and k are parameters according to the time stepping scheme. The parameter α is set to 1 in the nonstationary case, and $\alpha = 0$ for stationary problems in which case θ_i and k can be set to 1. The matrix M represents the *mass matrix* which might be a diagonal matrix in many applications, and L is the discretized Laplacian. Furthermore, K corresponds to the convective term and may depend on a given vector $\tilde{\mathbf{U}}$ (in the linear case of Oseen problems) or on the unknown vector $\mathbf{u}$ itself (in the fully nonlinear case).

Proposition 1 *If we assume that the operator S^{-1} exists, then the original problem (2.16) in primary formulation is equivalent to the following scalar pressure Schur complement formulation*

$$B^T S^{-1} (Bp - \frac{1}{k}\mathbf{g}) = 0. \tag{2.18}$$

Once the pressure p is known, the corresponding velocity vector $\mathbf{u}$ satisfies

$$\mathbf{u} = S^{-1}(\mathbf{g} - kBp). \tag{2.19}$$

Let us remark, that we restrict to the homogeneous case, $B^T\mathbf{u} = 0$, by technical reasons only. The corresponding modifications for inhomogeneous data are easily performed. Furthermore, the same techniques can be applied to other mixed equations, especially to "harder" coupled problems as the incompressible *low Reynolds number $k - \epsilon$* turbulence model, for example. In fact, *Schur complement* techniques are widely used in numerics, particularly if one part of unknowns can be formally eliminated. For instance, these approaches are currently applied for the algorithmic design of *domain decomposition* techniques on parallel computers, if coupling conditions at processor subdomain interfaces have to be derived. However, their range of applications is very general and almost infinite.

We will concentrate in the following on the case of linear problems. Then, the corresponding (linear) *pressure Schur complement* equation reads

$$B^T S^{-1} Bp = \frac{1}{k} B^T S^{-1} \mathbf{g}. \tag{2.20}$$

We have formally changed from the solution process of a coupled system of equations to a scalar problem in p only, in which case the velocity vector u can be achieved from the pressure p. However, from a computational point of view, this does not mean a big saving since in general the corresponding *Schur complement* matrix cannot be calculated explicitly (S^{-1} is a full matrix!). And moreover, the application of S^{-1} enforces the solution of velocity subproblems. However, in combination with iterative techniques it is often sufficient to apply the matrix–vector multiplication. This leads in our context to applications of S^{-1} to a given vector, that means to solutions of auxiliary problems involving S.

What we have gained so far is a starting point for numerical analysts who can now apply their knowledge about efficient iterative schemes for scalar problems. We forget all the difficulties behind B, S^{-1} and B^T, and we simply define

$$A := B^T S^{-1} B \quad , \quad f := \frac{1}{k} B^T S^{-1} \mathbf{g} . \tag{2.21}$$

Then, we are faced with solving a (standard) scalar problem of the following type.

$$\boxed{Ap = f}$$

One of the most general possibilities to solve such problems is to perform a *preconditioned Richardson* iteration. We call it in the following the **basic iteration** for the *pressure Schur complement* equation (2.20), with C^{-1} being an appropriate preconditioner for A, resp., for $B^T S^{-1} B$.

Basic iteration for the pressure Schur complement equation:

Given: Iterate p^{l-1}

Perform: One relaxation step to obtain p^l

$$
\begin{aligned}
p^l &= p^{l-1} - C^{-1}(Ap^{l-1} - f) \\
    &= p^{l-1} - C^{-1}\left(B^T S^{-1} B p^{l-1} - \frac{1}{k} B^T S^{-1} \mathbf{g}\right)
\end{aligned}
$$

Our task will be to analyse this *basic iteration* and – as usual in Numerical Linear Algebra – to accelerate it significantly. But first, let us stop for a moment and ask the question:

> '*What have we gained? Are there (well-known) methods which can be interpreted as* pressure Schur complement *iteration schemes? The answer is YES.*'

We will show that (at least) the following well-known solution schemes (plus variants) belong to the class of *pressure Schur complement* (PSC) methods:

- Projection methods

- Pressure correction schemes

- SIMPLE

- Uzawa iterations

- Vanka smoother

We start with the *projection schemes*. Following the articles of Chorin [22], Van Kan [115] or Gresho [41], they usually prescribe the following substeps in each time step:

Projection methods in classical notation:

Given: $\mathbf{u}(t_n), p(t_n)$ at time level t_n

Perform: 3 substeps to obtain $\mathbf{u}(t_{n+1})$ and $p(t_{n+1})$

1. Solve nonlinear Burgers- or linearized convection–diffusion/Helmholtz equations for an intermediate velocity $\tilde{\mathbf{u}}$. The right hand side $\tilde{\mathbf{g}}$ contains the original load vector (or modified related to a chosen linearization technique) and an "old" pressure approximation (with $\nabla_h \sim B$ denoting the discrete gradient).

$$S\tilde{\mathbf{u}} = \tilde{\mathbf{g}} - k\nabla_h p_{old}$$

2. Solve an "update" equation for the pressure ("Pressure–Poisson problem") with the (discrete) divergence of the intermediate velocity $\tilde{\mathbf{u}}$ on the right hand side. We denote by Δ_h the "Pressure–Poisson" operator which is in general a discrete Laplacian, and by $\nabla_h \cdot \sim B^T$ the discrete divergence operator.

$$\Delta_h q = \frac{1}{k}\nabla_h \cdot \tilde{\mathbf{u}}$$

3. Update the new pressure $p(t_{n+1})$ and the new "discretely divergence–free" velocity $\mathbf{u}(t_{n+1})$. The parameter σ is in the range $(0, 2]$, typically $\sigma = 1$.

$$p(t_{n+1}) = p_{old} + \sigma q \quad , \quad \mathbf{u}(t_{n+1}) = \tilde{\mathbf{u}} - k\nabla_h q$$

If we choose for each time step the initial pressure $p_{old} = 0$ we end up with a (discrete) analogue to the scheme first proposed by Chorin [22]. If we select $p_{old} = p(t_n)$, we obtain a 2nd order scheme due to Van Kan [115]. Even higher order extrapolation backwards in time is possible, for instance the choice $p_{old} = 2p(t_n) - p(t_{n-1})$. However our numerical experience with such potentially cubic schemes is very disadvantageous, as soon as varying time steps and especially non-equidistant spatial meshes are utilized (see the numerical section). This may contradict to some other papers which can be found in the literature, but those results mostly refer to uniform spatial discretizations with equidistantly small time steps due to explicit techniques. In our experience they are not working well for complex CFD configurations.

This classical projection approach can be reformulated in the following way. We write the same algorithm for time level t_{n+1}, but now beginning at the end, with step 3. Then, we obtain the following formulation for the pressure $p(t_{n+1})$ only:

Projection methods in pressure Schur complement notation:

$$
\begin{aligned}
p(t_{n+1}) &= p_{old} + \sigma q \\
&= p_{old} + \sigma \Delta_h^{-1} \frac{1}{k} \nabla_h \cdot \tilde{\mathbf{u}} \\
&= p_{old} + \sigma \Delta_h^{-1} [\nabla_h \cdot S^{-1} (\frac{1}{k} \tilde{\mathbf{g}} - \nabla_h p_{old})] \\
&= p_{old} - \sigma \Delta_h^{-1} [\nabla_h \cdot S^{-1} (\nabla_h p_{old} - \frac{1}{k} \tilde{\mathbf{g}})]
\end{aligned}
$$

These reformulations show that indeed one projection step can be interpreted as being exactly 1 (!) iteration of our *basic iteration* with the special preconditioner Δ_h^{-1}. Further, the corresponding velocity vector $\mathbf{u}(t_{n+1})$ is chosen to satisfy the discrete continuity equation, hereby neglecting the momentum equation with respect to $p(t_{n+1})$ from (2.19).

In a very similar way, "segregated" solution approaches involving SIMPLE–like or other *pressure correction* techniques (see for an overview [29] and the literature cited therein) can be reformulated as *pressure Schur complement* schemes. The main difference to the previous *projection* methods is that these schemes are directly applied as iterative solvers for the *Schur complement* equation, related to stationary <u>and</u> nonstationary problems. In contrast to the projection schemes which apply exactly 1 basic iteration in each time step, they are repeated several times. Additionally the corresponding velocity field can be adjusted to satisfy the momentum equation via (2.19), or an analogous treatment as performed by the *projection methods* is possible.

The choice of the preconditioner C^{-1} may lead to another difference. For many *pressure correction* and also for "discrete" projection schemes (Gresho [41]) the appropriate preconditioner for the *Schur complement* operator $B^T S^{-1} B$ is constructed on discrete level, as

$$C := B^T \tilde{S}^{-1} B . \tag{2.22}$$

In this case, $\tilde{S}$ is often chosen as diagonal matrix related to the (linear) operator S. Possible choices for $\tilde{S}$ are the exact diagonal of S or of a part of S only, or the row sums from S may be taken,

$$\tilde{S} = \text{diag}\,(S) \quad \text{or} \quad \tilde{S} = \text{diag}\,(\text{part}\,(S)) \quad \text{or} \quad \tilde{S} = \text{diag}\,(\text{row sum of } S) . \tag{2.23}$$

All these modifications (with damping strategies) result in different versions of the original SIMPLE scheme, but all of them can be interpreted as *pressure Schur complement* schemes.

Another class of schemes which falls into the class of *pressure Schur complement* schemes are Uzawa-like iterations. The corresponding preconditioner is the (damped) identity matrix or, in finite element schemes, the (lumped) mass matrix M_p in the pressure space is chosen,

$$C := \sigma I \quad or \quad C := \sigma M_p . \tag{2.24}$$

All these different approaches can be interpreted as special *pressure Schur complement* methods. To be more precise, they all belong to the *global pressure Schur complement* schemes (global MPSC) since – on discrete or continuous level – the major idea is to construct globally defined operators as preconditioners for the complete *Schur complement* operator $B^T S^{-1} B$.

Further variants are the *local pressure Schur complement* schemes (local MPSC) which follow the strategy of approximating locally the *pressure Schur complement* operator. On certain "parts" Ω_i of the domain Ω, or more general for certain subsets Ω_i of the complete number of unknowns, we invert exactly the complete operators $B^T_{|\Omega_i} S^{-1}_{|\Omega_i} B_{|\Omega_i}$, restricted onto Ω_i,

$$C^{-1} := \sum_i (B^T_{|\Omega_i} S^{-1}_{|\Omega_i} B_{|\Omega_i})^{-1} . \tag{2.25}$$

These subdomains – better called "patches" – are usually related to the underlying mesh and can consist of single elments or clusters of elements. These "local solvers" are embedded in an outer Jacobi–like or Gauß–Seidel–like process to obtain a global relaxation scheme. This technique comes very near to *domain decomposition* approaches, but is more flexible concerning boundary conditions since their construction is performed on discrete level. A typical example for these *local Schur complement* techniques is the mentioned *Vanka* smoother [114] and related schemes.

We can generalize all these seemingly different schemes by the subsequent approach. The essential point is to construct *additive preconditioners* of the following type,

$$C^{-1} := \sum_i \alpha_i C_i^{-1}$$

applying techniques of constructing "operator splitting" (global MPSC) or "domain decomposition" (local MPSC) operators. The strategy in the subsequent sections can be described as follows: We start with the derivation of several preconditioners and analyse their efficiency and robustness with respect to different parameter settings and other circumstances.

This step will provide us with improved basic iterations for solving the *pressure Schur complement* equations. If we compare with standard numerical approaches for solving iteratively systems of equations, this corresponds to the step of deriving simple relaxation schemes analogously to Jacobi, Gauß–Seidel, SOR or ILU methods.

The next step concerns the "classical" problem in Numerical Linear Algebra how to improve the convergence behaviour of simple basic iterations. It is well-known that the "right" choice of the preconditioner C and corresponding relaxation parameters leads already to an improvement, but usually this success in efficiency and robustness is not satisfying in practise. The essential idea is to accelerate these basic iteration schemes by applying them as:

- *preconditioners* in Krylov space methods, for instance in the conjugate gradient (CG) method, the BiCGSTAB [113] or the GMRES [86],

- *smoothers* in standard multigrid schemes.

We prefer the multigrid approach, since theoretical and particularly numerical results demonstrate that multigrid methods are the most efficient and robust "solver engines" in Numerical Linear Algebra, if they are applied "in the right way". There is often some misunderstanding with multigrid schemes: *multigrid* is <u>not</u> a "Black Box" tool which automatically performs best for all problems. If the optimal performance shall be obtained, one has to adopt all components (*smoothing, grid transfer, control of coarse grid solution*, etc., see Chapter 3) to the current problem. There are certain "Black Box" multigrid techniques for Poisson–type problems for example, but these are usually not optimal for other problems as Stokes- or elasticity equations, particularly if non-standard finite elements spaces are involved.

In the following, we will explain how "optimal" multigrid techniques for solving the derived *pressure Schur complement* problems may look like. The corresponding smoothers are the *pressure Schur complement* iterations, while all other components are standard with respect to the multigrid solution of (velocity) transport–diffusion or Pressure–Poisson-like problems and their corresponding discretizations. They will contain some new mathematical ideas and in fact they differ in a small but significant way from some other well-known approaches, if for instance SIMPLE is used inside a multigrid approach.

Beside the mathematical success that a generalization and classification of many existing and even new solution schemes can be provided, there will be another more "practical" result. We can offer the following explicit advices at least to those code developers who perform *projection* or *pressure correction* schemes:

> *'Take your "projection/pressure correction ingredients" and add some slight modifications to increase significantly the performance of your CFD–tool. The cost for these additional implementations are minimal.'*

2.3   "Global Multilevel Pressure Schur Complement"

The major idea behind the global *multilevel pressure Schur complement* schemes (global MPSC) is the construction of globally defined operators as preconditioners for the *pressure Schur complement* formulation related to the matrix

$$A := B^T S^{-1} B\,. \tag{2.26}$$

We explicitly assume that the matrix S has the following form

$$S := \alpha M + \theta_1 \nu k L + \theta_2 k K, \tag{2.27}$$

with parameters θ_i, α, k (depending on the spatial and temporal discretization) and ν (viscosity), and matrices M (representing the *mass matrix*), L (the discretized *Laplacian*) and K (corresponding to the convective term).

There are two possibilties to approximate this *Schur complement* operator A: by a *factorization* or by an *additive* approach. Let us start with the additive technique. Since in general one single operator C can be hardly constructed which takes into account <u>all</u> three components of A, resp., of S, with respect to the underlying diffusive (L), convective (K) and reactive (M) terms, we first develop preconditioners which are related to each of the three subproblems. That means, we construct "optimal" operators for the limit c ases of dominating diffusive (Stokes problem), convective (incompressible Euler problem) or reactive character (divergence–free L^2 projection, resp., time step k being "infinitely small"):

$$C^{-1} = \alpha_R A_R^{-1} + \alpha_D A_D^{-1} + \alpha_K A_K^{-1} \tag{2.28}$$

We claim that

A_R is an "optimal" (reactive) preconditioner for $B^T M^{-1} B$,

A_D is an "optimal" (diffusive) preconditioner for $B^T L^{-1} B$,

A_K is an "optimal" (convective) preconditioner for $B^T K^{-1} B$.

The meaning of "optimality" has to be defined more precisely: The ideal state would be if the partial preconditioners were direct solvers with respect to the underlying subproblem. In fact, this may be true for the fully "reactive" case, with $A = B^T M^{-1} B$. However, if these preconditioners are applied as smoothers in a multigrid context and if the resulting convergence behaviour is independent of outer parameters and the underlying mesh, then this is already sufficient as criterion for optimality in this context.

If we obtain that for the corresponding subproblem the preconditioner, resp., the smoother in the multigrid framework leads to bounded and "good" convergence rates, for instance $\rho_{MG} \leq 0.5$ on meshes containing arbitrarily large *aspect ratios*, then this scheme is called "optimal". Indeed, this seems to be satisfied by the diffusive preconditioner A_D. The reason for this weaker optimality criterion is caused by the construction process of the separate preconditioners: while the strict application of transformations on <u>discrete</u> level leads to such exact solvers, the use of properties of the underlying <u>continuous</u> partial differential operator misses this result, but we still can guarantee the weakened optimality. Thus, the way of designing *global MPSC* schemes can be characterized as

'Construction of **globally** defined **additive** preconditioning operators by **discrete** and **continuous** arguments.'

For the following derivation of separate preconditioners for $B^T S^{-1} B$ we repeat that the velocity matrix S is assumed to have the structure

$$S := \alpha M + \theta_1 \nu k L + \theta_2 k K .$$

Then, we construct preconditioning operators in the meaning of above for the distinct parts of S, that means,

$B^T M^{-1} B$ as preconditioner for the reactive part (M) in S

$B^T L^{-1} B$ as preconditioner for the diffusive part (L) in S

$B^T K^{-1} B$ as preconditioner for the convective part (K) in S

and we finish with an approximate preconditioning operator which is given in additive form, corresponding to the three different parts of the complete matrix S,

$$[B^T S^{-1} B]^{-1} \approx \alpha [B^T M^{-1} B]^{-1} + \theta_1 \nu k [B^T L^{-1} B]^{-1} + \theta_2 k [B^T K^{-1} B]^{-1} .$$
$$(2.29)$$

This final operator is thought to act as preconditioner or smoother in a multigrid approach, but we do not claim to view it as appropriate inverse operator.

2.3.1   The "reactive" preconditioner A_R

In many applications, the spatial discretization process of the zero-order term $\mathbf{u}_t$ leads to a (velocity) mass matrix M which is already diagonal by construction. This is satisfied by finite difference approaches or by using the nonconforming triangular finite elements [24]. Otherwise, the procedure of *lumping* may lead to diagonal mass matrices, that means a corresponding diagonal mass matrix can be reached by the following two possiblities:

- by using **special quadrature points** which have to be related to the degrees of freedom of the finite element space used. For instance, the trapezoidal quadrature rule is taken for conforming linear/bilinear finite elements with degrees of freedom at the vertices, or a modified midpoint/midface quadrature rule must be applied in the case of the edge/face–oriented nonconforming spaces.

- by **summing up** the row sum of the mass matrix M to obtain the diagonal matrix.

We prefer the first approach which often leads to sufficient accuracy in the usual finite element context. As a consequence we can consider the following "reactive" operator

$$B^T M_l^{-1} B.$$

Here, M_l indicates that the (velocity) mass matrix is diagonal, if necessary after *lumping*. In the case that the velocity matrix S contains a non–diagonal matrix M, the approach with $B^T M_l^{-1} B$ can be viewed as optimal preconditioner in the previously defined sense, but it is not necessarily an exact solver. The reason for the explicit demand for a diagonal mass matrix M_l is that then we can directly calculate the matrix $P = P(M_l)$,

$$P := B^T M_l^{-1} B. \tag{2.30}$$

This explicit calculation leads to several advantages, namely

1. In many situations – depending on the spatial discretization of pressure and velocity – the storage of P and its application in a matrix–vector multiplication is much cheaper than using the product form $B^T M_l^{-1} B$ (see the following section 2.5), particularly for the nonconforming finite elements.

2. Since P has been explicitly computed, standard multigrid solvers can be applied, even with robust smoothers of SOR or ILU–type. This is not the case for the implicit product form $B^T M_l^{-1} B$ which allows diagonal preconditioning/smoothing only. Hence, a robust behaviour on anisotropic meshes is possible only if the explicit form P is utilized.

This compact matrix P which usually has to be calculated only once in a preprocessing step (or if the spatial mesh has changed) has some other nice features. First of all, P can be interpreted as a discretization matrix steeming from a mixed formulation of the (continuous) Poisson problem

$$-\Delta q = rhs, \tag{2.31}$$

with appropriate boundary conditions. The corresponding continuous mixed formulation reads after introducing the quantity $\mathbf{v} = -\nabla q$, and then using $rhs = -\nabla\cdot\nabla q = \nabla\cdot\mathbf{v}$,

$$\begin{bmatrix} I & \nabla \\ \nabla\cdot & 0 \end{bmatrix} \begin{bmatrix} \mathbf{v} \\ q \end{bmatrix} = \begin{bmatrix} 0 \\ rhs \end{bmatrix}. \tag{2.32}$$

The resulting discretized stiffness matrix reads

$$\begin{bmatrix} M_l & B \\ B^T & 0 \end{bmatrix}. \tag{2.33}$$

This explains the expression "Pressure–Poisson" problem since this technique corresponds to a finite element formulation for the Poisson problem which can be even derived for piecewise constant pressure ansatz functions. If we calculate the stiffness matrix entries based on an equidistant mesh by using the introduced nonconforming velocity and the piecewise constant pressure ansatz, we obtain a "good old friend", the well–known 5–point stencil,

$$Stencil(P) = \begin{bmatrix} 0 & -1 & 0 \\ -1 & 4 & -1 \\ 0 & -1 & 0 \end{bmatrix}. \tag{2.34}$$

However on discrete level, in contrast to the continuous case, there may arise differences in comparison to the formulation (2.32) if we ask for boundary conditions involved in M_l. While in the continuous case, formulations requiring only $\mathbf{v}\cdot\mathbf{n} = 0$ (with $\mathbf{n}$ denoting the outer normal vector to $\partial\Omega$) lead to stable formulations, we have more freedom on discrete level due to the demand for preconditioning only. Since $B^T M_l^{-1} B$ is thought to approximate the full *pressure Schur complement* operator $B^T S^{-1} B$, we use for M_l the same boundary conditions as for S, for instance Dirichlet conditions in both components for $\mathbf{v}$ on rigid walls or at the inflow.

What are the resulting boundary conditions for the new (pressure) matrix $P = B^T M_l^{-1} B$? In fact, the same happens as with continuous projection methods: pressure cells attaching rigid walls lead to natural boundary conditions from type $\partial q / \partial \mathbf{n} = g$. This is demonstrated by the following matrix stencil which is derived on equidistant meshes with our preferred Stokes element $\tilde{Q}1/Q0$ (we assume full Dirichlet boundary conditions at the "right side"),

$$Stencil(P) = \begin{bmatrix} 0 & -0.5 \\ -1 & 2 \\ 0 & -0.5 \end{bmatrix}. \tag{2.35}$$

Other possiblities exist for prescribing boundary conditions in M_l and hence in P, namely by requiring $\mathbf{v} \cdot \mathbf{n} = g$ only (as in the continuous case) or by prescribing no boundary conditions at all. These variations lead to different resulting matrices P. We have made some first numerical attempts with these approaches, but they were not satisfactory at all. However, these tests were spontaneous only and have to be accompanied with a rigid numerical analysis.

There is another remarkable point concerning boundary conditions of the "Pressure–Poisson" matrix P. What happens on boundary parts where the velocity (mostly coupled with the pressure) satisfies a natural boundary condition? If we use the given M_l without any boundary modification, this corresponds to the *do nothing* techniques proposed in [56] for realizing natural boundary conditions in M_l. This approach leads to the following matrix stencil, with Neumann-type boundary conditions for the velocity, again at the "right side",

$$Stencil(P) = \begin{bmatrix} 0 & -1 \\ -1 & 4 \\ 0 & -1 \end{bmatrix}. \tag{2.36}$$

This shows that the resulting matrix is definite, instead of being semi–definite only, and the corresponding pressure boundary conditions can be expressed as Dirichlet values on a fictitious mesh which is enlarged by one layer of cells. Up to now, there is no rigorous analysis of this boundary condition available, however all numerical experiences are excellent.

Let us summarize these considerations concerning the reactive *pressure Schur complement* preconditioner P:

- The pressure matrix $P = B^T M_l^{-1} B$ behaves like a discrete Poisson operator but there are some differences to continuously derived stiffness matrices with respect to boundary conditions.

- Since P arises from a mixed formulation, even piecewise constant ansatz functions are allowed which is in contrast to the standard discretization procedure.

- Another difference lies in the implementation of boundary conditions related to P: since we perform all transformations on discrete level we have a much richer variety of possiblities.

The essential idea for this approach is to perform all modifications on discrete level with already discretized partial differential operators and this is due to the concept of building *preconditioners* for the discrete *Schur complement* matrix $P = B^T S^{-1} B$. The construction process of preconditioners is much more variable since in the context of defect correction principles, many possibilities are allowed which are hard to perform on continuous level! Since for very small time steps k the velocity matrix S behaves like

$$S = M_l + \theta_1 \nu k L + \theta_2 k K \to M_l \quad \text{for} \quad k \to 0, \tag{2.37}$$

we have constructed an almost exact solver/preconditioner $C^{-1} := P^{-1}$ for $B^T S^{-1} B \approx B^T M_l^{-1} B$ if we apply the same boundary conditions for M_l as for S. Consequently, our numerical scheme involving P as preconditioner leads to "optimal" convergence rates in the case of small time steps k. In that case that M_l is explicitly part of S, the convergence rates tend to 0 for $k \to 0$, while in the other case of M being not diagonal the preconditioning operator P, resp., applied as smoother in a multigrid context, may lead at least to excellent multigrid rates independent of parameter and mesh variations. This optimality is very important since this Pressure–Poisson problem due to the incompressibility constraint is often dominating the numerical cost for time step $k \to 0$. In the following Section 2.5, we will additionally show that our preferred Stokes element $\tilde{Q}1/Q0$ – nonconforming rotated multilinear velocity and piecewise constant pressure finite elements – leads to matrices P of minimal numerical complexity which explains why our approach is one of the most efficient for the fully nonstationary incompressible Navier–Stokes equations.

Unfortunately, this "discrete level" approach of calculating explicitly $P = B^T M_l^{-1} B$ does not always lead to satisfying results. We will show in Section 2.5 that the use of conforming bilinear velocity and pressure functions leads to matrices P which are large in the sense of nonzero matrix entries in each row. Moreover, P has a nontrivial kernel due to the stability failure of this element combination (the *Babuška–Brezzi–condition* is not satisfied, see [36],[58]). However, since the pressure is bilinear, a direct discretization of the Laplacian can be performed instead, with appropriate (?) boundary conditions such that we obtain

$$P := \Delta_h .$$ (2.38)

This technique is widely used and often discussed (see [41]), sometimes called *approximate projection* scheme. The advantage is that fast numerical solution techniques, as multigrid or conjugate gradient solvers, can be easily applied. However, we loose the property of having an (almost) exact preconditioner for the *Schur complement* formulation for small time step k. Additionally, in combination with pressure correction or projection methods, the final step of modifying the intermediate velocity $\tilde{u}$ to satisfy exactly the discrete continuity equation cannot be guaranteed. This explains the expression *approximate projection* scheme.

In contrast to the former approach which calculates directly on discrete level the appropriate "Pressure–Poisson" operator, the second access uses the knowledge that for small k the corresponding *Schur complement* equation behaves like a Laplacian. Hence, we choose as preconditioner a matrix P which arises from the discretization of a continuously given Poisson problem. We recommend, if possible at all, to perform the first "discrete" approach which may lead to exact solvers. Since in some cases, for the widely used $Q1/Q1$ element pair for instance, this technique fails, the second "continuous" approach has to be applied. However, so far is it not clear if "optimal" convergence rates as in the following "diffusive" case, meaning $\rho_{MPSC} \leq C < 1$, can be obtained independent of all parameters and grids.

Let us finish with some conclusions. First of all, we have constructed "reactive" preconditioners P for the term $B^T M^{-1} B$ which have been based on discrete or continuous formulations. In the later Section 'Resulting schemes and relation to other existing methods' we will demonstrate some of their numerical features more in detail for our preferred finite element combination. In any case, we define for the following considerations the "reactive" preconditioner A_R as

$$A_R := P = B^T M_l^{-1} B$$

with the following advantages if this discrete approach is applied:

- An (almost) exact solver/preconditioner A_R for time steps $k \to 0$

- A flexible treatment of pressure boundary conditions on discrete level

- Highly efficient multigrid solvers for applying A_R^{-1}

- Very compact matrices A_R in dependence of the spatial discretization

2.3.2 The "diffusive" preconditioner A_D

Next, we construct an "optimal" preconditioner A_D for the diffusive part of the *Schur complement* operator, $B^T L^{-1} B$, in which case L is a discrete (velocity) Laplacian from the momentum equations. Hence, we are looking for an optimal Stokes solver in the *Schur complement* formulation.

Unfortunately, the same transformations as for the "reactive" part, namely on discrete level only, are impossible in the diffusive case. The reason is simply that L^{-1} ("inverse discrete Laplacian") and therefore $B^T L^{-1} B$ are full (!) matrices, at least for all finite difference/element/volume approaches. Consequently, we have to concentrate on transformations on the continuous level only. The key is that the continuous *pressure Schur complement* operator for the Stokes problem is spectrally equivalent to the identity, i.e.,

$$\nabla \cdot \Delta^{-1} \nabla \sim I \quad \text{resp.} \quad \exists c_1, c_2 : c_1 I \leq \nabla \cdot \Delta^{-1} \nabla \leq c_2 I. \qquad (2.39)$$

In the finite element context, this relations reads:

$$B^T L^{-1} B \sim M_p \quad \text{resp.} \quad \exists c_1, c_2 : \text{cond}(M_p^{-1}[B^T L^{-1} B]) \leq c_2/c_1. \qquad (2.40)$$

The matrix M_p denotes the (pressure) mass matrix corresponding to the discrete ansatz functions for the pressure, and c_1, c_2 are constants which are independent of the mesh width. However, the shape of the mesh, i.e., uniform grid cells or very anisotropic elements with large *aspect ratios*, can determine the size of the quotient c_2/c_1.

The idea is that the operator $B^T L^{-1} B$ which is in fact a full matrix, can be preconditioned by a much simpler sparse mass matrix M_p in the pressure space. In most applications, M_p is even a diagonal matrix, for instance if piecewise constant pressure functions or if appropriate *lumping* procedures for linear elements are used, such that the numerical cost for preconditioning are neglectable. The numerical examples will show that indeed $A_D := M_p$ is an "optimal" preconditioner. If M_p is used as smoother in a multigrid context, the resulting convergence rates are independent of the mesh size parameter h and – at least for the nonconforming $\tilde{Q}1/Q0$ spaces – they even are independent of the shape of the underlying triangulation. However, in contrast to the "reactive" preconditioner A_R, the diffusive counterpart A_D is not an exact solver for $B^T L^{-1} B$. This is the meaning of "optimal" in this context.

$$A_D := M_p$$

In fact, this preconditioner for the Stokes equations is not new at all since it is nothing than classical Uzawa iteration, formulated on continuous and discrete level. The only difference will be how to use this simple preconditioner in the multigrid context, as part of the additive preconditioner for Stokes equations as well as for the full Navier–Stokes equations.

We finish this section with a remark concerning those relaxation schemes which follow the idea of SIMPLE or related *pressure correction* techniques: We propose to use exclusively the much simpler (diagonal) preconditioner $A_D := M_p$ in the Stokes case. This is in contrast to the variant of solving (second order) Pressure–Poisson problems, as often proposed in the literature. Although the preconditioner M_p looks so simple, all numerical tests show that indeed $A_D := M_p$ is sufficient. However, one has to keep in mind that in the *global MPSC* approach the defect calculation requires the multiplication with $B^T S^{-1} B$, and consequently the application of S^{-1} is the expensive part. In this context, the defect calculations may be more costly than the preconditioning step, in particular for stationary configurations. Hence, this solution scheme applied to stationary Navier–Stokes equations seems not to be the fastest one in all cases, especially in comparison to the following *local MPSC* schemes, but nevertheless it is very efficient as the numerical tests show. And in particular, it is absolutely robust against all variations of parameters and the shape of the mesh in the pure Stokes case.

2.3.3   The "convective" preconditioner A_K

We still have to treat the convective part $B^T K^{-1} B$ as the last part of the complete *pressure Schur complement* operator $B^T S^{-1} B$, with the "velocity" matrix S given as

$$S = M + \theta_1 \nu k L + \theta_2 k K \,. \tag{2.41}$$

This desired scalar operator is the equivalent *Schur complement* formulation (modulo boundary conditions) of a linearized incompressible Euler equation (with given $\mathbf{U}$),

$$\begin{bmatrix} \mathbf{U} \cdot \nabla & \nabla \\ \nabla \cdot & 0 \end{bmatrix} \begin{bmatrix} \mathbf{v} \\ q \end{bmatrix} = \begin{bmatrix} \mathbf{f} \\ g \end{bmatrix} \,. \tag{2.42}$$

The related discretized stiffness matrix reads, with $K\mathbf{v} = K(\mathbf{U})\mathbf{v} \approx \mathbf{U} \cdot \nabla \mathbf{v}$,

$$\begin{bmatrix} K & B \\ B^T & 0 \end{bmatrix} \,. \tag{2.43}$$

As in the previous diffusive case, the explicit construction of the *pressure Schur complement* operator $B^T K^{-1} B$ is impossible since in general K^{-1} is not a sparse matrix. In the case of dominating transport, the use of appropriate *upwind* techniques may lead to sparse matrices K of lower triangle type. Nevertheless, the corresponding inverse K^{-1} is a lower triangular but full matrix which has lost the desired sparsity pattern. And hence, $B^T K^{-1} B$ cannot be a sparse matrix, too. However, in contrast to the diffusive preconditioner A_D, we are not able to apply corresponding continuous techniques. At least to our knowledge, nothing is known about which (continuous) partial differential operator $\tilde{K}$ is spectrally equivalent to $\nabla \cdot (\mathbf{U} \cdot \nabla)^{-1} \nabla$. There is an open problem how to derive an operator $\tilde{K}$ such that

$$\nabla \cdot (\mathbf{U} \cdot \nabla)^{-1} \nabla \sim \tilde{K} \quad , \text{ resp.,} \quad \exists\, c_1, c_2 : c_1 \tilde{K} \leq \nabla \cdot (\mathbf{U} \cdot \nabla)^{-1} \nabla \leq c_2 \tilde{K} \,. \tag{2.44}$$

If this operator was known, a corresponding pressure matrix $\tilde{K}_p$ might be obtained with

$$B^T K^{-1} B \sim \tilde{K}_p \quad , \text{ resp.,} \quad \exists\, c_1, c_2 : \text{cond}(\tilde{K}_p^{-1}[B^T K^{-1} B]) \leq c_2 / c_1 \,. \tag{2.45}$$

However, the problem is how to construct this special partial differential operator? If we apply the *Schur complement* transformation $\nabla \cdot (\cdot)^{-1} \nabla$ onto the identity operator I or the Poisson operator Δ we obtain

$$\nabla \cdot I^{-1} \nabla \sim \Delta \quad , \text{resp.,} \quad \nabla \cdot \Delta^{-1} \nabla \sim I. \tag{2.46}$$

Roughly spoken, the *Schur complement* transformation $\nabla \cdot (\cdot)^{-1} \nabla$ applied to a zero–order operator (identity I) leads to a second–order operator (Poisson operator Δ) while in contrast the application to Δ leads to the identity operator. This gives hope that a *Schur complement* transformed first order operator $\mathbf{U} \cdot \nabla$ is again a first order operator,

$$\nabla \cdot (\mathbf{U} \cdot \nabla)^{-1} \nabla \sim (\tilde{\mathbf{U}} \cdot \nabla). \tag{2.47}$$

However, even if this relation is correct and the transformed operator corresponds to a transport equation: how does the advective direction $\tilde{\mathbf{U}}$ explicitly look like?

Coming back to the discrete approach, another successful way may be to work with "double" ILU–decompositions. It is well–known from the numerical treatment of convection–diffusion problems that an ILU–decomposition K_{ILU} of the convective operator K (and of the complete velocity matrix S, too) leads to efficient and robust convergence results if applied as smoother in a multigrid context. However, we have the problem in our *Schur complement* approach that the inverse matrix K_{ILU}^{-1} has to be calculated which is in general a dense matrix. The solution might be to apply a second ILU–decomposition: instead of explicitly calculating $B^T K_{ILU}^{-1} B$ we restrict to $ILU(B^T K_{ILU}^{-1} B)$ to generate a sparse matrix. Up to now, the research is not completed but we hope to be able to perform efficiently these calculations, based on a clever implementation. However, even then it is not guaranteed that this double ILU–factorization leads to a robust and efficient smoother in the *global MPSC* multigrid approach.

Since up to now neither discrete nor continuous approaches were successful, we cancel the convective *pressure Schur complement* preconditioner from our subsequently applied iteration schemes. Consequently, theoretical and numerical examples show that we failed to obtain the desired "discrete Black Box" solver based on these *global MPSC* techniques, at least in some cases of midrange Reynolds numbers. And it is exactly the case of medium Reynolds number flows since for higher Reynolds numbers the flow is in general fully nonstationary by physical reasons and can be captured by the "reactive" preconditioner.

However, this "missing range" is fortunately the case in which our subsequent *local MPSC* solvers show to work perfectly, concerning robustness and efficiency. Both together lead to optimized solution schemes which are applicable for the full range of Reynolds numbers, from Stokes problems to stationary Navier–Stokes equations over slightly time dependent flows up to highly nonstationary frameworks. Nevertheless, the work has to go on to find the "missing" preconditioner in the *global MPSC* context.

It is obvious that the "trick" to replace the velocity matrix S by its ILU–decomposition $ILU(S)$ or other appropriate factorizations,

$$\begin{bmatrix} ILU(S) & B \\ B^T & 0 \end{bmatrix} \quad \text{instead of} \quad \begin{bmatrix} S & B \\ B^T & 0 \end{bmatrix}, \qquad (2.48)$$

decreases the numerical cost to apply the corresponding *pressure Schur complement* operator, due to the easier application of $ILU(S)^{-1}$. Nevertheless, the total gain in inverting $[B^T ILU(S)^{-1}B]^{-1}$ is minimal since still a mixed (nonsymmetric) problem has to be solved which can be characterized as follows:

- poor **condition number**: $O(h^{-1}) - O(h^{-2})$,

- no **explicit construction** of an "optimal" preconditioner is known,

- sensitive to **mesh anisotropies** (no direct ILU–decomposition of $B^T ILU(S)^{-1}B$ is possible in general).

In fact, for the "easier" system

$$\begin{bmatrix} ILU(S) & B \\ B^T & 0 \end{bmatrix} \quad , \text{resp.,} \quad B^T ILU(S)^{-1}B \qquad (2.49)$$

it may be cheaper to perform a matrix-vector multiplication involving $ILU(S)^{-1}$ instead of S^{-1}, but the complete solution process is still almost so expensive as for the original system as long as no optimal preconditioner can be explicitely constructed. So, the statement of "replacing the velocity matrix S" by an "easier" preconditioner $\tilde{S}$ (in the sense that $\tilde{S}^{-1}$ can be easily applied) is worthless in many cases; most the difficulties of the original mixed problem are preserved. That is the reason why we encourage so strongly the search for the explicit form of the corresponding scalar *pressure Schur complement* preconditioner. If its explicit construction is possible then the condition number and the sensitivity with respect to mesh anisotropies stay the same, but:

- the preconditioning problem is a (definite) scalar problem.

- "optimal" (robust and efficient) multigrid tools can be used for "inverting" it since its matrix structure is now explicitly given.

- the number of total unknowns which is, i.e., the number of pressure unknowns only, is much smaller than the complete set of degrees of freedom.

2.3.4   The "factorized" preconditioner

In the previous three subsections we have shown how *pressure Schur complement* preconditioners can (or cannot) be constructed, individually related to the three components of the complete Navier–Stokes operator, namely the reactive, diffusive and convective part. Furthermore, the complete preconditioner has been derived in additive form consisting of these three (in fact, two!) partial preconditioners. In contrast to this splitting approach one can directly try to approximate the complete *Schur complement* operator $B^T S^{-1} B$ by constructing a spectrally equivalent - continuous or discrete - preconditioning operator. Previously, we have shown for which types of S, being the Laplacian or the identity matrix, this process may be successful on discrete or continuous level. Additionally, as indicated above, another approach may be to replace (at least on discrete level) the matrix S by an approximate $part(S)$ in which case S is assumed to be factorized as

$$S = part(S) + \text{rest} . \tag{2.50}$$

Possible choices are, for instance, the incomplete LU–decomposition (ILU), the lower triangle part (analogous to the Gauß–Seidel scheme, resp., the SOR and SSOR schemes) or the diagonal of S (Jacobi method). Here, we mention the papers of Wittum (see for instance [118]) about the *Distributive Smoothers* in combination with ILU, and Braess's paper [14] on variants of the SIMPLE scheme. In particular, Braess applies approximate *Pressure Schur complement* preconditioners of the form

$$\begin{bmatrix} part(S) & B \\ B^T & 0 \end{bmatrix}^{-1} \tag{2.51}$$

which have been analyzed as smoothers in multigrid. However, the main difference in comparison to our approach is that Braess still works in both primitive variables $\mathbf{u}$ <u>and</u> p, for which he solves the given approximate *Pressure Schur complement* problem as divergence–free preconditioner. Roughly spoken, his version which has been analyzed in [14] is a hybrid form between the original SIMPLE (working in both primitive variables) and our approach which strictly acts in the pressure variables only. While for fully nonstationary problems all variants still look like similar, there arise large differences for stationary problems (Stokes, for instance) in which case our reactive preconditioner leads to completely different results, especially in combination with the MPSC approach.

As explained in the previous section for the convective preconditioner, we claim that it is desirable by numerical and computational reasons to calculate explicitly the corresponding approximate *Schur complement* matrix

$$P_{fact} := B^T part(S)^{-1} B \,. \tag{2.52}$$

Up to now, we were only successful with its practical realization if we applied

$$part(S) = diag(S) \,. \tag{2.53}$$

Here, $diag(S)$ does not mean to perform the exact diagonal of S, but a diagonal related to the matrix S has to be taken, analogously defined as for the previous *lumping process*. This corresponds to a kind of modified (and damped) Jacobi–scheme for this *pressure Schur complement* approach with all its disadvantages concerning efficiency and robustness on anisotropic meshes (see Section 3.4). Therefore, our study has to be continued to derive more powerful approximations, for instance of modified ILU–type as shown before, but up to now we set:

$$\boxed{C := P_{fact} = B^T diag(S)^{-1} B}$$

This strategy is well-known and widely used, too. The corresponding schemes are often related to the SIMPLE relaxation and its derivates, and they are mainly utilized in pressure correction, pressure update, pressure projection schemes, etc. (see the papers [14] and [29] for a detailed overview and the literature cited therein). We made our own experiences with this approach but the results did not convince us to use this technique. Theoretical and numerical considerations show:

1. Only in the fully nonstationary case, we achieved satisfying convergence properties, but these were comparable with those obtained by our "reactive" preconditioner A_R which simply uses $part(S) = M_l$.

2. In the stationary case, or correspondingly for low Reynods numbers and large time steps, the results are in general worse than in comparison to our additive preconditioner. In fact, for the stationary case we can show (see also the numerical examples) that the simple diagonal pressure mass matrix M_p is already sufficient and leads to much better convergence results.

3. There are still problems to determine the 'optimal' relaxation parameter if we choose $P_{fact} := B^T diag(S)^{-1}B$, in particular on anisotropic meshes with strongly varying mesh sizes. Up to now, we could not solve this problem, neither theoretically nor practically.

4. In contrast to the components of our "additive" preconditioner which have to be assembled only once in a preprocessing step (assuming that the spatial mesh does not change), the matrix $P_{fact} = B^T diag(S)^{-1}B$ has to be calculated in every time step and in each iteration of the nonlinear solver, as soon as S is changing.

Due to these inherent problems, we continue our following work with the additive preconditioner only which finally reads

$$
\begin{aligned}
C^{-1} &= \alpha_R A_R^{-1} + \alpha_D A_D^{-1} & (2.54) \\
&= \alpha_R P^{-1} + \alpha_D M_P^{-1} . & (2.55)
\end{aligned}
$$

Hereby, the matrix P is obtained by an explicit calculation of $P := B^T M_l^{-1} B$, and M_l, resp., M_p, are the diagonal (lumped) mass matrices for the velocity, resp., the pressure ansatz functions. The parameters α_R, α_D are damping parameters which have to be chosen appropriately, with

$$
\begin{aligned}
\alpha_R \leq 1 \,, \alpha_D \leq \theta k \nu \quad &\text{(nonstationary calculations)} & (2.56) \\
\alpha_R = 0 \,, \alpha_D \leq \nu \quad &\text{(stationary calculations)} . & (2.57)
\end{aligned}
$$

On the basis of these settings, one *basic iteration* step for the (linear) *pressure Schur complement* equation can be easily formulated as follows:

Global Pressure Schur Complement iteration scheme:

Given: Iterate p^{l-1}
Perform: One relaxation step to obtain p^l

$$
\begin{aligned}
p^l &= p^{l-1} - [\alpha_R A_R^{-1} + \alpha_D A_D^{-1}](B^T S^{-1} B p^{l-1} - \frac{1}{k} B^T S^{-1} \mathbf{g}) \\
&= p^{l-1} - [\alpha_R P^{-1} + \alpha_D M_p^{-1}](B^T S^{-1} B p^{l-1} - \frac{1}{k} B^T S^{-1} \mathbf{g})
\end{aligned}
$$

This scheme has been written in the compact mathematical form as solver for the *pressure Schur complement* equation, but it can be rewritten in terms of classical *projection steps*. This formulation is dedicated to all users who are more familiar with this kind of "segregated" (see [29],[41]) or operator splitting approaches.

Global Pressure Schur Complement iteration in projection methods notation:

Given: Iterate p^{l-1}
Perform: 4 substeps to obtain p^l

1. Solve for an intermediate velocity $\tilde{\mathbf{u}}^l$ the corresponding nonlinear Burgers- or linearized convection–diffusion/Helmholtz equations. The right hand side contains the original load vector (or modified related to a chosen linearization) and the old pressure iterate p^{l-1}.

$$
S\tilde{\mathbf{u}}^l = \mathbf{g} - k B p^{l-1}
$$

---

2. Calculate the right hand side f_p for the preconditioning step in the *Schur complement* formulation which is just the *pressure Schur complement* residual for p^{l-1}.

$$
\begin{aligned}
f_p &= \frac{1}{k} B^T \tilde{\mathbf{u}}^l \\
&= \frac{1}{k} B^T S^{-1}[\mathbf{g} - kBp^{l-1}] \quad = \quad \text{Residual}\,(p^{l-1})
\end{aligned}
$$

3. Solve an update–equation for the pressure ("Pressure–Poisson problem") with the reactive preconditioning matrix P and the *pressure Schur complement* residual f_p on the right hand side.

$$
Pq = f_p
$$

4. Update the new pressure p^l by additionally applying the diffusive preconditioner M_p^{-1} which is a diagonal matrix.

$$
p^l = p^{l-1} + \alpha_R q + \alpha_D M_p^{-1} f_p
$$

The main differences in comparison to the classical pressure correction and projection–type methods are the following items:

- If the new pressure p^l is updated, not only the solution q of the "Pressure–Poisson problem" is involved. As an additional second preconditioner, the solution of the auxiliary diffusive step is added which involves only scaling by the diagonal matrix M_p^{-1}.

- Up to now, we did not explain how the corresponding velocity vector $\mathbf{u}^l$ can be calculated.

The introduction of the second preconditioner as additional scaling of the right hand side term $f_p = \frac{1}{k} B^T \tilde{\mathbf{u}}^l$ by the inverse diagonal pressure mass matrix M_p^{-1} is not new. What is new is the derivation on discrete level as separate optimal preconditioner for the Stokes part of the *Schur complement* equation. To our knowledge, this additional term is also introduced by Timmermans ([70],[71]), but his motivation is based on modifications of partial differential operators on continuous level which are mathematically not so clear. However, these continuous reformulations lead finally to the same methods as we proposed, without our interpretation as preconditioners in a solution process of linear systems of equations.

A different but more rigorous derivation of this diffusive preconditioner M_p can be found by Glowinski [39]. In this paper, the same splitting techniques of the *Schur complement* operator A for (nonsteady) Stokes equations are used,

$$A := -\nabla \cdot (\alpha I - \nu \Delta)^{-1} \nabla \,, \qquad (2.58)$$

to derive a reactive and a diffusive preconditioner C^{-1},

$$C^{-1} := -\alpha \Delta^{-1} + \nu I \,. \qquad (2.59)$$

In contrast to our approach this construction process is done on continuous level only such that the discrete preconditioners are obtained by applying appropriate discretization techniques. The advantage of this technique is a more rigorous analytical understanding of this preconditioning process. In fact, Cahouet [20] has shown by Fourier Analysis the spectral equivalence and hence the optimality of this preconditioner in the case of periodic boundary conditions. On the basis of this observation, Glowinski has embedded these optimal Stokes preconditioners in a CG–method and operator splitting approach which work very efficient and robust for a wide range of Reynolds numbers (see [39]). In contrast, we explicitly define the reactive preconditioner, resp., the Pressure–Poisson operator on discrete level such that our approach seems to have two advantages:

1. The additive *PSC* preconditioner for the discrete *Schur complement* equation,

$$C^{-1} := \alpha_R P^{-1} + \alpha_D M_P^{-1} \qquad (2.60)$$

   is an exact solver in the case of decreasing time steps k, $k \to 0$. In the multigrid context, this leads to convergence rates ρ_{MPSC} with $\rho_{MPSC} \to 0$. Hence, our approach is optimal in the case of small time steps k.

2. Since we apply our techniques on discrete level for deriving the "Pressure–Poisson" operator, this approach can be performed very easily for discontinuous and particularly for piecewise constant pressure discretizations. In fact, we recommend our approach for those Stokes

elements which include discontinuous or constant pressure approximations while the continuous approach by Glowinski has clear advantages for conforming linear or even higher order pressure approximations. Nevertheless, our techniques of embedding *PSC* schemes as smoothers in multigrid can be analogously applied.

The last remark in this section concerns the calculation of velocity approximations which are related to the new pressure iterate p^l. First of all, there is always an intermediate velocity $\tilde{\mathbf{u}}^l$ computed which results from the calculation of the *pressure Schur complement* residual for the old iterate p^{l-1} (see step 1 and 2 in the description of *"Global Pressure Schur Complement iteration in projection methods notation"*). Then, the following relation is satisfied:

$$\tilde{\mathbf{u}}^l = \mathbf{u}(p^{l-1}) = S^{-1}[\mathbf{g} - kBp^{l-1}]. \tag{2.61}$$

This shows that $\tilde{\mathbf{u}}^l = \mathbf{u}(p^{l-1})$ is the velocity approximation which for given p^{l-1} satisfies exactly the momentum equation. And in addition, if for a certain iterate p^L the *Schur complement* equations is satisfied,

$$B^T S^{-1}[kBp^L - \mathbf{g}] = 0, \tag{2.62}$$

it follows as a direct consequence that $\mathbf{u}^L := \mathbf{u}(p^L)$ which is given by

$$\mathbf{u}^L = S^{-1}[\mathbf{g} - kBp^L], \tag{2.63}$$

satisfies the momentum <u>and</u> the continuity equation, that means

$$\begin{bmatrix} S & kB \\ B^T & 0 \end{bmatrix} \begin{bmatrix} \mathbf{u}^L \\ p^L \end{bmatrix} = \begin{bmatrix} \mathbf{g} \\ 0 \end{bmatrix}. \tag{2.64}$$

Hence, for given pressure p, the corresponding "momentum-free" velocity approximation can be always found by solving the transport-diffusion problem $S\mathbf{u} = \mathbf{g} - kBp$.

This approach renders $\mathbf{u}$ to be a "slave variable" from the pressure p and leads to "momentum–free" velocity approximations, that means $\mathbf{u}$ satisfies the discrete momentum equations for arbitrary p. However, there is another possiblity to calculate a velocity field which now is stressed to satisfy the continuity equation. After each step of the *global pressure Schur complement* iteration we keep as result the intermediate velocity $\tilde{\mathbf{u}}^l$ (which indeed is $\tilde{\mathbf{u}}^l = \mathbf{u}(p^{l-1})$ as seen before) and the pressure update q. Based on these two quantities and the necessary condition that the reactive preconditioner $A_R = P$ is explicitly calculated from

$$P := B^T M_l^{-1} B, \tag{2.65}$$

then the following setting

$$\mathbf{u}^l = \tilde{\mathbf{u}}^l - k M_l^{-1} B q \tag{2.66}$$

satisfies the discrete continuity equation exactly, $B^T \mathbf{u}^l = 0$. A simple calculation shows

$$
\begin{aligned}
B^T \mathbf{u}^l &= B^T(\tilde{\mathbf{u}}^l - k M_l^{-1} B q) \\
&= B^T S^{-1}[\mathbf{g} - k B p^L] - k B^T M_l^{-1} B q \\
&= k f_p - k P q \\
&= 0.
\end{aligned}
$$

In fact, since the quantity q can be interpreted as approximate term for "kp_t", this velocity update is a correction of a "momentum-free" velocity $\tilde{\mathbf{u}}$ via a second order term "$k^2 \nabla p_t$". For more details about rigorous mathematical aspects of these projection techniques see the book of Prohl [79]. If we summarize these both approaches we can state the following two possibilities to derive velocity approximations $\mathbf{u}$ for a given pressure p:

Velocity calculations in global Pressure Schur Complement schemes:

Given: Iterate p, $\tilde{\mathbf{u}}$ and q
Perform: Corresponding iterate $\mathbf{u}$ which satisfies:

- **Momentum equation:** $\mathbf{u} = \mathbf{u}(p) := S^{-1}[\mathbf{g} - k B p]$

- **Continuity equation:** $\mathbf{u} = \mathbf{u}(q) := \tilde{\mathbf{u}} - k M_l^{-1} B q$

2.4 "Local Multilevel Pressure Schur Complement"

The *global MPSC* approach leads to preconditioning matrices which are in general very high-dimensional, but sparse. As usual in the context of partial differential equations, the bandwidth of the assembled matrices is very small (typically 5 – 11 in our applications) such that only few matrix entries have to be stored in comparison to the huge dimension of up to several millions of unknowns. If appropriate storage techniques are consequently applied which take into account this locally regular band structure, we can perform solution schemes which explicitly exploit the pipelining and vectorization facilities of modern processors. If the vector length is large enough (what is automatically satisfied), a high percentage of the large floating point Peak performance can be reached. In particular, if subroutines in LAPACK and BLAS-style can be applied, this high performance is achievable by large parts of the code.

In contrast, the following approach is influenced by "very local" techniques which are related to *domain decomposition* approaches. This local *pressure Schur complement* approach is oriented at the idea of solving "small" problems which thereby exploit the fast cache of processors, in contrast to the vectorization facilities of the *global MPSC* approach. Consequently, the way of designing *local MPSC* schemes can be characterized as follows:

> 'Solve **exactly** on subsets and perform an outer **Block–Gauß-Seidel/Jacobi** iteration.'

2.4.1 The local "Pressure Schur Complement" preconditioner

We define *patches* Ω_i in a very general manner as set of unknowns which can be viewed as "subdomains" of the complete set of unknowns. In our context of the $\tilde{Q}1/Q0$ element, one patch consists of one or several <u>neighboured</u> mesh cells,

$$\Omega_i = \bigcup_{j \in I(i)} T_j \quad , \quad \Omega = \bigcup_{i=1}^{NP} \Omega_i . \tag{2.67}$$

Here, NP is the number of patches and $I(l)$ is the index set of elements which belong to the patch l. The construction of the subsets will be performed in such a way that the following relations hold for the application of the resulting *local MPSC* schemes (with NEL the total number of elements):

- $1 \leq NP \leq NEL$ (NP number of patches)

- $\sum_{i=1}^{NP} I(i) \geq NEL$

- if $NP = 1$, then we solve directly the original problem on the complete domain Ω, resp., for the complete set of unknowns.

- if NP increases, $NP \rightarrow NEL$, the convergence rates shall stay bounded independent (!) of NEL and hence the mesh size.

Before we go more into detail, we explain shortly the idea of the following algorithmic constructions. On the patches Ω_i which still have to be appropriately defined, we will explictly solve local subproblems involving the locally defined coupled stiffness matrix A_i,

$$A_i := \begin{bmatrix} \tilde{S}_{|\Omega_i} & kB_{|\Omega_i} \\ B_{|\Omega_i}^T & 0 \end{bmatrix}. \qquad (2.68)$$

The matrix entries – and hence the corresponding "boundary conditions" on the "subdomains" – are taken from the global matrices. $\tilde{S}$ means that the complete velocity matrix S or only parts of S are taken, for instance the diagonal part $diag(S)$. As will be explained later, these local subproblems may be solved by a direct solver ("Gaussian elimination"), that means we explicitly apply the inverse of A_i^{-1} to a given (short) right hand side vector. Since this elimination process leads to fill-up of the matrix and increases dramatically the storage requirements, we solve the equivalent local *pressure Schur complement* problem with the very compact matrix

$$P_i := B_{|\Omega_i}^T \tilde{S}_{|\Omega_i}^{-1} B_{|\Omega_i}. \qquad (2.69)$$

This matrix is in general a full matrix, but the number of unknowns – which is the number of pressure variables defined on the patch Ω_i – is much smaller than the number of local pressure <u>and</u> velocity degrees of freedom. For a moderate number $I(i)$ of elements belonging to the i-th patch Ω_i, this full *pressure Schur complement* matrix is moderately small and fits mostly into the processor cache. Then, having solved this local subproblem for the pressure, the corresponding velocity field can be obtained as described in the previous section.

Summing up, a corresponding local subproblem is solved on the subset Ω_i and approximations for $\mathbf{u}$ and p restricted to the patch Ω_i have been gained. Finally, the overall convergence towards the solution of the complete problem

$$S\mathbf{u} + kBp = \mathbf{g} \quad , \quad B^T\mathbf{u} = 0 \qquad (2.70)$$

is obtained by embedding this local solution step into an outer blockwise Jacobi- or Gauß-Seidel iteration. The difference between both iteration schemes lies in the different evaluation of the right hand sides for the local subproblems. In the Jacobi case, we can simply formulate the basic iteration in suggestive notation as

---

Basic iteration for the local Pressure Schur
Complement approach:

Given: Iterates $\mathbf{u}^{l-1}$, p^{l-1}
Perform: One relaxation step to obtain $\mathbf{u}^l$, p^l

$$\begin{bmatrix} \mathbf{u}^l \\ p^l \end{bmatrix} = \begin{bmatrix} \mathbf{u}^{l-1} \\ p^{l-1} \end{bmatrix} - \omega^l \sum_{i=1}^{NP} \begin{bmatrix} \tilde{S}_{|\Omega_i} & kB_{|\Omega_i} \\ B^T_{|\Omega_i} & 0 \end{bmatrix}^{-1}$$

$$\times \left(\begin{bmatrix} S & kB \\ B^T & 0 \end{bmatrix} \begin{bmatrix} \mathbf{u}^{l-1} \\ p^{l-1} \end{bmatrix} - \begin{bmatrix} \mathbf{g} \\ 0 \end{bmatrix} \right) .$$

In practical, the iteration step is split into several substeps. First, we successively calculate the defect for $\mathbf{u}^{l-1}$ and p^{l-1} with respect to the patch Ω_i,

$$\begin{bmatrix} \mathbf{defu}_i^{l-1} \\ defp_i^{l-1} \end{bmatrix} = \left(\begin{bmatrix} S & kB \\ B^T & 0 \end{bmatrix} \begin{bmatrix} \mathbf{u}^{l-1} \\ p^{l-1} \end{bmatrix} - \begin{bmatrix} \mathbf{g} \\ 0 \end{bmatrix} \right)_{|\Omega_i} \qquad (2.71)$$

and solve the corresponding local problem

$$\left[\begin{array}{cc} \tilde{S}_{|\Omega_i} & kB_{|\Omega_i} \\ B_{|\Omega_i}^T & 0 \end{array} \right] \left[\begin{array}{c} \mathbf{v}_i^l \\ q_i^l \end{array} \right] = \left[\begin{array}{c} \mathbf{defu}_i^{l-1} \\ defp_i^{l-1} \end{array} \right]. \qquad (2.72)$$

Finally, we obtain the new iterates $\mathbf{u}_{|\Omega_i}^l$ and $p_{|\Omega_i}^l$ (with relaxation parameters ω^l) via

$$\left[\begin{array}{c} \mathbf{u}_{|\Omega_i}^l \\ p_{|\Omega_i}^l \end{array} \right] = \left[\begin{array}{c} \mathbf{u}_{|\Omega_i}^{l-1} \\ p_{|\Omega_i}^{l-1} \end{array} \right] - \omega^l \left[\begin{array}{c} \mathbf{v}_i^l \\ q_i^l \end{array} \right]. \qquad (2.73)$$

Then, the same procedure has to be repeated for the next patch. Usually, the problem may arise that due to subdomain boundaries certain velocity or pressure components are relaxed several times. The easiest way to get globally defined values with respect to iteration loop l is to take the resulting values from the last treated patch, or to calculate an averaged value over all results coming from the different patches. This variant is a simple block–Jacobi iteration applied to the mixed problem

$$Su + kBp = \mathbf{g} \quad, \quad B^T \mathbf{u} = 0. \qquad (2.74)$$

As usual in Numerical Linear Algebra, this scheme can be easily improved by modifying the defect calculation in the first step. Instead of using the old values $\mathbf{u}_{|\Omega_i}^{l-1}$ and $p_{|\Omega_i}^{l-1}$ only, we incoorporate into the update procedure the new iterates $\mathbf{u}_{|\Omega_i}^l$ and $p_{|\Omega_i}^l$ which belong to patches which have already been treated. Then, we employ already these new values if the local defects are calculated on the next patch. This strategy is well-known as block-Gauß-Seidel iteration and has usually a better efficiency and robustness behaviour than the block–Jacobi scheme, involving the same numerical effort on standard sequential processors.

The main philosophy behind this approach is the following: It is well-known that Jacobi/Gauß-Seidel–like iterations work very efficiently as long as anisotropies in the matrix entries are small. However, on triangulations containing elements with large *aspect ratios*, for instance very flat cells to resolve boundary layers or complex geometrical details, or with large jumps of the size of two neighboured cells, the convergence rates deteriorate dramatically. One remedy is to employ ILU–techniques which is a hard job for systems of equations, or to apply block versions resulting in isotropic subdomains. Hence, the essential idea is to collect elements of the triangulation in such a way that the resulting patches have about the same "shape and size". In contrast, all the "bad" anisotropies are hidden inside of the patches. Then, the outer global convergence behaviour will be satisfactory while the small local subproblems only are very ill-conditioned. However, the size of these local problems is usually small, and the complete inverse of the matrix fits into the RAM and sometimes even into the cache of the processor such that fast direct solvers can be applied. Consequently, we should obtain convergence rates which are independent of such grid distortions and which are of the same quality as those for very regular structured meshes. Additionally, if machine–optimized routines as BLAS subroutines are employed, the computing times for solving moderately sized problems do not differ much from the pointwise relaxation techniques, as long as the local matrix fits into the cache. Hence, excellent convergence rates and performance results should be expected since the highly tuned processor facilities can be exploited.

A further essential assumption for the expected numerical behaviour is that the convergence rates improve with increasing the local problem size, but stay bounded for increasing the number of patches. Particularly the second point is in contrast to some *domain decomposition* approaches in which the convergence rate often depends on the number of subdomains and additionally on the size of the overlap of the subdomains. Our approach, at least applied to the $\tilde{Q}1/Q0$ Stokes element, works formally without any overlap since our partitioning due to elements leads to an direct sum in (2.67). However, since we take the matrix entries from the global matrices, restricted only to the patch Ω_i, we are in the same position as before when we constructed our reactive preconditioner P in the previous section. In particular, we perform the same boundary treatment, now applied to interior subdomain boundaries, but the corresponding stencils behave analogously. For instance, for an interior boundary part on the "right side" we obtain

$$Stencil(P) = \begin{bmatrix} 0 & -1 \\ -1 & 4 \\ 0 & -1 \end{bmatrix}. \tag{2.75}$$

This strategy can be interpreted again as prescribing homogeneous Dirichlet values on a fictitious mesh which is enlarged by one additional layer of cells around the subdomain. So, we automatically obtain definite problems. Further, this approach can be interpreted as a *domain decomposition* scheme with overlapping size of 1 element of (local) mesh width h.

It is obvious that the convergence rates of our basic iteration tend to 0 if the number of patches tends to 1, since in this case we perform an exact solver (if $\tilde{S} = S$). However, if we apply this modified discrete *domain decomposition* ansatz as smoother in a typical multigrid context, we can additionally guarantee that our convergence rates stay bounded below from 1, although

the overlapping size depends on the actual mesh width. The explanation
is that in the extreme case of $NP = NEL$, that means if the patch size
is 1, we simply apply a pointwisely working smoothing operator. Hence, as
"worst case" we obtain the typical multigrid rates of pointwise smoothers
which further improve for increasing patch sizes. Again, the essential key is
to perform all modifications on discrete level if possible, and then to apply
acceleration techniques via multigrid. This approach, practiced in the *local*
as well as the *global MPSC* context, is one of the most important steps in all
our mathematical work.

2.4.2   Blocking strategies for building patches Ω_i

The remaining question is for the strategy concerning the construction of the
patches Ω_i. Essentially, there are 2 possibilities:

- **Blocking by hand:** we explicitly prescribe which elements have to
  be collected to build a common patch Ω_i. This requires a complete
  knowledge of the shape of the elements and the grid structure. Then, a
  list containing the selected elements for each patch Ω_i can be prescribed.
  This information must be redefined if the mesh has changed, and the
  building of the patches has to be repeated.

- **Adaptive blocking:** The analogous lists of elements for the definition
  of patches Ω_i can be generated in an adaptive way during the solution
  process itself.

Our implemented procedure to construct the corresponding patches is based
on such a list/array containing the information which elements have to be
collected together to build a patch Ω_i. Then, inside of the smoothing procedure, resp., of our local *Schur complement* basic iteration, we successively
build on every patch Ω_i the corresponding local *Schur complement* matrix A_i.
For the local solution process, we apply the inverse A_i^{-1} which currently is
generated by LAPACK-like subroutines, and obtain first the (local) pressure
subproblem solution and subsequently the corresponding velocity field by ap-
plying $S_{|\Omega_i}^{-1}$. The major part of the CPU time is spent by building up the
corresponding local numbering vectors and local matrix entries which have
to extracted from the global matrices. This is a memory access intensive
process which is difficult to optimize! Then, the local *pressure Schur com-
plement* matrix A_i (which involves $S_{|\Omega_i}^{-1}$) and its inverse are calculated. All
computations of these inverse matrices, resp., the solution of corresponding
local subproblems, should be implemented by the support of LAPACK-like
subroutines. Then, even for local subproblem sizes of 100 – 1000 elements

the corresponding linear algebra tasks can be performed in a very efficient way if the processor facilities are appropriately exploited.

We still have to be more precise in giving instructions how to generate the lists of elements which determine the size and shape of the patches. This is the crucial point for these block iteration schemes and determines essentially the resulting convergence rates and hence the total elapsed CPU time. This field is still under research and we can only provide recent numerical experiences with blocking strategies which are implemented up to now.

Our current blocking strategy follows two *geometrical* criterions and works elementwisely. We need the following information which has to be calculated for each element i, resp., for each edge j:

- the **aspect ratio** $AR(i)$ which is the quotient of the length of the two lines connecting opposite midpoints. If $AR(i) < 1$, we set $AR(i) = 1/AR(i)$.

- the **volume ratio** $VR(j)$ which is the quotient of the area/volume of the two element attached with edge j. If $VR(j) < 1$, we set $VR(j) = 1/VR(j)$.

Adaptive blocking strategy based on geometrical criterions:

Given: $AR(i)$ for all element i and $VR(j)$ for all edges j
Perform: Run through all elements i of the mesh

1. Let i_0 be the actual element number and check $AR(i_0)$ and $VR(k)$ for all edges k belonging to i_0.

2. If both are smaller than a given tolerance TOL, for instance with $TOL = 10$ or 100, take the next element and denote it by i_0.

3. If $AR(i_0)$ and/or $VR(k)$ are larger than TOL, then start the following recursive procedure and create a new patch:

   - If $AR(i_0) > TOL$, add the 2 elements which are neighboured along the both longer edges to the current patch. Perform recursively step 3 with these 2 elements i until $AR(i)$ and $VR(l)$ are smaller than TOL for the recursively defined list of elements i with edges l.

> - if $VR(k) > TOL$, add the 1 element which is neighboured along the edge k to the current patch. Perform recursively step 3 with this element until $AR(i)$ and $VR(l)$ are smaller than TOL for the recursively defined list of elements i with edges l.
>
> All elements which are not collected to patches by the previous steps build their own patches which consequently consist of exactly 1 element.

The *blocking strategy by hand* follows these geometrical criterions, too. However, in our applications we explicitly exploit one possible grid generation process in *FEATFLOW* ("uniform refinement"). We start with prescribing a coarse mesh which can be arbitrarily complex and unstructured, and obtain the refined meshes by simply connecting opposite midpoints. As a consequence, the number of elements increases by a factor of 4 from level to level, and the shape of the mesh, that means *aspect ratios* and *volume ratios*, is preserved. Our implemented *blocking strategy by hand* uses the explicit knowledge about the coarse mesh structure and constructs patches by an analogous strategy as in the adaptive case, but collecting macro-elements (these are the elements of the coarse mesh) only. Since all *aspect ratios* and *volume ratios* of the coarse mesh are well-known, the lists containing the macro-elements which build a patch can be easily defined. Then, in the refinement process, refined elements belong recursively to the same patch if their "father–element" is part of this patch. Consequently, the number of elements belonging to one patch increases from level to level by a factor 4 (in 2D), and this strategy may become inefficient due to large storage and CPU requirements. However, due to the easy implementation this rough approach is a good starting point. In the following section containing numerical results we demonstrate both strategies and give corresponding convergence rates.

It is obvious that the adapative strategy should be the performed part of future software packages. However, up to now the derivation and optimization of such strategies is not finished and in fact under heavy research. Since we could not finish all implementations in the most efficient way with regard to exploit the fast caches of modern processors we actually provide only convergence rates. The reader may believe our statements about the resulting CPU times and we hope to finish soon the needed implementations. Up to now, we are very optimistic since so far all computational results show no contradiction to the theoretical propositions. As indicated, the crucial point is the optimal strategy for clustering elements to build a patch Ω_i. If the number of elements per patch $I(i)$ is too large, the CPU-cost will increase since direct solvers involving cubic run-time behaviour are applied. However, if the patch size is too small and if the "bad" anisotropies are not inside of the patches, the outer convergence rates will decrease and again the CPU cost may increase dramatically.

We discussed so far only anisotropies due to geometrical effects, as large *aspect ratios* and *volume ratios*. These deficiencies can be detected by checking carefully the underlying triangulation and are essentially initiated by complex geometries or by accuracy requirements if boundary layers have to be resolved. However, even on very regular meshes the standard Jacobi/Gauß–Seidel–like iterations may fail, depending on the underlying partial differential equation which has been discretized. If for instance an anisotropic diffusion problem or an anisotropic transport–diffusion equation, stabilized by streamline–diffusion techniques, is treated,

$$-\epsilon u_{xx} - u_{yy} = f \quad , \text{ resp.}, \quad -\epsilon \Delta u + \mathbf{b} \cdot \nabla u = f,$$

then these standard methods may fail due to anisotropies which can be detected by a careful examination of the matrix entries (see Section 3.1.4 for numerical properties of streamline–diffusion discretizations). In fact, also geometrical effects as the introduced quantities *aspect ratio* and *volume ratio* can always be detected in corresponding matrix entries which may show large differences in size. We expect for the future that all blocking strategies can be reduced to very general *algebraic blocking techniques* which then will enclose our geometrically motivated approaches, too.

Up to now, we only have performed numerical tests for the nonconforming Stokes element $\tilde{Q}1/Q0$. In this case, the *domain decomposition* character is easily motivated since the discrete pressure ansatz functions have local support on separate cells only. These patchwise techniques can be generalized to other finite elements, as for instance for conforming spaces involving bilinear $Q1$ ansatz functions for both velocity and pressure. For this case, Becker/Rannacher [7] have performed similar techniques ("stringwise Vanka") and have come to convincing results. Moreover, these general blocking approaches are recently discussed by Lötzbeyer/Rüde [65] and Douglas [26] who applied similar techniques for solving Poisson problems. We expect this approach to be very efficient in future (see the FEAST project [1]), becoming a standard tool for solving large linear systems, in particular on fast computer platforms. Due to the splitting into regular outer blocks and small patches containing the bad anisotropies, the resulting convergence rates show a very efficient and robust behaviour and the resulting CPU times are small compared to classical pointwise relaxation schemes if the fast superscalar architecture including efficient cache handling is exploited. Moreover, in combination with the *pressure Schur complement* approach, these techniques are easily applicable to systems of equations and hence, they provide an alternative approach to classical ILU–schemes.

The last point in this section concerning *local MPSC* schemes is the question for related existing methods. We have already introduced the general formulation involving $\tilde{S}$ as approximation for the complete velocity matrix S in the preconditioning step. If we choose

$$\tilde{S} = diag(S) \quad , \quad I(IEL) = 1 \quad \text{for all elements } IEL \qquad (2.76)$$

we obtain exactly a variant of the classical *Vanka* scheme [114]. This scheme successively treats pressure and velocity for each element separately such that for the $\tilde{Q}1/Q0$ elements the resulting local *pressure Schur complement* problems are one-dimensional. More recent information about this relaxation scheme can be found in the work by Schieweck [90]. The efficiency of this scheme if applied as smoother in a standard multigrid approach is quite good, especially in combination with a direct fully coupled solution approach for low or medium Reynolds numbers. However, the problems with regard to mesh deformations and triangulations containing large *aspect ratios* are well-known (see [90]). We demonstrate some of these deficiencies in our subsequent numerical examples.

One easy remedy is to replace $\tilde{S}$ by S itself which requires the solution of small 4×4 velocity subproblems on each element while the size of the *pressure Schur complement* subproblems is still one. The robustness behaviour is improved in comparison to the classical *Vanka* scheme, but it is not satisfying at all. The final solution step is to increase the number of elements in certain patches j, $I(j) \geq 1$, if necessary by applying the introduced blocking strategies.

Let us shortly summarize the advantages and disadvantages of the *local MPSC* schemes which were introduced in this section:

1. In our present numerical tests the proposed blocking strategies lead to very robust and efficient convergence behaviour if the *local pressure Schur complement* basic iteration is applied as a smoother in a standard multigrid approach. Our theoretical considerations let expect multigrid convergence rates which should be (almost) independent of the domain Ω, of the underlying shape and size of the mesh and of the viscosity parameter if appropriate stabilization techniques for the convective term are applied. Hence, the *local MPSC* techniques provide very robust solvers for linear problems of the type

$$S\mathbf{u} + kBp = \mathbf{g} \quad , \quad B^T\mathbf{u} = 0. \tag{2.77}$$

2. If the patch size is sufficiently small (100 – 1,000 elements) such that
   the complete inverse *local pressure Schur complement* matrix A_i^{-1} fits
   into the cache, the resulting run-time behaviour in combination with
   machine optimized linar algebra tools (BLAS) is excellent.

3. The proposed adaptive blocking strategy is very flexible and robust. In
   fact, for every time step or nonlinear iteration and even on every grid
   level, the optimal blocking structure may be different.

4. In combination with a *Schur complement* approach this technique is
   easily applicable to systems of equations and can be an alternative
   to ILU–schemes. However, it must be stated that it is a hard job to
   implement these techniques, particularly if computational efficiency is
   desired.

5. The advantageous robustness behaviour can turn into a disadvantage
   since this solver is additionally robust against the time step parameter k. This is nice for low Reynolds number flow if large time steps are
   possible, but for highly nonstationary flows there are often physical reasons for small values k. However, the resulting multigrid convergence
   rates stay bounded away from 0 due to the incompressibility constraint.
   This seems to be in contrast to the *global MPSC* schemes which lead to
   convergence rates $\rho_{gMPSC} \to 0$ for decreasing time step k. This is not
   exactly true since still the Pressure–Poisson problem with the matrix
   $P = B^T M_l^{-1} B$ has to be inverted. In fact, the convergence rates for
   inverting P are the same as those obtained by the *local MPSC* for very
   small time steps, $\rho_{gMPSC} \approx \rho_{lMPSC}$. But the major difference is that
   the size of the pure pressure subproblems involved with P in the *global
   MPSC* is much smaller. This essential point concerning the numerical complexity of both *MPSC* approaches is highlighted in the next
   section which explains the consequences for optimized Navier–Stokes
   solvers. Further, the numerical examples show that these *local MPSC*
   schemes are the right counterparts for the *global MPSC* techniques in
   the range of low or medium Reynolds numbers. Both together lead to
   an optimized Navier–Stokes solver and both are absolutely necessary.

2.5   Resulting schemes
and relation to other existing methods

We have explained so far that the characterization of most existing solution approaches can be reduced to the following two major issues:

1. The kind of performed **ODE-solver** for the discretization of the time derivative (Explicit/Implicit Euler, Crank–Nicolson, Fractional–step–θ–scheme, etc.) and the applied **spatial discretization scheme** (finite differences, finite elements, finite volumes, etc.)

2. The strategy of solving (approximately) generalized **discretized Navier-Stokes problems** in each time step,

$$S\mathbf{u} + kBp = \mathbf{g} \quad , \quad B^T\mathbf{u} = 0 . \tag{2.78}$$

Again,we explicitly assume the following structure of the velocity matrix S,

$$S := \alpha M + \theta_1 \nu k L + \theta_2 k K , \tag{2.79}$$

with parameters $\alpha, \theta_1, \theta_2$ and k and right hand side $\mathbf{g}$ depending on the performed discretizations in space and time.

Furthermore, the second aspect of the solution of such highly dimensional systems of equations – linear or nonlinear – can be reduced to the following three major topics, **treatment of the nonlinearity, treatment of the incompressibility** and **complete outer control**.

According to them, we can specify subtrees in the *Navier–Stokes tree* all of which represent different solution schemes. However, due to this splitting between discretization and solution part there is one common issue for all of the resulting schemes: If the discretization is fixed (for instance the Crank–Nicolson scheme together with a finite element discretization in space) they all result in exactly the same solution. Even if there seem to be completely different approaches, they all will provide the same velocity and pressure approximates if they obtain a solution at all. This can be enforced by requiring that the time step k is small enough or some iteration counter N (for the nonlinear solver) or L (for the linear coupled solver) are sufficiently large. Since all of them can be interpreted as special **solution methods** for the discrete problem (2.78), they are now comparable with respect to robustness and efficiency and hence with respect to the accuracy of the resulting time stepping scheme, too.

This fact can also be read from the *Navier–Stokes tree* (see Section 2.1). As can be seen, we have at the top of the tree two different possible approaches:

1. First we perform an outer nonlinear scheme of *fixed point defect correction type* (see [101]) which will be explained more in detail in Chapter 3. As a result, this approach requires the solution of linearized discrete Navier–Stokes, resp., Oseen equations in each nonlinear step. These linear coupled equations which all are of the same structure

$$Su + kBp = \mathbf{g} \quad , \quad B^T \mathbf{u} = 0, \qquad (2.80)$$

can be treated by one of the proposed *local* or *global pressure Schur complement* schemes. Here we have two additional free parameters: When do we stop the nonlinear iteration, resp., what is the upper limit N_{max} of nonlinear iteration steps? How many *PSC* iteration steps L are allowed to "solve" the resulting linear coupled problems via *pressure Schur complement* techniques? How is the corresponding velocity approximation determined if the solution process for the coupled systems is interrupted at L_{max}? All these techniques span the left subtree in the *Navier–Stokes tree* and are called **Galerkin** schemes in our context.

2. The other approach corresponds to the right subtree and denotes the methods of **projection** type. First, we perform as outer iteration the *global pressure Schur complement* approach, applied to the fully nonlinear or already linearized equations (2.80). In the inner loop we first have to solve the corresponding velocity equations for $\tilde{\mathbf{u}}$ involving the solution of linear or nonlinear transport–diffusion problems with operator S. This nonlinear problem can be solved again with a quasi-Newton approach. Then, the pressure is updated and a corresponding velocity field is adjusted. This procedure corresponds to $L = L_{max} = 1$ (nonlinear) *pressure Schur complement* steps and can be repeated until the equation (2.80) is solved, or if $L \geq L_{max}$ for a prescribed L_{max}. If the iteration counter L_{max} is sufficiently large, we obtain exactly the same velocity field and pressure as calculated by the *Galerkin* approach. Since every iteration step requires the solution of nonlinear Burgers equations in the inner loop (or at least of linearized versions) we omitted this approach with $L > 1$ up to now. However, we have to comment that this technique might be a good candidate for a nonlinear multigrid approach.

   We always apply in our numerical tests the case of exactly $L = 1$ outer decoupling steps, with corresponding linear or nonlinear inner loops for S^{-1}. This approach is related very much to the classical *projection* methods and, in fact, analogous discrete versions of schemes proposed by Chorin [22] or Van Kan [115] can be easily derived (*discrete projection*, see [103]).

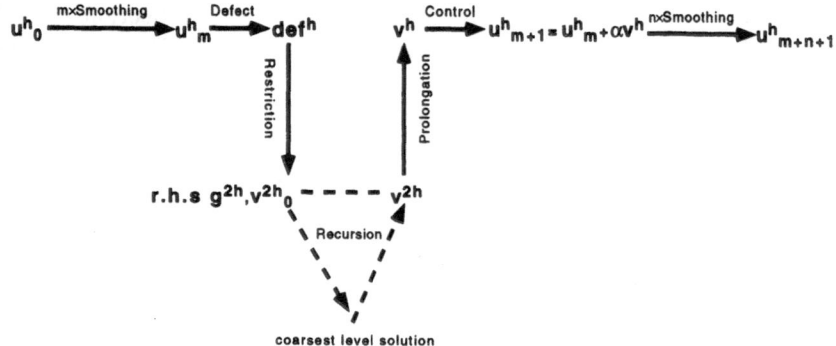

Figure 2.3: Standard multigrid cycle $MPSC(N, \cdot, \cdot)$

Additionally, we have the possibility to apply the *pressure Schur complement* schemes as simple basic iterations on the finest mesh only – this is denoted by SPSC schemes ('S' for 'single grid') and corresponds in the *global PSC* context to the standard *projection*, resp., *pressure correction* schemes – or as smoothers in a multigrid approach (called MPSC schemes, 'M' for 'multigrid').

The standard multigrid approach is addressed in Chapter 3. Nevertheless, we want to explain carefully the typical run of a standard multigrid MPSC algorithm to highlight the characteristics of the *MPSC* schemes.

We assume that we want to solve a (discrete) linear system of equations

$$A_N u_N = f_N \,. \tag{2.81}$$

Further, there exists a *hierarchy of levels* $i, i = 1, \ldots, N$, which may be connected with a *mesh size* parameter h_i, for instance. On each of these levels, we have to be able to define the discrete problem matrix A_i and the corresponding right hand side f_i. f_N is given a priori on the finest level N only, while all other terms f_i are generated during the multigrid run. Then, the N-level multigrid algorithm $MPSC(N, \cdot, \cdot)$ for the solution of (2.81) reads as follows:

Performing one iteration step $MPSC(N, u_N^0, f_N)$ yields, for a given initial guess u_N^0, the new approximate u_N^{m+n+1}, which we may write as

$$MPSC(N, u_N^0, f_N) = u_N^{m+n+1} \,. \tag{2.82}$$

Each run of $MPSC(k, \cdot, \cdot)$ is called one *cycle* of the multigrid iteration, and
the application of sufficiently many *cycles* on level N ensures the (approximate) solution of the problem (2.81). The most important components of
such a typical *MPSC* multigrid algorithm are:

- matrix–vector multiplication with matrix A_k

- smoothing operator, resp., performed *basic iteration*

- grid transfer routines (*prolongation* operator I^k_{k-1} and *restriction* operator I^{k-1}_k)

The k-level iteration $MPSC(k, u^0_k, f_k)$:

The k-level iteration with initial guess u^0_k yields an approximation to
u_k, the solution of

$$A_k u_k = f_k . \qquad (2.83)$$

One step can be described in the following way:

For $k = 1$, $MPSC(1, u^0_1, f_1)$ is the exact solution:
$MPSC(1, u^0_1, f_1) = A^{-1}_1 f_1$.

For $k > 1$, there are four steps:

1) m-Presmoothing steps

Apply m smoothing steps to u^0_k with a *basic iteration* to obtain u^m_k.

2) Correction step

Calculate the restricted residual (with *restriction operator* I^{k-1}_k)

$$f_{k-1} = I^{k-1}_k (f_k - A_k u^m_k) , \qquad (2.84)$$

and let u^i_{k-1} ($1 \le i \le p$, $p \ge 1$) be defined recursively by

$$u^i_{k-1} = MPSC(k - 1, u^{i-1}_{k-1}, f_{k-1}), \, 1 \le i \le p, \, u^0_{k-1} = 0 . \qquad (2.85)$$

3) Step size control of the correction

Calculate u_k^{m+1} (with *prolongation operator* I_{k-1}^k) via

$$u_k^{m+1} = u_k^m + \alpha_k I_{k-1}^k u_{k-1}^p, \qquad (2.86)$$

where α_k may be a fixed value or chosen adaptively so as to minimize the error $u_k^{m+1} - u_k$ in an appropriate norm, for instance in the discrete energy norm by

$$\alpha_k = \frac{(f_k - A_k u_k^m, I_{k-1}^k u_{k-1}^p)_k}{(A_k I_{k-1}^k u_{k-1}^p, I_{k-1}^k u_{k-1}^p)_k}. \qquad (2.87)$$

4) n-Postsmoothing steps

Analogously to step 1), apply n smoothing steps to u_k^{m+1} and obtain u_k^{m+n+1}.

In our *multilevel pressure Schur complement* framework we have two possibilities to perform this described multigrid approach:

1. **Local MPSC approach in primary formulation for velocity and pressure:**

   The problem matrix A and the unknown u read

   $$A := \begin{bmatrix} S & kB \\ B^T & 0 \end{bmatrix}, \quad u := \begin{bmatrix} \mathbf{u} \\ p \end{bmatrix},$$

   while the basic iteration may be written (in the Jacobi-like approach) as

   $$\begin{bmatrix} \mathbf{u}^l \\ p^l \end{bmatrix} = \begin{bmatrix} \mathbf{u}^{l-1} \\ p^{l-1} \end{bmatrix} - \omega^l \sum_{i=1}^{NP} \begin{bmatrix} \tilde{S}_{|\Omega_i} & kB_{|\Omega_i} \\ B_{|\Omega_i}^T & 0 \end{bmatrix}^{-1}$$
   $$\times \left(\begin{bmatrix} S & kB \\ B^T & 0 \end{bmatrix} \begin{bmatrix} \mathbf{u}^{l-1} \\ p^{l-1} \end{bmatrix} - \begin{bmatrix} \mathbf{g} \\ 0 \end{bmatrix} \right).$$

The cost intensive part is the smoothing step, analogously to standard multigrid techniques for typical scalar problems, i.e., Poisson- or convection-diffusion equations. We apply the *local pressure Schur complement* basic iteration only in this multigrid context, as *local MPSC*.

2. **Global MPSC approach in scalar formulation for the pressure only:**

In this formulation, the problem matrix A and the unknown u have the form

$$A := B^T S^{-1} B \quad , \quad u := p,$$

while the basic iteration reads

$$p^l = p^{l-1} - [\alpha_R P^{-1} + \alpha_D M_p^{-1}](B^T S^{-1} B p^{l-1} - \frac{1}{k} B^T S^{-1} \mathbf{g}).$$

The cost intensive part beside smoothing is also the matrix-vector multiplication with $A = B^T S^{-1} B$ which is needed for smoothing, defect calculation and adaptive coarse grid correction. In contrast to standard multigrid for typical scalar problems, i.e., Poisson- or convection-diffusion equations, the global MPSC schemes require the solution of such Poisson- or convection-diffusion problems S^{-1} even in each matrix-vector multiplication step. As we will show later, the corresponding numerical cost are "neglegible" in the case of highly nonstationary flows, but for low or medium Reynolds numbers the effort increases. Nevertheless, the numerical tests show that the resulting multigrid solvers are optimal in that sense that the convergence rates are excellent and particularly independent of domain and mesh anisotropies. And we claim that these *global MPSC* schemes are not only robust but also very efficient solution tools. Especially if a careful implementation is performed, some of these expensive matrix-vector multiplications can be saved.

More details about the other components, *adaptive coarse grid correction* and *grid transfer*, are presented in Chapter 3. Additionally, we give there some background information how to perform efficiently the subproblems involved with S^{-1} and P^{-1}.

If we come back to the components of the *Navier–Stokes tree* we see immediately that the *Galerkin* schemes (involving N_{max} and L_{max} sufficiently large) represent the most robust schemes which should provide the largest time steps k in nonstationary embeddings. In fact, the time step size has to be adjusted only by accuracy reasons, but not due to robustness requirements. For instance, special conditions which impose restrictions onto time step k with respect to the mesh width h (of *CFL condition* type) are not necessary in this fully implicit approach. However, this strong accuracy and robustness behaviour has to be payed by larger numerical cost due to (many) nonlinear steps requiring coupled Oseen problems. In contrast, certain linearization techniques for the nonlinearity including even a fully explicit treatment of the convective term decrease the cost for each iteration step, but the strong robustness and accuracy are lost. The same may happen if the incompressibilty condition is not satisfied sufficiently accurate by setting L_{max} too small.

The simplest scheme that we can derive with our approach includes a fully explicit treatment of the convective term (however being formally of 2nd order), inside of a *projection scheme* approach (which is also of 2nd order accuracy in time). In combination with the Crank–Nicolson scheme we obtain a well-know variant of *projection* schemes which is 2nd order accurate in time and which involves – in 2D – the solution of exactly two Helmholtz- and one Pressure–Poisson problem in each time step. In contrast, if we combine the Crank–Nicolson with the solution of generalized nonlinear discrete Navier–Stokes problems in each time step, we also obtain scheme which is formally of exactly the same order. What is the resulting time step depending on the type of problem and the underlying spatial mesh, and what are the corresponding CPU cost to solve the complete problem? These questions cannot be answered by theoretical considerations but we are able to examine them by computational experiments. These will serve as *numerical proofs* which are needed to allow ratings of resulting Navier–Stokes solvers. In fact, these results show large differences in the numerical behaviour of these seemingly different approaches which nevertheless are formally of second order accuracy. In addition, we can even show that on very anisotropic spatial meshes which are typically designed to improve the boundary layer resolution, the fully nonlinear approach is the only robust and efficient candidate for a "discrete Black Box" solver.

As representants for many approaches which typically are performed by the CFD community we have introduced in [105] some concretely specified schemes. The used notation consists of two parts: first the outer iteration and secondly the (approximate) solver for the resulting inner problems. This notation explains what is treated first as outer iteration: the nonlinearity or the incompressibility. And then, which of both problems is approximated in the inner loop, correspondingly to the *Navier–Stokes tree*.

In the following, we restrict to some typical schemes of exemplary character which will be examined more carefully in the later numerical tests. Here, they are ordered with respect to their expected robustness and implicitness.

The schemes on the "highest level" are the *Galerkin* schemes which solve "exactly" the discrete nonlinear coupled problems

$$Su + kBp = \mathbf{g} \quad , \quad B^T\mathbf{u} = 0, \tag{2.88}$$

while on the "lowest level" the *1-step projection* scheme with explicit treatment of the nonlinearity are positioned.

Typical example for the Galerkin schemes:

Problem: $Su + kBp = \mathbf{g} \quad , \quad B^T\mathbf{u} = 0$

with: $Su := [\alpha M + \theta_1 \nu kL + \theta_2 kK(\mathbf{u})]\mathbf{u}$ (nonlinear operator)

Given: Iterates $\mathbf{u}^{l-1}$, p^{l-1}

Perform: One *fixed point defect correction* step to obtain $\mathbf{u}^l$, p^l

$$\begin{bmatrix} \mathbf{u}^l \\ p^l \end{bmatrix} = \begin{bmatrix} \mathbf{u}^{l-1} \\ p^{l-1} \end{bmatrix} - \omega^l \begin{bmatrix} \tilde{S}(\mathbf{u}^{l-1}) & kB \\ B^T & 0 \end{bmatrix}^{-1}$$
$$\times \left(\begin{bmatrix} S & kB \\ B^T & 0 \end{bmatrix} \begin{bmatrix} \mathbf{u}^{l-1} \\ p^{l-1} \end{bmatrix} - \begin{bmatrix} \mathbf{g} \\ 0 \end{bmatrix} \right)$$

with: linear preconditioner $\tilde{S}(\mathbf{u}^{l-1}) := \alpha M + \theta_1 \nu kL + \theta_2 kK(\mathbf{u}^{l-1})$

$\implies$ Solve a linear Oseen equation in each nonlinear step.
$\implies$ This can be done either by *local* or *global MPSC* schemes.
$\implies$ The (linear) convection term depends on the last iterate $\mathbf{u}^{l-1}$.
$\implies$ Iterate more nonlinear steps l if necessary.

These methods provide "direct solvers" for stationary (generalized) Navier–Stokes equations and due to their fully implicit character they are the most robust and accurate time stepping schemes. In fact, the resulting time steps to reach a desired accuracy are the largest ones. However, the cost for one time step are most, too. Additionally, due to their fully implicit *Galerkin* character they are the only variants which seem to allow a rigorous a posteriori error control (see Chapter 3).

The cost can be diminished by weakening the threshold parameters or by applying only a fixed small number of nonlinear and/or linear steps. However, the accuracy and robustness behaviour may be weakened, too.

Typical example for the nonlinear "1-step projection" schemes:

Given: Iterates $\mathbf{u}^{n-1} = \mathbf{u}(t_{n-1})$ and $p^{n-1} = p(t_{n-1})$
from time level $n - 1$.

Perform: 5 substeps to obtain $\mathbf{u}^n$ and p^n on the new time level n

1. Solve for an intermediate velocity $\tilde{\mathbf{u}}^n$ the corresponding nonlinear Burgers equations. The right hand side contains the original load vector from time level t_n and the last pressure iterate p^{n-1}.

$$S\tilde{\mathbf{u}}^n = S(\tilde{\mathbf{u}}^n)\tilde{\mathbf{u}}^n = \mathbf{g} - kBp^{n-1}$$

2. Calculate the right hand side f_p for the "Pressure–Poisson problem" involving the discrete divergence of $\tilde{\mathbf{u}}^n$.

$$f_p = \frac{1}{k}B^T\tilde{\mathbf{u}}^n$$

3. Solve an update–equation for the pressure ("Pressure–Poisson problem") with the reactive preconditioning matrix R $[= P = B^T M_l^{-1} B]$ and the right hand side f_p.

$$Rq = f_p$$

4. Update the new pressure p^n by additionally applying the diffusive preconditioner D $[= M_p]$ which is a diagonal matrix in general.

$$p^n = p^{n-1} + \alpha_R q + \alpha_D D^{-1} f_p$$

5. Update the new velocity $\mathbf{u}^n$ to satisfy the incompressibility constraint.
$$\mathbf{u}^n = \tilde{\mathbf{u}}^n - kM_l^{-1}Bq$$

These schemes are applied to nonstationary flows only. They provide an "exact" treatment of the nonlinearity, but only depending on the last pressure iterate. We can expect smaller time steps than compared to the *Galerkin* schemes, but the cost for one time step are cheaper, too. In fact, we applied these schemes (as *Discrete Projection Method*) in the *1995 DFG Benchmark* (see [87]). For fully nonstationary flows with dominating convective term and on complex domains, this approach is up to now our favourized one, even if a rigorous error control in time is not clear. The resulting solutions satisfy the continuity equation in (2.88), but the discrete momentum equation only approximately.

Typical example for semi–implicit "1-step projection" schemes:

Given: Iterates $\mathbf{u}^{n-1(2)} = \mathbf{u}(t_{n-1(2)})$ and $p^{n-1} = p(t_{n-1})$ from time level $n - 1(2)$.

Perform: 5 substeps to obtain $\mathbf{u}^n$ and p^n on the new time level n

1. Solve for an intermediate velocity $\tilde{\mathbf{u}}^n$ the corresponding **linear** convection–diffusion equation involving the linearized operator $\hat{S}$. The right hand side $\hat{\mathbf{g}}$ contains the modified load vector from time level t_n (due to the performed linearization strategy) and the last pressure iterate p^{n-1}.

$$\hat{S}\tilde{\mathbf{u}}^n = \hat{\mathbf{g}} - kBp^{n-1}$$

2. Calculate the right hand side f_p for the "Pressure–Poisson problem" involving the discrete divergence of $\tilde{\mathbf{u}}^n$.

$$f_p \;=\; \frac{1}{k}B^T\tilde{\mathbf{u}}^n$$

3. Solve an update–equation for the pressure ("Pressure–Poisson problem") with the reactive preconditioning matrix R $[= P = B^T M_l^{-1} B]$ and the right hand side f_p.

$$Rq = f_p$$

---

4. Update the new pressure p^n by additionally applying the diffusive preconditioner D $[= M_p]$ which is a diagonal matrix in general.

$$p^n = p^{n-1} + \alpha_R q + \alpha_D D^{-1} f_p$$

5. Update the new velocity $\mathbf{u}^n$ to satisfy the incompressibility constraint.

$$\mathbf{u}^n = \tilde{\mathbf{u}}^n - k M_l^{-1} B q$$

These schemes are discrete counterparts of the classical schemes proposed for instance by Chorin [22] or Van Kan [115]. Moreover, the *discrete projection-2* schemes by Gresho [41] are included, too. In every time step, 1 Pressure–Poisson problem and 2 linear (nonsymmetric) transport–diffusion problems, one in in each velocity component, have to be solved. Hence, the cost per time step are comparatively small. The resulting solutions satisfy "exactly" the continuity equation in (2.88), but the corresponding discrete momentum equation only approximately.

Typical example for semi–explicit "1-step projection" schemes:

Given: Iterates $\mathbf{u}^{n-1(2)} = \mathbf{u}(t_{n-1(2)})$ and $p^{n-1} = p(t_{n-1})$ from time level $n - 1(2)$.

Perform: 5 substeps to obtain $\mathbf{u}^n$ and p^n on the new time level n

1. Solve for an intermediate velocity $\tilde{\mathbf{u}}^n$ the corresponding **linear** Helmholtz equations involving the **symmetric positiv definite** operator $\hat{S}$. The right hand side $\hat{\mathbf{g}}$ contains the modified load vector from time level t_n, the explicitly evaluated convective term and the last pressure iterate p^{n-1}.

$$\hat{S}\tilde{\mathbf{u}}^n = \hat{\mathbf{g}} - k B p^{n-1}$$

2. Calculate the right hand side f_p for the "Pressure–Poisson problem" involving the discrete divergence of $\tilde{\mathbf{u}}^n$.

$$f_p \;=\; \frac{1}{k} B^T \tilde{\mathbf{u}}^n$$

---

3. Solve an update–equation for the pressure ("Pressure–Poisson problem") with the reactive preconditioning matrix R $[= P = B^T M_l^{-1} B]$ and the right hand side f_p.

$$Rq = f_p$$

4. Update the new pressure p^n by additionally applying the diffusive preconditioner D $[= M_p]$ which is a diagonal matrix in general.

$$p^n = p^{n-1} + \alpha_R q + \alpha_D D^{-1} f_p$$

5. Update the new velocity $\mathbf{u}^n$ to satisfy the incompressibility constraint.

$$\mathbf{u}^n = \tilde{\mathbf{u}}^n - k M_l^{-1} B q$$

These schemes are also discrete counterparts of the classical projection schemes, but now with an explicit treatment of the complete convective term. Hence, in every time step 1 Pressure–Poisson problem and 2 linear Helmholtz equations, one in each velocity component, have to be solved. The cost per time step seem to be the smallest since only the same symmetric matrices are involved. However, we may have to satisfy a "hidden" *CFL condition* due to stability reasons. These semi–explicit methods are often used in codes acting with higher order spectral elements or finite differences. In fact, they can provide very high computer performance results on regular meshes due to possible regular data structures. However, we have to analyse how they perform in general domains, in comparison to the more expensive implicit schemes. The resulting solutions satisfy again the continuity equation in (2.88).

In the numerical section we will give more details about the schemes actually performed. In addition, we will specify necessary parameters as for instance threshold parameters for stopping criterions, maximum number of linear or nonlinear iteration steps, number of basic iterations, resp., number of performed smoothing steps, type of multigrid cycle and many more. By purely theoretical considerations we have arranged all proposed schemes in a scale which is mainly oriented at the stability and robustness behaviour. However, since we fix the spatial mesh, all schemes potentially lead to the same solution due to the implicit defect-correction principle, if the time step k is sufficiently small or if the threshold parameters are appropriately adjusted.

Therefore, the stability and robustness behaviour is not so decisive. In fact, the more important question is:

> 'What is the required total numerical work to obtain a certain accuracy? This involves the measurement of number of time steps, nonlinear iteration steps, linear multigrid sweeps, but the final measure is the elapsed CPU time to achieve a defined accuracy (for a given spatial mesh)!'

The answer to this question which involves mathematical aspects as accuracy, robustness and efficiency as well as computational issues, is the main topic of our subsequent numerical tests.

We finish this chapter about the mathematical derivation of general Navier–Stokes solvers with several additional remarks concerning the characteristic behaviour of our preferred $\tilde{Q}1/Q0$ Stokes element. It is a quadrilateral element which (in 2D) uses piecewise *rotated bilinear* shape functions for the velocities, spanned by $\langle x^2 - y^2, x, y, 1 \rangle$, and piecewise constant pressure approximations (see Figure 2.4).

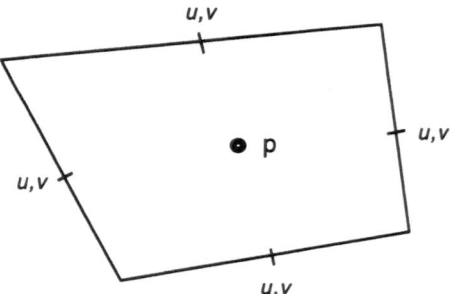

Figure 2.4: Nodal points of the nonconforming finite element pair $\tilde{Q}1/Q0$

The nodal values are the mean values of the velocity vector over the element edges (or the midpoint values), and the mean values of the pressure over the elements rendering this approach "nonconforming". An analogous approach on hexahedral elements is easily derived and implemented in the FEATFLOW software package.

We show that the spatial discretization performed by these finite elements together with the Crank–Nicolson- or the Fractional-step-θ-scheme leads to an excellent solver for nonstationary problems in 2D and especially in 3D. The essential point is hereby the application of the derived global PSC preconditioners as basic iterations. Moreover, the *domain decomposition–multigrid* ansatz in the corresponding *local MPSC* guarantees excellent performance for stationary problems, particularly in combination with data-splitting parallel approaches (see Schieweck [90]).

For a better understanding of the computational performance of the *PSC* schemes in combination with the $\tilde{Q}1/Q0$ Stokes element, the following two considerations are essential.

I) Complexity of the coupled solution methods:

Since mainly iterative methods are considered, a natural measure for the numerical effort are the cost for a matrix–vector multiplication (MV) with the (coupled) matrix A,

$$A = \begin{bmatrix} S & kB \\ B^T & 0 \end{bmatrix}. \tag{2.89}$$

This is one of the most time–consuming operations in a usual multigrid or conjugate gradient iteration since it is needed in defect calculations or applications of smoothing operators. In the following we express the cost in terms of the number of elements (NEL), for three common finite element discretizations which seem to have about the same accuracy and which use asymptotically the same number of elements NEL in 2D and 3D. The Stokes elements are:

a) $Q1/Q1$ bi/trilinear velocity and pressure approximations at vertices ([36], [58])

b) $Q1/Q0$ bi/trilinear velocity at vertices, constant pressure in cells ([36], [58])

c) $\tilde{Q}1/Q0$ nonconforming *rotated multilinear* velocity, constant pressure in cells ([84])

We get the following complexity estimates for one matrix–vector multiplication on tensor product meshes (*d.o.f.* stands for degrees of freedom). Similar results – being even identical for the nonconforming approach – can be obtained on general grids.

Case $Q1/Q1$:

d.o.f.(each velocity component) = *d.o.f.*(pressure) $\approx$ number of elements = NEL
stencil(S) = 9 (27 in 3D), stencil(B) = stencil(S)

- 1 MV in 2D: $2 \cdot (9 \, (\sim S u_i) + 9 \, (\sim B_i p)) + 2 \cdot 9 \, (\sim B_i^T u_i)$ $\approx$ **54**
NEL

- 1 MV in 3D: $3 \cdot (27 \, (\sim S u_i) + 27 \, (\sim B_i p)) + 3 \cdot 27 \, (\sim B_i^T u_i)$ $\approx$ **243**
NEL

Case $Q1/Q0$:

d.o.f.(each velocity component) $\approx$ *d.o.f.*(pressure) = NEL
stencil(S) = 9 (27 in 3D), stencil(B) = 4 (8 in 3D)

- 1 MV in 2D: $2 \cdot (9 \, (\sim S u_i) + 4 \, (\sim B_i p)) + 2 \cdot 4 \, (\sim B_i^T u_i)$ $\approx$ **34**
NEL

- 1 MV in 3D: $3 \cdot (27 \, (\sim S u_i) + 8 \, (\sim B_i p)) + 3 \cdot 8 \, (\sim B_i^T u_i)$ $\approx$ **129**
NEL

Case $\tilde{Q}1/Q0$:

d.o.f.(velocity component) $\approx$ 2 (3 in 3D) *d.o.f.*(pressure), *d.o.f.*(pressure) =
NEL
stencil(S) = 7 (11 in 3D), stencil(B) = 4 (6 in 3D)

- 1 MV in 2D: $2 \cdot (2 \cdot 7 \, (\sim S u_i) + 4 \, (\sim B_i p)) + 2 \cdot 4 \, (\sim B_i^T u_i)$ $\approx$
44 NEL

- 1 MV in 3D: $3 \cdot (3 \cdot 11 \, (\sim S u_i) + 6 \, (\sim B_i p)) + 3 \cdot 6 \, (\sim B_i^T u_i)$ $\approx$
135 NEL

We notice that in 2D the linear algebra work is about the same while in 3D the
Stokes elements which use constant pressure approximations are cheaper by a
factor 2 (assuming the same discretization and convergence rates). However,
the element pairs $Q1/Q0$ and $Q1/Q1$ may lead to unstable pressure approximations ("checkerboard modes") and hence may need additional stabilization
while the nonconforming one is always stable (see [7],[36],[58],[84]).

II) Time invariability of the projection operator:

Assuming large Reynolds numbers, the corresponding time steps k are in
general small due the physical scales of motion. Therefore, we get the following relation for the operator S (with S being linear or nonlinear) from
the discretization of the nonstationary equations (with M mass matrix, L
discretized Laplacian, K discretized convection part, h mesh width)

$$S := M + \theta_1 \nu k L + \theta_2 k K \approx M + O(k) \quad \text{for } k \text{ being sufficiently small.} \quad (2.90)$$

Hence, S can be interpreted as a nonsymmetric (and nonlinear), but well conditioned perturbation of the mass matrix M for small values k. On the other hand we derive for the Schur complement equation and equivalently for the matrix A in (2.89) (with Δ_h being a discretized Laplacian)

$$\text{cond}(B^T S^{-1} B) = \text{cond}(B^T [M + O(k)]^{-1} B) \approx \text{cond}(\Delta_h) = O(h^{-2}). \quad (2.91)$$

This shows that if we assume that $\nu << 1$ and $k << 1$, e.g., large Reynolds number and hence small time steps k, the condition number of the coupled system is bounded from below by $O(h^{-2})$. This limit steems from a second order problem due to the incompressibility constraint, and no further improvement may be gained for even smaller k. Hence, we see that the coupled matrix A in (2.89) consists of a time dependent large part S (acting on the velocity components) which gets "better" for small k, and a much smaller projection part $B^T S^{-1} B \approx B^T M^{-1} B \approx B^T M_l^{-1} B$ (M_l the lumped mass matrix) acting on the pressure, which is (almost) time step invariant. However, this small part, corresponding to a discrete Laplacian, determines the overall convergence rate of the coupled solvers for small k!

The velocity matrix S can be usually "inverted" in a very robust and efficient way by standard solvers of multigrid or conjugate gradient type, if necessary in connection with a standard nonlinear iteration. Even on highly anisotropic grids very robust solvers are available if appropriate smoothers/preconditioners of modified ILU–type with renumbering strategies are used (see also [106],[118]). In many applications even one multigrid sweep is enough.

Therefore if we neglect the application of S^{-1}, the time consuming part is the application of $[B^T M_l^{-1} B]^{-1}$ or of another "Pressure–Poisson" operator Δ_h^{-1}. We consider the approach of constructing *pressure Schur complement* preconditioners on discrete level only, since in this case optimal (since exact) PSC preconditioning operators are available for $k \to 0$. Possible solution strategies are multigrid or conjugate gradient type methods, with diagonal smoothing or preconditioning, since the iteration matrix $[B^T M_l^{-1} B]^{-1}$ is only implicitly given.

Again the cost for performing one matrix–vector multiplication (MV) are a natural measure, now for the implicitly given matrix $\sum_{i=1}^d B_i^T M_l^{-1} B_i$ (or for other diagonal matrices S_{diag} instead of M_l) with space dimension $d = 2$ or $d = 3$. For comparison see the results corresponding to (2.89) which were also shown for tensor product meshes.

Case $Q1/Q1$:

- 1 MV in 2D: $2 \cdot (9 \, (\sim B_i^T) + 1 \, (\sim M_l^{-1}) + 9 \, (\sim B_i))$ $\approx$ **38 NEL**

- 1 MV in 3D: $3 \cdot (27 \, (\sim B_i^T) + 1 \, (\sim M_l^{-1}) + 27 \, (\sim B_i))$ $\approx$ **165 NEL**

Case $Q1/Q0$:

- 1 MV in 2D: $2 \cdot (4 \, (\sim B_i^T) + 1 \, (\sim M_l^{-1}) + 4 \, (\sim B_i))$ $\approx$ **18 NEL**

- 1 MV in 3D: $3 \cdot (8 \, (\sim B_i^T) + 1 \, (\sim M_l^{-1}) + 8 \, (\sim B_i))$ $\approx$ **51 NEL**

Case $\tilde{Q}1/Q0$:

- 1 MV in 2D: $2 \cdot (4 \, (\sim B_i^T) + 2 \, (\sim M_l^{-1}) + 4 \, (\sim B_i))$ $\approx$ **20 NEL**

- 1 MV in 3D: $3 \cdot (6 \, (\sim B_i^T) + 3 \, (\sim M_l^{-1}) + 6 \, (\sim B_i))$ $\approx$ **45 NEL**

This is an improvement over the previous results for the fully coupled solver, but it is not all that satisfactory, especially observing that exclusively diagonal preconditioners/smoothers can be used since the matrix is given only implicitly in product form. Hence, on anisotropic grids there may be no gain in robustness. The final step towards a robust and efficient solution technique is to evaluate (only once!) the matrix product and to store the result in compact form as

$$P := \sum_{i=1}^{d} B_i^T M_l^{-1} B_i \,. \tag{2.92}$$

The resulting storage and numerical effort for one matrix–vector multiplication with P is given in the following Proposition. The results are shown for tensor product meshes, but they are analogous for general meshes, and even identical in the case of the nonconforming finite elements.

Proposition 2 *(Storage and numerical work for the application of $P = B^T M_l^{-1} B$)*
We derive the following sizes for the pressure matrix P (that means, the number of non–zero matrix elements), and equivalently the cost for one matrix–

vector multiplication. The units are the numbers of elements (NEL) of the used mesh, and the improvement (given as factor) is related to the previous results for the coupled solution algorithm:

For Q1/Q1: **25 NEL** *in 2D (factor of 2) and* **125 NEL** *in 3D (factor 2).*

For Q1/Q0: **9 NEL** *in 2D (factor of 4) and* **27 NEL** *in 3D (factor 5).*

For Q̃1/Q0: **5 NEL** *in 2D (factor of 9) and* **7 NEL** *in 3D (factor 19).*

For the element pair $Q1/Q1$ we obtain a very large stencil for matrix P which corresponds to a discretization of the Laplacian using quadratic finite elements. In this case, continuous projection methods, leading to much smaller stencils with 9 or 27 entries, seem to be preferrable if we accept the spurious pressure boundary layers. Additionally, we should not forget that this element pair requires some additional stabilization techniques (see [7],[58]). The same is true for $Q1/Q0$ (see [36],[41]). The checkerboard instabilities can be detected studying the matrix structure of P,

$$Stencil(P) = \begin{bmatrix} \star & 0 & \star \\ 0 & \star & 0 \\ \star & 0 & \star \end{bmatrix}. \tag{2.93}$$

As a consequence, certain filtering techniques have to be used. Among the three proposed Stokes elements the only always stable pair of elements is the nonconforming one which, as we have seen, requires the least numerical effort. Evaluating the matrix entries on cartesian grids we obtain in 2D the well–known 5–point, resp., the 7–point stencil in 3D (we get analogous 5–point, resp., 7–point stencils on general meshes) which correspond in 2D to the most common finite difference discretization of the Laplacian with cell centered values:

$$Stencil(P) = \begin{bmatrix} 0 & -1 & 0 \\ -1 & 4 & -1 \\ 0 & -1 & 0 \end{bmatrix}. \tag{2.94}$$

This 5–point (or 7–point) discretization is the simplest admissible discretization in 2D (or 3D) and the corresponding numerical effort is minimal. Additionally, we are able to construct very robust solvers since well known ILU–techniques (see [118]) are applicable in this case.

Moreover, this Stokes element $\tilde{Q}1/Q0$ is similar to some well known "good old" discretization techniques, at least on cartesian grids. Almost the same matrix stencils have been used since years in connection with some finite difference or staggered grid discretizations. However, at the same time this finite element scheme can also be applied on general meshes, and allows a rigorous error analysis based on variational arguments. This clearly is a further advantage. In our opinion, this Stokes element together with the proposed discretization and solution techniques is the best choice among all other possibilities with regard to efficiency and robustness in complex frameworks. The point which is still unsolved is the question of accuracy, resp., the problem of reliable and efficient a posteriori error control in combination with fully adaptive grid refinement techniques. This is a major point of further research.

These considerations concerning numerical efficiency have shown that the global MPSC approach in combination with the proposed nonconforming finite elements leads to one of the most efficient solvers for nonstationary flow problems. This finally explains the high efficiency of our FEATFLOW results with respect to the *1995 DFG Benchmark* (see [87] and Section 1.1) in comparison to other codes. In contrast to many other approaches, we do <u>not</u> spent 90% or even more of the CPU time in solving the pressure subproblems. In fact, our multigrid-based solvers are optimized in such a way that the elapsed time for assembling matrices and right hand sides in every time step is of about the same magnitude as for the solution of the (linear) velocity or the pressure subproblems. And this is true on complex domains even for the fully nonlinear as well as the semi–implicit cases which involve the solution of highly nonsymmetric equations.

The proposed MPSC techniques can be analogously applied to other finite element pairs or even to different spatial discretization techniques, however the final computational efficiency of these combinations is still unknown to us and should be a point of further research. We do expect a similar behaviour, and particularly for higher order elements (as for instance $Q2/P1$) we recommend the MPSC approach to construct corresponding fast and robust solution schemes.

Chapter 3

Other mathematical components

In addition to Chapter 2, 'Derivation of Navier–Stokes solvers', we will explain other important mathematical components and tools which are necessary in the framework of numerical solution techniques for the incompressible Navier–Stokes equations. While Chapter 2 mainly concentrated on the aspect of 'how to solve efficiently the resulting discretized (coupled) systems of equations', we provide in the following sections supplementary components concerning discretization techniques, or error control mechanisms, and other necessary solver tools. It is self-evident that we cannot present these components with <u>all</u> their important and interesting details. Items like 'multigrid', 'numerical linear algebra', 'finite elements for Navier–Stokes equations' or 'time discretization techniques' fill many books separately. Therefore, we restrict to some special features only among these voluminous topics which are

important in the context of the incompressible Navier–Stokes equations.

necessary to understand the FEATFLOW software package.

new in the sense that they correspond to numerical experiences which are non-standard or which even contradict to some widely accepted views.

For the following discussion we lay our particular emphasis on the practical aspects of the considered components. Consequently, all items are explained not only theoretically, but they are also confirmed by accompanying numerical calculations if possible. In addition, we always try to support the reader with references with regard to further explanation of some specific topics. The major aspects we will highlight are:

1. **Finite element** spaces (including approximation and stability properties of Stokes elements, the nonconforming $\tilde{Q}1/Q0$ finite elements, stabilization techniques for convective terms via upwind or streamline–diffusion techniques, explicit construction of discretely divergence–free subspaces, a posteriori error control mechanisms).

2. **Time discretization** techniques (including the Fractional–step–θ–scheme and other One–step θ–schemes, adaptive time step control).

3. **Nonlinear iteration** schemes (including adaptive fixed point defect correction techniques, quasi–Newton schemes, stopping criterions, linearization techniques for nonstationary problems, least square CG methods).

4. **Multigrid** tools (including properties of simple basic iterations as smoothers or as preconditioners in Krylov–space methods, construction of grid transfer operators and coarse grid matrices, adaptive step-length control of the correction).

5. **Boundary conditions** (including natural *do nothing* conditions, pressure drop and flux settings, iterative implementation techniques, treatment of moving boundaries).

3.1   Finite element spaces

As a model problem, we consider first the following stationary Stokes system,

$$-\Delta \mathbf{u} + \nabla p = \mathbf{f} \quad , \quad \nabla \cdot \mathbf{u} = 0 \quad \text{in } \Omega \quad , \quad \mathbf{u} = \mathbf{g} \quad \text{on } \partial\Omega . \qquad (3.1)$$

The pair $\{\mathbf{u}, p\}$ represents the velocity and the pressure in a bounded region $\Omega \subset \mathbf{R}^2$ with sufficiently regular boundary $\partial\Omega$. The boundary values $\mathbf{g}$ on $\partial\Omega$ and the force $\mathbf{f}$ are given. For simplicity, we assume in the following that Ω is a convex polygon and that the boundary values $\mathbf{g}$ are homogeneous.

Defining the bilinear forms $a(\mathbf{u}, \mathbf{v}) := (\nabla \mathbf{u}, \nabla \mathbf{v})$ and $b(p, \mathbf{v}) := -(p, \nabla \cdot \mathbf{v})$, a usual weak formulation of the Stokes problem reads as follows:

Find a pair $\{\mathbf{u}, p\} \in H := \mathbf{H}_0^1(\Omega) \times L := L_0^2(\Omega)$, *such that*

$$a(\mathbf{u}, \mathbf{v}) + b(p, \mathbf{v}) \quad = \quad (\mathbf{f}, \mathbf{v}) \qquad \forall \, \mathbf{v} \in H \qquad (3.2)$$

$$b(q, \mathbf{u}) \quad = \quad 0 \qquad \forall \, q \in L. \qquad (3.3)$$

Here, $L_0^2(\Omega)$ and $\mathbf{H}_0^1(\Omega)$ are the usual Lebesgue and Sobolev spaces. Let $(\cdot, \cdot)_\Omega$ denote the inner product of $L_0^2(\Omega)$, with corresponding norm $\|\cdot\|_\Omega = \|\cdot\|_{0,\Omega}$. In the following, the subscript Ω is usually omitted. The notation is simultaneously used for spaces and norms of scalar as well as vector–valued functions in which case we often use boldfaced typesetting. The partial integration involved in the definition of the bilinear form $b(\cdot, \cdot)$ is necessary by two reasons: Since the partial differential operator is now applied onto a velocity field, lower order approximates for the pressure are possible. In fact, even piecewise constant pressure functions being in $L_0^2(\Omega)$ only are allowed. Further, this partial integration is important for the derivation of associated "stable" natural boundary conditions as can be seen later.

An equivalent "shorter" formulation with $\mathbf{V}(\Omega) = \{\mathbf{v} \in \mathbf{H}_0^1(\Omega) : \nabla \cdot \mathbf{v} = 0\}$ reads:

Find $\mathbf{u} \in \mathbf{V}(\Omega)$, *such that*

$$a(\mathbf{u}, \mathbf{v}) = (\mathbf{f}, \mathbf{v}) \quad \forall \, \mathbf{v} \in \mathbf{V}(\Omega). \qquad (3.4)$$

Problem (3.2) has a unique solution for any force $\mathbf{f} \in \mathbf{H}^{-1}(\Omega)$ (see, e.g., [36]), which is a consequence of the well known stability estimate (with $\beta > 0$)

$$\sup_{\mathbf{v} \in \mathbf{H}_0^1(\Omega)} \frac{(q, \nabla \cdot \mathbf{v})}{\|\nabla \mathbf{v}\|_0} \geq \beta \|q\|_0 > 0 \quad, \forall \, q \in L_0^2(\Omega), \, q \neq 0, \qquad (3.5)$$

and if $\mathbf{f} \in \mathbf{L}^2(\Omega)$, then the solution is in $\mathbf{H}^2(\Omega) \times \mathrm{H}^1(\Omega)$ and satisfies the a priori estimate

$$\|\mathbf{u}\|_2 + \|p\|_1 \leq c \|\mathbf{f}\|_0. \qquad (3.6)$$

For the approximation of problem (3.2), resp., (3.4), one chooses finite-dimensional spaces "$H_h \subset H$" and $L_h \subset L$ consisting of functions which are piecewise polynomial with respect to a (regular) decomposition $\mathbf{T}_h = \bigcup\{T\}$ into simple elements T (triangles, quadrilaterals, etc.). The symbol h is used as a parameter characterizing the maximum width of the elements of

$\mathbf{T}_h$. $\partial\mathbf{T}_h$ denotes the set of all boundary edges (faces) Γ of the elements $T \in \mathbf{T}_h$. Additionally, the family $\{\mathbf{T}_h\}_h$ is often assumed to satisfy the usual *uniform shape condition* (see [23]). The common edge between two elements $T_i, T_j \in \mathbf{T}_h$ is denoted by Γ_{ij} with corresponding midpoint m_{ij}. Analogously, we define the boundary edges $\Gamma_{i0} \subset (\partial\mathbf{T}_h \cap \partial\Omega)$ with the midpoints m_{i0}. To obtain the *fine* mesh $\mathbf{T}_h$ from a *coarse* mesh $\mathbf{T}_{2h}$ we simply connect opposing midpoints ("regular refinement"). Domain boundaries are respected, this means, all boundary nodes are on the true domain boundary. In the new grid $\mathbf{T}_h$, the old midpoints become vertices. For technical and programming reasons, we do not use adaptive techniques at this moment.

Since the spaces H_h are allowed to be nonconforming, i.e., $H_h \not\subset \mathbf{H}_0^1(\Omega)$, we have to work with elementwisely defined bilinear forms and corresponding energy norms:

$$a_h(\mathbf{u}_h, \mathbf{v}_h) := \sum_{T \in \mathbf{T}_h} \int_T \nabla\mathbf{u}_h \cdot \nabla\mathbf{v}_h \, dx, \quad \|\mathbf{v}_h\|_h := (a_h(\mathbf{v}_h, \mathbf{v}_h))^{1/2} \quad (3.7)$$

$$b_h(q_h, \mathbf{v}_h) := -\sum_{T \in \mathbf{T}_h} \int_T q_h \nabla\cdot\mathbf{v}_h \, dx. \tag{3.8}$$

Then, the corresponding discrete formulation of the Stokes problem in primary variables reads:

Find a pair $\{\mathbf{u}_h, p_h\} \in H_h \times L_h$, such that

$$a_h(\mathbf{u}_h, \mathbf{v}_h) + b_h(p_h, \mathbf{v}_h) = (\mathbf{f}, \mathbf{v}_h) \qquad \forall \mathbf{v}_h \in H_h \qquad (3.9)$$
$$b_h(q_h, \mathbf{u}_h) = 0 \qquad \forall q_h \in L_h, \quad (3.10)$$

where, in the nonconforming case, the subscript h refers to differentation and integration in the elementwise sense with respect to the mesh $\mathbf{T}_h$.

The necessary and sufficient conditions (see [17],[36]) for the existence and convergence of the approximations $\{\mathbf{u}_h, p_h\}$ are the

- **approximability property**

$$\inf_{\mathbf{v}_h \in H_h} \|\mathbf{v} - \mathbf{v}_h\|_h \leq ch^{m-1}\|\mathbf{v}\|_{\mathbf{H}^m} \quad \forall \mathbf{v} \in H \cap \mathbf{H}^m(\Omega) \quad (3.11)$$
$$\inf_{q_h \in L_h} \|q - q_h\|_0 \leq ch^{m-1}\|q\|_{H^{m-1}} \quad \forall q \in L \cap H^{m-1}(\Omega) \quad (3.12)$$

for some integer $m \geq 2$.

- **stability estimate** (*Babuška–Brezzi condition*)

$$\min_{q_h \in L_h/\mathbf{R}} \max_{\mathbf{v}_h \in H_h} \frac{b_h(q_h, \mathbf{v}_h)}{\|\mathbf{v}_h\|_h \|q_h\|_0} \geq \tilde{\beta} \qquad (3.13)$$

for some fixed constant $\tilde{\beta} > 0$ independent of h.

Under these conditions one has the following **approximation** result

$$\|\mathbf{u} - \mathbf{u}_h\|_0 + h \|\mathbf{u} - \mathbf{u}_h\|_h + h \|p - p_h\|_0 \leq ch^{m-1}\{\|\mathbf{u}\|_{\mathbf{H}^m} + \|p\|_{\mathbf{H}^{m-1}}\}. \quad (3.14)$$

There are numerous admissible pairs of finite element spaces ("Stokes elements") or, more general, of discrete spaces H_h and L_h proposed in the literature. They also include finite differences, finite volumes or spectral elements and even other approaches. However, there is one common interest: they have to satisfy the *approximability property*, the *stability estimate* and the corresponding linear systems of equations have to be efficiently solved. We want to discuss these aspects in the next section.

3.1.1 Criterions for the comparison of various Stokes elements

Stokes elements and the Babuška–Brezzi (BB) condition

There are several Stokes elements, for instance nonconforming linear or rotated multilinear finite elements for the velocity and piecewise constant pressure functions, which automatically satisfy the BB-condition. In addition, also certain higher order elements, as the Taylor-Hood element $(P2/P1)$ with quadratic ansatz function or the biquadratic/linear version $(Q2/P1)$, satisfy directly the stability estimate. There is a complete mathematical framework available (see for instance [17]) such that the rigorous mathematical derivation of stable element pairs is sufficiently solved. One major result is that in general the approximate velocity space H_h has to have at least a higher polynomial degree than the corresponding pressure space L_h.

However for several theoretical and practical reasons, it is sometimes desirable to work with equal order degree trial functions for velocity and pressure, particularly allowing the use of continuous pressure approximations. This ansatz would be unstable in general if used with the standard weak formulation in (3.9). However, due to a *least square* ansatz (see [58] or [17] and the literature cited therein) which adds certain "correction terms" in the variational formulation of the continuity equation, a modified stability condition can be satisfied.

These correction terms, for instance for the $Q1/Q1$ element (conforming bilinear velocity and pressure), are mostly chosen in such a way that the solution of the continuous problem satisfies exactly the discrete equation. Thus, this formulation introduces an additional stabilizing term in a consistent way, at least in the case of the linear Stokes equations. Similar constructions are possible for the $Q1/Q0$ element (conforming bilinear velocity and piecewise constant pressure) which will not be discussed here in more detail.

Summing up, there have been developed many stable Stokes elements in the last 30 years. And also certain "unstable" pairs (in a mathematical sense), for instance the widely-used $Q1/Q0$ element without any stabilization as filtering out *checkerboard modes*, show often a "stable" approximation behaviour in contrast to the analysed instabilities. Nevertheless, in many applications leads the failure of the BB-condition to numerical instabilities as spurious pressure oscillations or bad multigrid convergence behaviour, such that the mathematically rigorous demand for the *Babuška–Brezzi condition* has proven to be necessary for practical calculations.

However, even if the mathematical proof of such a stability condition has shown to be successful for many element pairs, there still remain problems which may occur in realistic applications:

- **Mesh dependence** of the *Babuška–Brezzi condition*

  Does the constant $\tilde{\beta}$ depend on the shape of the elements? For some stable elements this constant seems to be related to the *aspect ratio* such that the approximation property decreases dramatically on meshes with very flat cells. Moreover, in many stabilization approaches, a mesh-dependent term of order $O(h^{\alpha})$ occurs. What is h on a very anisotropic mesh? Results by Becker/Rannacher [7] show that the quantitatively incorrect choice of a local mesh width parameter "h" leads to over- or understabilization and deteriorates the resulting solution and approximation behaviour.

- **Problem dependence** of the stabilization technique

  How does the "right" stabilization term look like for the fully nonlinear Navier–Stokes equations since a consistent stabilization must be nonlinear? Further, how sensitively does the computational behaviour depend on additional constants occuring in the stabilization term, resp., can these quantities be fixed adaptively during the calculation?

- **Problem dependence** of the *Babuška–Brezzi* stability constant

  Is stability still guaranteed for the fully incompressible Navier–Stokes equations, resp., the linearized Oseen versions? And what happens for very small time steps k in which case the Stokes character can be neglected?

Stokes elements and resulting accuracy

The *approximability* property and the *approximation* result lead to the final accuracy estimates of order $O(h^m)$, resp., $O(h^{m-1})$ for the resulting discrete velocity $\mathbf{u}_h$ and pressure p_h, with respect to the (unknown) exact solutions of the Stokes system. Since m is the polynomial degree of the ansatz functions, consequently a higher order ansatz should lead to more accurate approximations and hence to a more efficient solution method since much less grid points are needed to achieve a prescribed accuracy – at least theoretically.

However, to exploit these higher order approximations, the *approximation* result requires a high degree of regularity, namely $\mathbf{H}^m$ for the velocity and H^{m-1} for the pressure. These regularity assumptions are often unrealistic in "real life" complex frameworks if boundary layers near geometrically complicated contours or large velocity and pressure gradients occur. From a computational point of view, there is no evidence that typical flow problems provide this needed higher degree of regularity. Therefore, what is necessary is a fair comparison of several Stokes elements for <u>realistic</u> flows to measure their accuracy and the corresponding total efficiency of the complete numerical scheme.

In fact, we started this project some years ago, but we never finished by multiple reasons. The most importing reason for never publishing the results was that we have compared certain element pairs for Stokes problems with a given (polynomial) exact solution only. These are not <u>realistic</u> flow problems which represent typical CFD problems!
However, with the knowledge from the previously described *1995 DFG Benchmark* we should perform at least Stokes flow for a similar framework (*'flow in a channel around a cylinder'*). And, the flow quantities which have to be measured should include the lift and drag coefficients. This project can (and will) be done. It is workable, at least on the basis of our program package FEAT which includes many possible classes of Stokes elements.

However, for comparing "real life" situations, we have to simulate the fully in-incompressible Navier–Stokes equations. Then, additional questions and problems may occur:

- What about the expected accuracy for the Navier–Stokes equations including the (nonlinear) convective term? How can stabilized discretizations be derived which also satisfy the required demand for higher order accuracy? There are (at least) two techniques proposed in the literature: upwind schemes and streamline–diffusion techniques. However, both are hard to optimize from a computational point of view, and it seems to be a very delicate task to adapt (by hand or during the calculation) the involved internal "free" parameters. Up to now, only few numerical experience with higher order approximations is available. Thus, we are at least sceptic about higher order ansatz spaces.

- How can the resulting discrete linear or nonlinear systems of equations be solved which arise from the discretization process of the Navier–Stokes equations? To do this in an efficient and robust way is not at all an easy job. In fact, it might be even impossible to derive "optimally" efficient solution approaches for these higher order discretizations.

If we take into account these theoretical approximation results which all are of a priori type, and if we compare them with numerical experience which follows directly from computational simulations, then the following question arises:

> 'Is the concept of achieving higher order accuracy by simply increasing the (local) polynomial degree of ansatz functions working at all for general CFD problems? Or are other concepts better suited if a certain accuracy shall be guaranteed?'

The background for this question is the following: Even if higher order approximations of order $O(h^m)$ are provided, there is nothing known about the size of the involved *approximation* and *stability* constants which may depend on the domain Ω, the size of derivatives of the solutions u and p, the mesh, the Reynolds number and many more. In fact, these constants can theoretically (and practically they do!) vary between 1 and 10^9. Consequently, a control mechanism, which can provide sharp estimates of a posteriori type for the resulting discrete solutions with respect to the (unknown) exact solution, is much more required: On the basis of the computational data only, a "sharp bound" for the resulting error is needed. And if these quantitative estimates are provided, a higher accuracy can be easily obtained by refining locally or globally the computational mesh in the regions where needed.

Thus, even with simple elements which are only linear and which provide by a priori estimates an $O(h^2)$ behaviour for the resulting error, arbitrarily accurate solutions can be obtained in general. This is the task of the recently developed a posteriori error control mechanisms which are shortly explained in a subsequent section. First numerical tests show their numerical reliability and efficiency such that on this basis a good compromise between "guaranteed accuracy" and "high efficiency" seems to be possible. We expect that this technique of error control by a posteriori mechanisms – on the basis of the calculated data only – will be part of many codes in the future. Software tools may use this error control in such a way that simple finite elements are performed in regions which require the accurate resolution of complex details via grid refinement, while in flow regions of greater smoothness higher order elements can be exploited. But, this has to be done always under the restriction that fast solvers are available and that the resulting error can be precisely controlled in a quantitative manner.

Stokes elements and corresponding solution tools

The third aspect which is important for the evaluation of the quality of certain Stokes elements is the availability of corresponding solvers. After discretizing the Navier–Stokes equations in space and time, very large nonlinear or linearized coupled systems of equations arise which have to be solved in every time step, resp., in each nonlinear iteration step even. Several issues have to be taken into consideration:

- First of all, it is wrong that a smaller number of total unknowns automatically leads to less storage demands. In addition to the vector length of the unknowns, the band width of the resulting stiffness matrices has to be respected. While in 3D the simple nonconforming rotated multilinear finite elements have a maximum band width of 11, in contrast the corresponding number for a conforming triquadratic ansatz is (at least) 125. If we take into account these band width results for all matrices involved – including also the gradient matrices B_i and the Pressure–Poisson matrix P – the resulting discrepancy can be very large. Consequently, many more grid points may be allowed for simpler elements to satisfy the same storage requirements as some higher order elements.

- Even for simple scalar problems, as Poisson- or transport–diffusion equations, it is a well-known fact that the numerical solution of discretized partial differential problems based on quadratic or even higher ansatz functions is disproportionately expensive. In contrast to simple constant or linear approaches, certain matrix properties as *strict diagonal dominance* or *M–matrix property* are not satisfied. Correspondingly, the derivation of fast and robust solution tools for coupled nonlinear

Navier–Stokes-like systems is still an open problem. This also shows the *1995 DFG Benchmark* and other comparative calculations.

- We expect that the derived solution tools of *global* or *local MPSC* type can lead to strong improvements with respect to efficiency and robustness. However, these are only theoretical considerations and have to be proven by corresponding numerical schemes and their implemented counterparts. And in particular, these tests have to be performed for "real life" CFD applications. Nevertheless, we are optimistic to claim that on the basis of the derived solution approaches the efficient computational solution of higher order discretized Navier–Stokes equations can be performed. One of our personal favourites is the $Q2/P1$ element (conforming bi/triquadratic velocities and piecewise discontinuous linear pressure) or the quadratic nonconforming relative (under progress by F. Schieweck) which seem to be well suited for many of our approaches.

If we take into account the three major requirements, stability, accuracy and efficiency, and if we compare these issues with our theoretical and numerical experience, our favourized Stokes element is clearly the $\tilde{Q}1/Q0$ element pair (nonconforming rotated multilinear velocity, constant pressure in cells [84]). The advantages in comparison to other element pairs are (at least with respect to our subjective experience) the following ones. Some will be explained more in detail in the next section.

- **Robustness:**

  This $\tilde{Q}1/Q0$ element pair satisfies the *Babuška–Brezzi condition* without any additional stabilization, and the stability constant seems to be independent of the shape and size of the used triangulation. In particular on meshes containing cells with very large *aspect ratios*, that means very long flat cells which are often needed to resolve boundary layers, the stability and hence the approximation property is always satisfied. In addition, robust upwind and streamline–diffusion mechanisms can be applied for higher Reynolds number flow.

- **Accuracy:**

  Looking at other first order elements ($Q1/Q0$, $Q1/Q1$, $P1/P1$) the accuracy seems to be comparable. On the other hand, $\tilde{Q}1/Q0$ is the only one among these which is unconditionally stable. Especially on highly anisotropic meshes, the approximation property seems to remain valid (see [7]). From technical and computational point of view, adaptive grid refinement can be performed in an analogous way. The only critical point which is not solved up to now is the rigorous derivation of a posteriori error estimators due to the nonconformity. This topic is just under research.

- **Efficiency:**

  As we have shown, the derived *local* and particularly the *global MPSC* solvers in the case of fully nonstationary problems seem to lead to "optimal" solution schemes. Our personal conclusion is that this finite element approach has the potential to be one important component of the desired "Black Box" solver for relevant CFD problems.

Since this $\tilde{Q}1/Q0$ Stokes element is our recently favourized discretization approach which is also implemented in the FEATFLOW package, we will discuss some of its specific properties more in detail.

3.1.2 Some properties of the nonconforming rotated multilinear spaces

The $\tilde{Q}1/Q0$ Stokes element uses *rotated bi/trilinear* shape functions for the velocity and piecewise constants for the pressure. It was introduced and analyzed in [84] and may be viewed as the natural quadrilateral analogue of the well-known triangular Crouzeix–Raviart element (see [36]). In numerical tests it has shown satisfactory stability and approximation properties and its 2D and 3D versions have been successfully implemented in the FEATFLOW code.

For the approximation of problem (3.2) by the finite element method we have introduced discrete spaces $H_h \approx \mathbf{H}_0^1(\Omega)$ and $L_h \approx L_0^2(\Omega)$. For this we use the reference element $\hat{T} = [-1, 1]^2$ and define for each $T \in \mathbf{T}_h$ the corresponding 1–1–transformation $\psi_T : \hat{T} \to T$. Then we set ("rotated bilinear elements", see [84])

$$\tilde{Q}_1(T) := \{ q \circ \psi_T^{-1} \mid q \in \text{span}\langle x^2 - y^2, x, y, 1 \rangle \} . \tag{3.15}$$

The degrees of freedom are determined by the nodal functionals $\{ F_\Gamma^{(a/b)}(\cdot), \Gamma \subset \partial \mathbf{T}_h \}$, with

$$F_\Gamma^{(a)}(v) := |\Gamma|^{-1} \oint_\Gamma v \, d\gamma \quad \text{or} \quad F_\Gamma^{(b)}(v) := v(m_\Gamma) . \tag{3.16}$$

Either choice is unisolvent with $\tilde{Q}_1(T)$, but each leads to different finite element spaces since the applied midpoint rule is only exact for linear functions.

The corresponding (*parametric*) finite element spaces $H_h = H_h^{(a/b)}$ and L_h are

$$L_h := \{q_h \in L_0^2(\Omega) \mid q_{h|T} = \text{const.}, \forall T \in \mathbf{T}_h\} \quad , \quad \mathbf{H}_h^{(a/b)} := S_h^{(a/b)} \times S_h^{(a/b)},$$
$$(3.17)$$

with

$$S_h^{(a/b)} := \left\{ \begin{array}{l} v_h \in L^2(\Omega) \mid v_{h|T} \in \tilde{Q}_1(T), \forall T \in \mathbf{T}_h, v_h \text{ continuous w.r.t.} \\[2mm] \text{all the nodal functionals } F_{\Gamma_{ij}}^{(a/b)}(\cdot), \forall \Gamma_{ij}, \\[2mm] \text{and } F_{\Gamma_{i0}}^{(a/b)}(v_h) = 0, \forall \Gamma_{i0} \end{array} \right\}.$$
$$(3.18)$$

Since the spaces $H_h^{(a/b)}$ are nonconforming, i.e., $H_h^{(a/b)} \not\subset \mathbf{H}_0^1(\Omega)$, we have to work with elementwise defined bilinear forms and corresponding energy norms

$$a_h(\mathbf{u}_h, \mathbf{v}_h) := \sum_{T \in \mathbf{T}_h} \int_T \nabla \mathbf{u}_h \cdot \nabla \mathbf{v}_h \, dx, \quad \|\mathbf{v}_h\|_h := (a_h(\mathbf{v}_h, \mathbf{v}_h))^{1/2} \quad (3.19)$$

$$b_h(q_h, \mathbf{v}_h) := - \sum_{T \in \mathbf{T}_h} q_{h|T} \int_T \nabla \cdot \mathbf{v}_h \, dx. \quad (3.20)$$

Let $j_h : L_0^2(\Omega) \to L_h$ be the operator of piecewise constant interpolation (modified to preserve the zero–mean value property) which satisfies for $q \in L_0^2(\Omega) \cap H^1(\Omega)$

$$\|q - j_h q\|_0 \le ch \|q\|_1 \quad , \forall q \in L_0^2(\Omega) \cap H^1(\Omega). \quad (3.21)$$

Further, let $i_h^{(a/b)} : \mathbf{H}_0^1(\Omega) \to H_h^{(a/b)}$ be the global interpolation operator in $H_h^{(a/b)}$, which is determined by

$$F_\Gamma(i_h^{(a/b)} \mathbf{v}) = F_\Gamma(\mathbf{v}) \quad , \forall \Gamma \subset \partial \mathbf{T}_h. \quad (3.22)$$

Unfortunately, on general non–uniform meshes the optimal order estimates do not hold for $i_h^{(a/b)}$. This is due to the fact that the spaces $H_h^{(a/b)}$ are

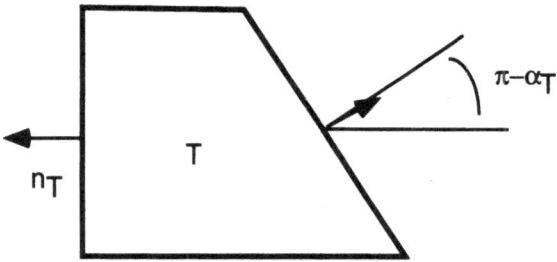

Figure 3.1: Deterioration of quadrilaterals

not *isoparametric*, i.e., the bilinear transformations $\psi_T : \hat{T} \to T$ are of another polynomial type than the shape functions on $\hat{T}$. In order to guarantee proper approximation properties for $H_h^{(a/b)}$, we have to impose a certain weak uniformity condition on the meshes $\mathbf{T}_h$. For each element $T \in \mathbf{T}_h$, let $\alpha_T \in (0, \pi]$ denote the maximum angle enclosed between the normal unit vectors corresponding to any opposite edges of T (see Figure 3.1). Then, the quantity

$$\sigma_h := \max\{|\pi - \alpha_T|, \forall T \in \mathbf{T}_h\} \qquad (3.23)$$

is a measure for the degeneration of the mesh $\mathbf{T}_h$.

Lemma 1 *For the interpolation operators* $i_h = i_h^{(a/b)}$ *the error estimate holds ([84]):*

$$\|v - i_h v\|_0 + h \|v - i_h v\|_h \le ch(h + \sigma_h) \|v\|_2, \quad \forall v \in \mathrm{H}_0^1(\Omega) \cap \mathrm{H}^2(\Omega).$$

The above element is the fully parametric version of the $\tilde{Q}1/Q0$ element as it is defined via transformation (bi- or trilinear) to a fixed reference element. It was shown in [84] that the stability and approximation properties of this ansatz deteriorate on strongly distorted meshes, i.e., depend very sensitively on the amount of deviation of the elements T from parallelogram shape and on the local *aspect ratio*. The first defect can be neglected if the computational mesh is generated through a systematic refinement process as described above. One always starts from a fixed (arbitrary) macro–decomposition and obtains the refined elements by connecting opposite midpoints. This approach is commonly used in standard multigrid tools.

Alternatively, one may also use a *non–parametric* version of the element where the reference space $\tilde{Q}_1(T) := \{q \in \mathrm{span}\langle \chi^2 - \eta^2, \chi, \eta, 1\rangle\}$ is defined for each element T independently with respect to the coordinate system (χ, η) spanned by the directions connecting the midpoints of sides of T. This ansatz turns out to be robust with respect to the shape of the elements T, and the following error estimate holds ([84]):

Lemma 2 *For the corresponding* non–parametric *counterparts, the optimal error estimate for $i_h = i_h^{(a/b)}$ holds without any dependency on σ_h:*

$$\|v - i_h v\|_0 + h \, \|v - i_h v\|_h \le ch^2 \, \|v\|_2 \quad, \forall v \in \mathrm{H}_0^1(\Omega) \cap \mathrm{H}^2(\Omega). \qquad (3.24)$$

Analogously as in the triangular case (see [36]) it is easy to see that together with the continuous stability estimate the following uniform stability result holds for the pair $(H_h^{(a)}, L_h)$. For the *non–parametric* counterparts it can be shown (see [7]) that the constant $\tilde{\beta}$ is even independent of the mesh *aspect ratio*,

$$\tilde{\beta} \, \|p_h\|_0 \le \max_{\mathbf{v}_h \in H_h^{(a)}} \frac{b_h(p_h, \mathbf{v}_h)}{\|\mathbf{v}_h\|_h} . \qquad (3.25)$$

This property seems to be very important, either for the resulting approximation property and for the convergence behaviour of the multigrid schemes. Only for the *non–parametric* counterparts the resulting multigrid rates stay bounded independent of the *aspect ratio*. The "midpoint oriented" space $H_h^{(b)}$ generally does not satisfy the stability property, we have to require the meshes $\mathbf{T}_h$ to be sufficiently uniform (see [84]).

Lemma 3 *Suppose that the quantity $\sigma = \sup_{h>0} \sigma_h$ is sufficiently small, then the stability estimate (3.25) holds true also for the pairing $(H_h^{(b)}, L_h)$.*

On the basis of our stability estimate (3.25) and the approximation property of Lemma 1 we can derive the following asymptotic error estimates for the *parametric* case (see [84]):

Lemma 4 *Suppose that the preceeding assumptions hold. Then, for $H_h = H_h^{(a)}$, and if the quantity $\sigma = \sup_{h>0} \sigma_h$ is sufficiently small also for $H_h = H_h^{(b)}$, the discrete Stokes problems have unique solutions $\{\mathbf{u}_h, p_h\} \in H_h^{(a/b)} \times L_h$, and further there holds:*

$$\|\mathbf{u} - \mathbf{u}_h\|_h + \|p - p_h\|_0 \;\leq\; c(h + \sigma_h)\left\{\|\mathbf{u}\|_2 + \|p\|_1\right\},$$

$$\|\mathbf{u} - \mathbf{u}_h\|_0 + \|p - p_h\|_{-1} \leq c(h + \sigma_h)^2\{\|\mathbf{u}\|_2 + \|p\|_1\}.$$

(3.26)

These results indicate that the convergence of the *parametric* rotated bilinear Stokes elements, for $h \to 0$, requires the underlying meshes $\{\mathbf{T}_h\}_h$ to be asymptotically uniform in the sense that $\sigma_h = \max\{|\pi - \alpha_T|, \forall T \in \mathbf{T}_h\} \to 0$, as $h \to 0$. This conclusion is supported by our numerical tests. In fact, the condition for convergence is often not very restrictive, since it is, for instance, automatically satisfied by weakly uniform meshes which we obtain when using our systematic grid refinement process described before. Therefore, also very complex domains are admitted.

When we work with the "midpoint oriented" finite element space $H_h^{(b)}$ it is convenient to replace the bilinear form $b_h(\cdot, \cdot)$ by its numerically integrated version

$$\tilde{b}_h(q_h, \mathbf{v}_h) \;:=\; -\sum_{T \in \mathbf{T}_h} q_{h|T} \sum_{\Gamma \subset \partial T} |\Gamma|\, F_\Gamma^{(b)}(\mathbf{v}_h) \cdot \mathbf{n}_\Gamma \qquad (3.27)$$

$$\approx\; -\sum_{T \in \mathbf{T}_h} q_{h|T} \sum_{\Gamma \subset \partial T} \oint_\Gamma \mathbf{v}_h \cdot \mathbf{n}_\Gamma\, d\gamma.$$

In this case the uniform stability condition is satisfied without any additional condition on the meshes $\{\mathbf{T}_h\}$. By some standard perturbation arguments (see [23]) the estimates of Lemma 4 carry over to this case without any condition on the size of σ_h. The *non–parametric* versions of the spaces $H_h^{(a/b)}$ have satisfactory approximation properties on general regular meshes. The stability properties are the same as those of their *parametric* counterparts, i.e., the optimal order convergence estimates for $H_h^{(b)}$ can be guaranteed only if using the modified bilinear form $\tilde{b}_h(\cdot, \cdot)$. As a conclusion, we finally get for $H_h^{(a)}$ and also for $H_h^{(b)}$ the optimal results:

Lemma 5 *Suppose that the preceeding assumptions are satisfied. Then, for the non–parametric versions of the spaces $H_h^{(a/b)}$, there holds (with the modifications of b_h if necessary):*

$$\|\mathbf{u} - \mathbf{u}_h\|_h + \|p - p_h\|_0 \;\leq\; ch\left\{\|\mathbf{u}\|_2 + \|p\|_1\right\},$$

$$\|\mathbf{u} - \mathbf{u}_h\|_0 + \|p - p_h\|_{-1} \leq ch^2\{\|\mathbf{u}\|_2 + \|p\|_1\}.$$

(3.28)

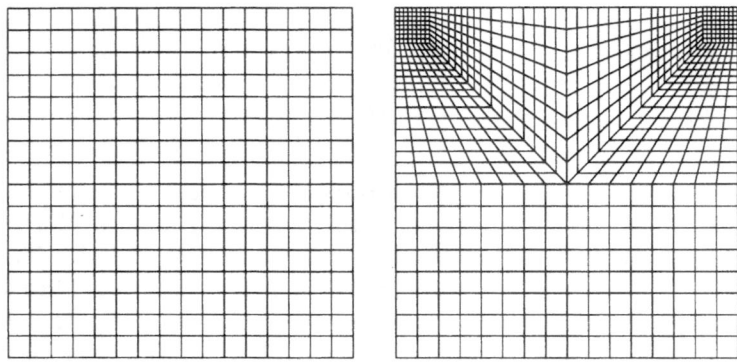

Figure 3.2: Uniform rectangular and locally refined mesh ("$h = 1/16$")

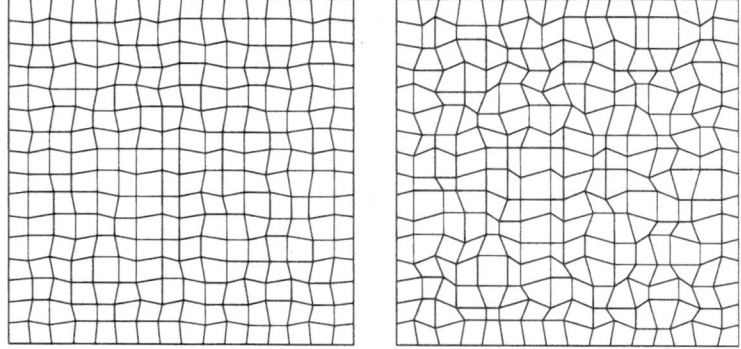

Figure 3.3: Perturbed meshes: 10% and 20% ("$h = 1/16$")

For the numerical demonstration of these theoretical results we have chosen one of the usual artificial test problems on the unit square $\Omega = (-1, 1) \times (-1, 1)$ with the exact solutions

$$
\begin{aligned}
u_1(x_1, x_2) &= -256\, x_1^2(x_1 - 1)^2 x_2(x_2 - 1)(2x_2 - 1)\,, \\
u_2(x_1, x_2) &= -u_1(x_2, x_1)\,, \\
p(x_1, x_2) &= 150\,(x_1 - 0.5)(x_2 - 0.5)\,.
\end{aligned}
$$

In the following Figures, four types of quadrilaterals meshes are shown (with mesh width $h = 1/16$) for which some calculations have been carried through. The stochastical perturbations have been performed on the corresponding finest level only.

h	$\varepsilon_u^{(a)}$	$\varepsilon_p^{(a)}$	$\varepsilon_u^{(b)}$	$\varepsilon_p^{(b)}$	h	$\varepsilon_u^{(a)}$	$\varepsilon_p^{(a)}$	$\varepsilon_u^{(b)}$	$\varepsilon_p^{(b)}$
1/8	0.0401	0.0137	0.0602	0.0162	1/8	0.0286	0.0113	0.0301	0.0125
1/16	0.0428	0.0130	0.0728	0.0145	1/16	0.0334	0.0106	0.0447	0.0122
1/32	0.0437	0.0127	0.0776	0.0133	1/32	0.0351	0.0102	0.0537	0.0112
1/64	0.0440	0.0125	0.0793	0.0128	1/64	0.0358	0.0100	0.0576	0.0104

Table 3.1: Convergence results on uniform (left) and locally refined (right) meshes

h	$\varepsilon_u^{(a)}$	$\varepsilon_p^{(a)}$	$\varepsilon_u^{(b)}$	$\varepsilon_p^{(b)}$	h	$\varepsilon_u^{(a')}$	$\varepsilon_p^{(a')}$	$\varepsilon_u^{(b')}$	$\varepsilon_p^{(b')}$
1/8	0.0420	0.0138	0.0604	0.0162	1/8	0.0431	0.0139	0.0646	0.0164
1/16	0.0501	0.0133	0.0798	0.0149	1/16	0.0493	0.0133	0.0987	0.0154
1/32	0.0810	0.0130	0.1151	0.0142	1/32	0.0515	0.0130	0.1741	0.0159
1/64	0.2348	0.0129	0.2753	0.0150	1/64	0.0519	0.0129	0.5022	0.0201

Table 3.2: Convergence results on nonuniform meshes (10% stochastical perturbation of the corresponding uniform mesh)

As measures for the quality of the various nonconforming versions we take the normalized relative L^2-errors for the velocity and pressure approximations

$$\varepsilon_{\mathbf{u}}(h) := \frac{\|\mathbf{u} - \mathbf{u}_h\|_0}{h^2 \|\mathbf{f}\|_0} \quad , \quad \varepsilon_p(h) := \frac{\|p - p_h\|_0}{h \, \|\mathbf{f}\|_0} . \tag{3.29}$$

The following Tables contain the results of a series of test calculations on the meshes shown. The two *parametric* versions (a) and (b) of the rotated bilinear element have about the same quantitative stability and convergence behaviour. In particular, both approximation schemes fail to provide satisfying results on strongly perturbed meshes. The results obtained for the corresponding *non-parametric* counterparts are indicated by superscripts (a') and (b'). As is predicted by the theory, only the *meanvalue oriented* nonparametric element (a') behaves well on strongly perturbed meshes (if $b_h(\cdot, \cdot)$ is not accordingly modified in case (b')).

Next, we show that large *aspect ratios* do not influence the approximation property if the nonparametric versions are used. The following Tables show results for the drag and lift coefficients for channel flow around a squared cylinder. Figure 3.4 shows the coarse mesh which can be modified by adapting the first inner line near to the square. The resulting *aspect ratios* are $AR = 10$ (this Figure, the mesh is called S1), $AR = 10^3$ (called S2) and $AR = 10^5$ (called S3). Obviously, the corresponding flow quantities (drag and lift) are not negatively influenced. The following configuration is identical with the framework in Section 4.1. While the examination of the resulting multigrid convergence rates for solving the incompressible Navier–Stokes equations is

%	$\varepsilon_u^{(a')}$	$\varepsilon_p^{(a')}$	$\varepsilon_u^{(b')}$	$\varepsilon_p^{(b')}$
0%	0.0437	0.0127	0.0776	0.0133
5%	0.0484	0.0128	0.1070	0.0139
10%	0.0515	0.0130	0.1741	0.0159
15%	0.0567	0.0134	0.2850	0.0191
20%	0.0638	0.0140	0.4405	0.0235
25%	0.0729	0.0148	0.6414	0.0292

Table 3.3: Convergence results on perturbed quadrilateral meshes (0 – 25% stochastical perturbation of the corresponding uniform mesh) for mesh width $h = 1/16$

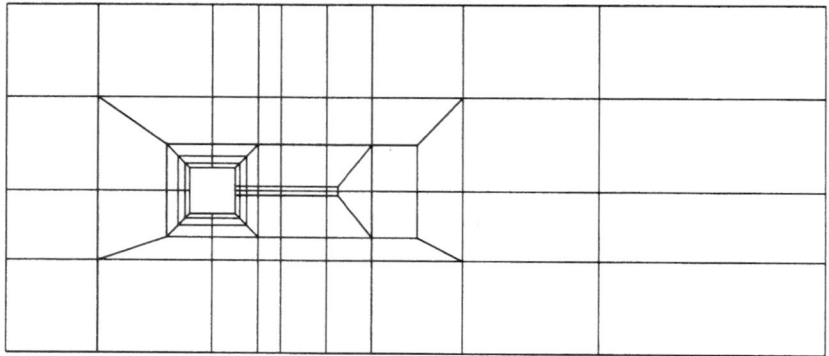

Figure 3.4: Typical (coarse) mesh S1 for approximation test

the aim in Section 4.1, here we show the corresponding approximation results. We perform simulations for the viscosity parameters $1/\nu \in \{5, 50, 500\}$, hereby calculating the solution of the corresponding stationary Navier–Stokes equations. For more details about the used flow configurations, see Section 4.1.

Obviously, the quality of the solution with respect to the underlying anisotropic mesh cells near the square is not significantly influenced. However, it must be stated that then the question arises whether such anisotropic grids are necessary at all for better approximations of these local quantities as lift and drag. The shown calculations do not answer this question, but it must be kept in mind that the performed Reynolds numbers are still far away from such calculations which might lead to boundary layers. Additionally, the influence of the performed upwind and streamline–diffusion stabilizations on the observed results is not clear at this moment, particularly for this kind of meshes. However, the tests show at least that very flat cells do not negatively disturb the approximation quality if utilized with the described nonconforming finite elements.

level	elements	vertices	midpoints	total unknowns
1	86	104	190	**466**
3	1,376	1,448	2,824	**7,024**
4	5,504	5,648	11,152	**27,808**
5	22,016	22,304	44,320	**110,656**
6	88,064	88,640	176,704	**441,472**

Table 3.4: Geometrical information and degrees of freedom for various levels

LEV	S1 ($AR = 10^1$) Drag / Lift		S2 ($AR = 10^3$) Drag / Lift		S3 ($AR = 10^5$) Drag / Lift	
	\multicolumn{6}{c}{$1/\nu = 5$}					
3	4.8239+1	0.1241+1	4.9155+1	0.1260+1	4.9164+1	0.1260+1
4	5.0098+1	0.1292+1	5.0696+1	0.1304+1	5.0705+1	0.1304+1
5	5.1009+1	0.1316+1	5.1399+1	0.1324+1	5.1408+1	0.1324+1
6	5.1507+1	0.1330+1	5.1761+1	0.1335+1	5.1773+1	0.1335+1
	\multicolumn{6}{c}{$1/\nu = 50$}					
3	6.0175+0	0.1904-0	6.1274+0	0.1943-0	6.1284+0	0.1943-0
4	6.0820+0	0.2000-0	6.1557+0	0.2024-0	6.1567+0	0.2025-0
5	6.1274+0	0.2047-0	6.1756+0	0.2064-0	6.1767+0	0.2064-0
6	6.1651+0	0.2072-0	6.1964+0	0.2082-0	6.1977+0	0.2083-0
	\multicolumn{6}{c}{$1/\nu = 500$}					
3	1.7795+0	-0.7125-2	1.8057+0	-0.8005-2	1.8058+0	-0.8085-2
4	1.7195+0	-0.5231-2	1.7576+0	-0.5261-2	1.7580+0	-0.5463-2
5	1.6843+0	-0.4375-2	1.7126+0	-0.4650-2	1.7129+0	-0.4290-2
6	1.6733+0	-0.4459-2	1.6899+0	-0.4306-2	1.6902+0	-0.4345-2

Table 3.5: Drag and lift values for different Reynolds numbers and aspect ratios

3.1.3 The discretely divergence–free subspaces

We demonstrate the explicit construction process of the *discretely divergence–free* subspaces corresponding to the proposed finite element spaces $H_h^{(a/b)}$. For this we introduce the modified discrete bilinear form $\tilde{b}_h(\cdot, \cdot)$ where

$$\tilde{b}_h(q_h, \mathbf{v}_h) := - \sum_{T \in \mathbf{T}_h} q_{h|T}\, Q_T(\mathbf{v}_h) \quad , \quad Q_T(\mathbf{v}_h) := \sum_{\Gamma \subset \partial T} |\Gamma|\, F_\Gamma(\mathbf{v}_h) \cdot \mathbf{n}_\Gamma,$$

(3.30)

which is for $H_h^{(b)}$ an $O(h^2)$ approximation to the original bilinear form $b_h(\cdot, \cdot)$ (exact for $H_h^{(a)}$!). Then, we call a function $\mathbf{v}_h \in H_h = H_h^{(a/b)}$ *discretely divergence-free*, if the condition

$$\tilde{b}_h(q_h, \mathbf{v}_h) = 0 \,, \, \forall\, q_h \in L_h \,,$$

(3.31)

is satisfied. Since we only use piecewise constant pressure approximations, an equivalent criterion is

$$Q_T(\mathbf{v}_h) = 0, \, \forall T \in \mathbf{T}_h \, . \qquad (3.32)$$

With these modifications we can introduce subspaces $H_h^d \subset H_h$ such that our discrete problem for the velocity is reduced to:

Find $\mathbf{u}_h^d \in H_h^d$, *such that*

$$a_h(\mathbf{u}_h^d, \mathbf{v}_h^d) = (\mathbf{f}, \mathbf{v}_h^d) \, , \, \forall \mathbf{v}_h^d \in H_h^d \, . \qquad (3.33)$$

Finally, the corresponding pressure $p_h \in L_h$ is determined by the condition

$$\tilde{b}_h(p_h, \mathbf{v}_h^r) = (\mathbf{f}, \mathbf{v}_h^r) - a_h(\mathbf{u}_h^d, \mathbf{v}_h^r) \, , \, \forall \mathbf{v}_h^r \in H_h^r \, , \qquad (3.34)$$

where the functions $\mathbf{v}_h^r$ span the curl–free part of the complete space H_h. In our configuration this is performed by a marching process from element to element, without solving any linear system of equations (see [25] and also [104]). Then, for the solution $\mathbf{u}_h^d$ of problem (3.33) and its corresponding pressure p_h, there hold again the optimal error estimates

$$\|\mathbf{u} - \mathbf{u}_h^d\|_h + \|p - p_h\|_0 \; \leq \; ch \{\|\mathbf{u}\|_2 + \|p\|_1\} \, ,$$
$$\|\mathbf{u} - \mathbf{u}_h^d\|_0 + \|p - p_h\|_{-1} \leq ch^2 \{\|\mathbf{u}\|_2 + \|p\|_1\} \, . \qquad (3.35)$$

Now, consider a general quadrilateral $T \in \mathbf{T}_h$ (see Figure 3.5) with vertices a^i, midpoints m^j, edges Γ^j, unit tangential vectors $\mathbf{t}^j$ and normal unit vectors $\mathbf{n}^j$. Let $\varphi_h^j \in S_h^{(a/b)}$ be the usual nodal basis functions of the finite element space $S_h = S_h^{(a/b)}$ (cf. (3.22)), restricted to the element T, satisfying $F_{\Gamma^i}(\varphi_h^j) = \delta_{ij}, \, i, j = 1, \ldots, 4$.

Then, the first group of basis functions $\{\mathbf{v}_h^{i,t}\}$ of H_h^d, corresponding to the edges of $\mathbf{T}_h$, is given by the local definition

$$\mathbf{v}_{h|T}^{i,t} \in \{\varphi_h^j \mathbf{t}^j \, , \, j = 1, \ldots, 4\} \, . \qquad (3.36)$$

The second group $\{\mathbf{v}_h^{i,\psi}\}$, corresponding to the vertices, is locally determined by

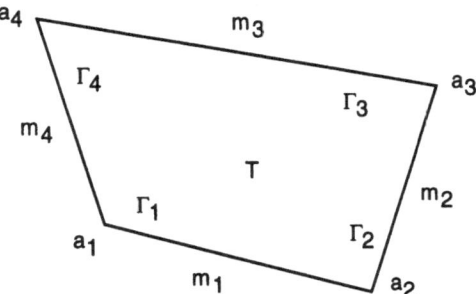

Figure 3.5: General quadrilateral T

$$\mathbf{v}_{h|T}^{i,\psi} \in \{\frac{\varphi_h^k \mathbf{n}^k}{|\Gamma^k|} - \frac{\varphi_h^j \mathbf{n}^j}{|\Gamma^j|}, \; j = 1,\ldots,4, \; k = (j+2) \bmod 4 + 1\}. \qquad (3.37)$$

Thus, we get approximations for the tangential velocities on the edges, and for the streamfunction values in the nodes (see [47],[104]). If we eliminate one of the functions $\{\mathbf{v}_h^{i,\psi}\}$ by prescribing the value in one (boundary) point or by the mean value zero condition, we get a basis for the discretely divergence–free subspace H_h^d, assuming that the problem has only one boundary component. For several boundary components (for instance, for flows around obstacles) we do not need any additional basis function, we only have to modify our linear solvers in a simple way: (by *iterative filtering techniques* (cf. [109] and see the following section 3.5). This technique of implementing boundary conditions in iterative solution methods – which is a very general one – allows to perform many boundary components without introducing additional "non-local" basis functions since only the matrix–vector multiplication subroutine has to be changed. This is a very essential aspect of the practicability of these divergence–free techniques which is in fact not published in many standard CFD books.

After introducing these new basis functions the size of our linear system is reduced from about $5\,NEL$ unknowns (NEL number of elements) for the usual primary formulation to about $3\,NEL$ in the divergence–free case. A disadvantage of the new formulation is that the corresponding stiffness matrix S_d with

$$S_d^{(i,j)} = a_h(\mathbf{v}_h^{i,d}, \mathbf{v}_h^{j,d}), \qquad (3.38)$$

has a condition number like $O(h^{-4})$ (cf. [104]), and the corresponding mass matrix like $O(h^{-2})$. However, since we have been able to derive a multigrid algorithm with typical convergence rates independent of h, this fact is negligible for our approach (see [104]).

The corresponding multigrid algorithm is developed for linear (nonsymmetric) problems of generalized Stokes or Oseen type, with a given coefficient function $\mathbf{U}$,

$$\alpha \mathbf{u} - \nu \Delta \mathbf{u} + \mathbf{U} \cdot \nabla \mathbf{u} + \nabla p = \mathbf{f} \quad , \quad \nabla \cdot \mathbf{u} = 0. \tag{3.39}$$

These equations typically arise from the linearization of the stationary or nonstationary Navier–Stokes equations. The components of the algorithm are the usual ones:

- An appropriate **smoothing operator** which can be the Gauß–Seidel-, the Jacobi- or even the ILU scheme since we have to handle scalar definite systems only.

- A **coarse grid correction** with appropriate grid transfer operators which are of second order accuracy and which have to be discretely divergence–free.

- A **step length control** of the correction as usual for nonconforming finite elements.

Then, for the positive definite case we can show that the convergence rates are independent of the mesh size h (see [109]). In all our computations (with an F–cycle) we use for the grid transfer a *macro–elementwise divergence–free interpolation*, which interpolates a given discretely divergence–free function from level $2h$ onto level h, again being discretely divergence–free and second order accurate (see [109]).

To explain this essential idea, we have to go more into detail. The following Figure 3.6 shows one element on level $2h$, and the corresponding refined elements on level h.

The *macro-elementwise interpolation*, which also works analogously for the scalar nonconforming and the Morley space (cf. [95],[109]), can be described as follows:

1. Transfer the divergence–free coefficient vector (Ψ_{2h}, Ut_{2h}) into the primary coefficient vector (U_{2h}, V_{2h}).

2. Interpolate "fully" on the macro element to get (U_h, V_h) (see Section 3.4).

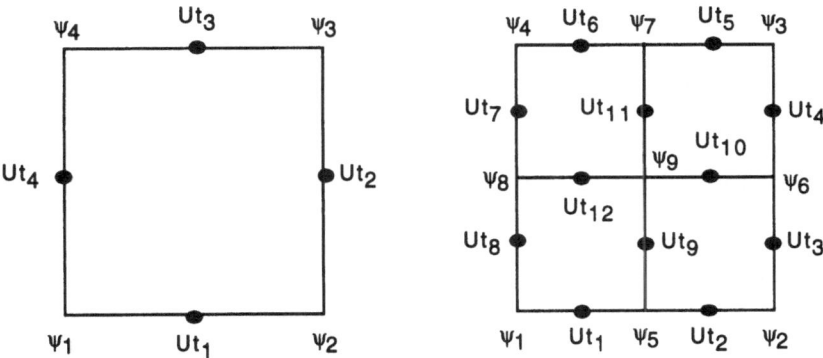

Figure 3.6: Macro element and refined elements

3. Compute Ut_h and Un_h on all fine grid edges.

4. Set $\Psi_h = \Psi_{2h}$ in the macro nodes and compute in the new vertices the values for Ψ_h by simply integrating Un_h.

5. Take the average of those Ψ_h and Ut_h, which lie on macro edges.

These operations, using *local transfer matrices*, can be performed very quickly and efficiently. In fact, solving a Stokes equation takes about the same amount of time as solving a Laplace equation. After all our extensive numerical tests (see for instance [109]) we can state that this proposed combination is a very efficient and robust solver for Stokes/Oseen problems of the above type, independent of all given data, domain and triangulation. So we can really claim to have a "Black Box" solver, at least for the linear problems arising in the discretization process for the fully stationary and nonstationary Navier–Stokes equations. Nevertheless, we have stopped the application of this special approach and prefer recently solution techniques which are directly derived for the primary formulation in velocity and pressure simultanously. Therefore, let us finish this section concerning the explicit construction of discretely divergence–free subspaces with some concluding remarks which also explain our changed preference during the last years (which may change again ...).

- **Advantages of discretely divergence–free subspaces:**

    The main advantage of the formal elimination of the pressure is not the seemingly smaller number of unknowns. In fact, the storage requirements are at least the same since the matrix stencils are increased, and additionally the resulting stiffness matrix is even worse ill–conditioned. Due to the introduction of streamfunction-like basis functions, the problem becomes a fourth order problem. However, if the derived multigrid components are applied, this ill-conditioness "vanishes" since the typical excellent multigrid rates independent of h can be provided. And moreover, since the corresponding discrete problems lead to definite stiffness matrices, all well-known techniques of SOR- or ILU–type or Krylov-space methods can be easily applied without any modification. Hence, the typical problem in mixed formulations that robust and efficient solvers on very anisotropic triangulations are hard to construct, has been cancelled by this approach. This is (or better: was) the major advantage of this approach. In combination with the proposed special prolongation and restriction operators which have to be carefully implemented, the resulting multigrid solver is very efficient and satisfies even the requirement that the convergence rates tend to 0 for time steps $k \to 0$. The explicit construction of the subspaces can be efficiently performed for the triangular as well as the quadrilateral elements and leads to exactly the same accuracy results as the primary formulation since the same solution is directly approximated in the subspace due to the incompressibility constraint. However, one difference is that if we stop the iteration process without having achieved the final convergence, we always satisfy at least the incompressibility constraint on discrete level. We mentioned triangles and quadrilaterals: In fact, we started with the triangular cases but we changed to the quadrilateral elements:

- Our experience is that quadrilateral elements lead to more accurate solutions with respect to the same number of unknowns, especially on orthogonal elements.

- Only in regions where grid refinement is needed, both are of the same quality. However, even then we prefer the quadrilateral elements since with "hanging node" techniques adaptive refinement processes can be easily applied, in particular with regard to the 3D cases.

- In 3D calculations, the approaches with hexaeders seem to be more efficient than with tetraeder elements, with respect to numerical but also computational aspects.

- **Problems of discretely divergence–free subspaces:**

  The main reason that we stopped our discretely divergence–free approach was due to the necessary step of solving 3D problems. It is not a big problem, and in deed it is solved by other authors (see for instance [97]), to construct and to implement 3D discretely divergence–free ansatz functions. The difference with regard to the 2D case is the fact that not all ansatz functions are automatically discretely divergence–free basis functions; some tools from graph theory are needed. However, also this task has been solved (see for an overview [97]) and the corresponding basis functions can be easily implemented. Then, the corresponding discrete problems are again scalar problems and can be treated by Krylov–schemes (CG, GMRES), for instance. However, the condition number behaves like $O(h^{-4})$ and can be improved by preconditioning techniques, but the resulting convergence rates are generally still depending on h. Thus, the expected convergence is too slow and tends to $\rho_{CG} \to 1$ for $h \to 0$. Consequently, multigrid is an absolutely necessary tool for an efficient solution process.

In this case, the construction of the corresponding grid transfer routines as described above is the main problem. It is not a mathematical problem since an algorithm which provides the desired results can be easily written down. In fact, it is a computational problem to implement this algorithm in a very efficient way by using *local transfer matrices*. Without them, prolongation and restriction in 2D require typically more than 95% of the CPU time in relation to the complete multigrid approach, while the application of the corresponding *local transfer matrices* can decrease this numerical amount to 10% and less which is the typical percentage for multigrid. However, this is a very hard process to derive and to implement these *local transfer matrices* which took us almost 2 months already in 2D, and what about the needed labour in 3D?

The fact that a "large-scale" implementation is absolutely necessary to obtain an efficient solution tool seems to be one of the major problems. Beside the 3D issues, the problems of implementing the indicated *iterative filtering techniques* for boundary conditions have to mentioned. Having done this task, the application to very complex domains involving many interior objects is very easy, but first of all these special iterative techniques have to be understood and then to be added to the code. And this job must be carefully done since otherwise only simple channel geometries can be treated.

- **Comparison with MPSC techniques for primary formulations:**

  On the basis of the developed *multilevel pressure Schur complement* techniques (MPSC) some of the "old" problems with regard to the solution of Stokes-like systems in velocity/pressure formulation have been successfully solved. For instance, if the *global MPSC* schemes are performed, we can analogously guarantee that the approximate solution satisfies the discrete continuity equation, but in addition also for inhomogeneous data $\nabla \cdot \mathbf{u} = g$. Very fast and robust solvers have become available, with respect to arbitrary time step sizes and anisotropies in the mesh. And, probably the essential point, these techniques can be applied without major additional work in the 3D case. Hence, on the basis of the recent results with respect to accuracy, efficiency and robustness of the *MPSC* schemes for 2D and 3D laminar incompressible flows, we favourize this approach for the mixed formulation, in particu-

$NEQ =$	5,184	20,352
GS2	0.282	0.276
GS4	0.153	0.141

Table 3.6: Multigrid convergence rates for flow around an ellipse

lar if we take into consideration future projects involving more complex flow models. However, it must be also mentioned that at least issues as prescribed *net flux* problems or *slip* boundary conditions can be treated much easier with the divergence–free techniques.

Let us finish this section about the explicit construction of discretely divergence–free subspaces with some concluding remarks. If we are exclusively interested in 2D incompressible flows, this ansatz leads to very promising results, with respect to numerical and algorithmic details but also in view of computational aspects. However, the advantages are less the traditionally claimed topics as "less unknowns and elimination of the pressure", but the resulting robustness and efficiency of this ansatz in combination with "optimal" multigrid tools. And it must be stated that the underlying work of deriving and implementing these technical tools is very hard, but it must be performed even if the task does not look like very mathematical! This construction process can be performed for our preferred $\tilde{Q}1/Q0$ Stokes element, but theoretically for other elements, too. For instance, the construction of the discretely divergence–free subspaces for the $Q2/P1$ is explicitly shown in [25] and [47]. But the construction process is not the crucial point: the derivation and implementation of the corresponding multigrid tools is the "hard" computational task, and this task has not been done as far as we know.

We end with showing two examples concerning the efficiency of this divergence–free technique in combination with multigrid. Many more examples for flow simulations with this divergence–free ansatz can be found in [101] and [104]. Table 3.1.3 shows the convergence rates for Stokes flow around an ellipse. The corresponding coarse mesh and after two *regular* refinements (this corresponds to $NEQ = 5.184$ unknowns) are given in Figure 3.7. We perform Gauß–Seidel as smoother, with 2 and 4 (total) smoothing steps.

As an additional example, we show results for the following class of problems (with $\alpha \geq 0$)

$$\alpha \mathbf{u} - \Delta \mathbf{u} + \nabla p = \mathbf{f} \quad , \quad \nabla \cdot \mathbf{u} = 0 \, . \tag{3.40}$$

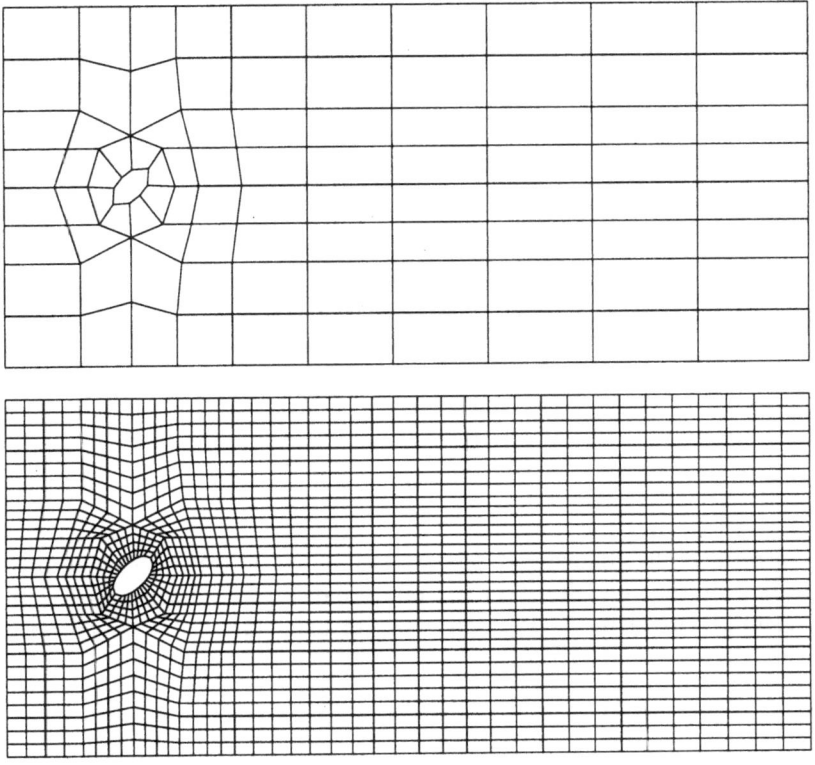

Figure 3.7: Coarse grid and refined mesh for ellipse configuration

n	$NEL =$	1,024	4,096	16,384
0	GS1	0.241	0.238	0.236
3	GS1	0.253	0.242	0.237
6	GS1	0.030	0.027	0.087
9	GS1	0.050	0.054	0.053

Table 3.7: Multigrid convergence rates for $1/k := \alpha = 10^n, n = 0, 3, 6, 9$; 1 smoothing step

This configuration is a typical example for a nonsteady calculation in which case the mass matrix is weighted with the inverse time step $O(\frac{1}{\Delta t})$. Here, the values $1/\Delta t := \alpha = 10^n$, $n \in \{0, 3, 6, 9\}$, are taken. These calculations are performed on an equidistant tensor product mesh for the unit square which is the simplest discretization for a typical *lid driven cavity* problem (see also the following subsection). They demonstrate the "optimal" behaviour of this solver for $\Delta t = 1/\alpha \rightarrow 0$, similar to the global MPSC case.

3.1.4   Stabilization techniques for convective terms

Our continuous formulation of the stationary Navier–Stokes equations reads

$$-\nu\Delta\mathbf{u} + \mathbf{u}\cdot\nabla\mathbf{u} + \nabla p = \mathbf{f}, \ \nabla\cdot\mathbf{u} = 0, \quad \text{in } \Omega \quad , \quad \mathbf{u} = \mathbf{g} \quad \text{on } \partial\Omega, \quad (3.41)$$

while the corresponding variational formulation (with the standard bilinear forms $a(\cdot,\cdot)$ and $b(\cdot,\cdot)$) can be written as:

Find a pair $\{\mathbf{u}, p\} \in H := \mathbf{H}_0^1(\Omega) \times L := \mathrm{L}_0^2(\Omega)$, such that

$$
\begin{aligned}
\nu a(\mathbf{u}, \mathbf{v}) + n(\mathbf{u}, \mathbf{u}, \mathbf{v}) + b(p, \mathbf{v}) &= (\mathbf{f}, \mathbf{v}) & \forall\, \mathbf{v} \in H & \quad (3.42) \\
b(q, \mathbf{u}) &= 0 & \forall\, q \in L. & \quad (3.43)
\end{aligned}
$$

We introduce the following trilinear form $n(\cdot,\cdot,\cdot)$ (other formulations are described in [56]) and its discrete counterpart $n_h(\cdot,\cdot,\cdot)$ which is related to the computational mesh $\mathbf{T}_h$, consisting of elements (here: quadrilaterals) $T \in \mathbf{T}_h$,

$$n(\mathbf{u}, \mathbf{v}, \mathbf{w}) := \int_\Omega u_i \frac{\partial v_j}{\partial x_i} w_j \, dx, \quad n_h(\mathbf{u}_h, \mathbf{v}_h, \mathbf{w}_h) := \sum_{T \in \mathbf{T}_h} \int_T u_{h,i} \frac{\partial v_{h,j}}{\partial x_i} w_{h,j} \, dx.$$

$$(3.44)$$

Next, we define the corresponding discrete spaces H_h, resp., L_h, and discrete bilinear forms $a_h(\cdot,\cdot)$, resp., $b_h(\cdot,\cdot)$, such that our discrete problem becomes:

Find a pair $\{\mathbf{u}_h, p_h\} \in H_h \times L_h$, such that

$$
\begin{aligned}
\nu a_h(\mathbf{u}_h, \mathbf{v}_h) + n_h(\mathbf{u}_h, \mathbf{u}_h, \mathbf{v}_h) + b_h(p_h, \mathbf{v}_h) &= (\mathbf{f}, \mathbf{v}_h) & \forall\, \mathbf{v}_h \in H_h & \quad (3.45) \\
b_h(q_h, \mathbf{u}_h) &= 0 & \forall\, q_h \in L_h. & \quad (3.46)
\end{aligned}
$$

This central discretization of the convection part may lead in principle to 2nd order accuracy, but only if the local mesh width h is "fine enough". In relevant applications the general result is to set h in relation to the Reynolds number which imposes impossible restrictions onto the mesh, especially in 3D. If this relation is not taken into account, this approach can lead to:

- stiffness matrices not of positive type (*M–matrices*) such that standard iterative schemes including multigrid cannot lead to efficient convergence results, if at all.

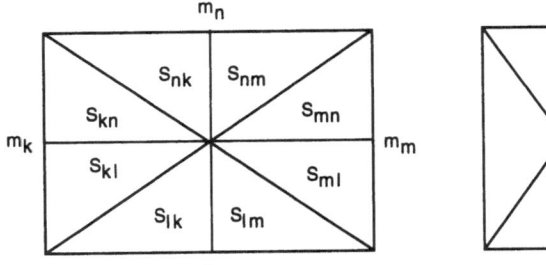 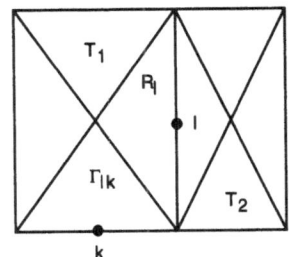

Figure 3.8: Barycentric fragments S_{ij} of T and lumping regions R_l around midpoint l

- oscillations and deterioration of the solution which have purely numerical character.

To stabilize the convection part for higher Reynolds numbers, there are two approaches which are widely used by the CFD community: **Upwind** schemes and **Streamline–diffusion** techniques.

Finite element upwind strategies for stabilizing the convective terms:

We describe an upwind discretization for the nonconforming rotated bilinear finite elements which is based on works of Ohmori/Ushijima [74] and Tobiska/Schieweck [91]. The main idea is to introduce new edge oriented lumping regions and lumping operators: We divide each quadrilateral $T \in \mathbf{T}_h$ into 8 barycentric fragments S_{ij} and define for each edge Γ_l, respectively midpoint m_l, the lumping region R_l by

$$R_l := \bigcup_{k \in \Lambda_l} S_{lk} , \qquad (3.47)$$

where Λ_l is the set of indices k, such that m_l and m_k are neighbouring midpoints.

Defining the edge $\Gamma_{lk} := \partial S_{lk} \cap \partial S_{kl}$, the boundary ∂R_l of the region R_l can be written as follows, and we achieve a new (edge oriented) partition of $\bar{\Omega} = \cup_{T \in \mathbf{T}_h} \bar{T}$ by

$$\partial R_l = \bigcup_{k \in \Lambda_l} \Gamma_{lk} \quad , \quad \bar{\Omega} = \bigcup_l \bar{R}_l . \qquad (3.48)$$

Next, we define the piecewise constant lumping operator L_h with

$$L_h \mathbf{v}_h(x) = \mathbf{v}_h(m_l) \,, \ \forall\, x \in R_l \,. \tag{3.49}$$

Writing $n_h(\mathbf{u}_h, \mathbf{v}_h, \mathbf{w}_h)$ as the sum $n_h = n_h^1 + n_h^2$ with

$$
\begin{aligned}
n_h^1(\mathbf{u}_h, \mathbf{v}_h, \mathbf{w}_h) &:= \sum_{T \in \mathbf{T}_h} \int_T \frac{\partial u_{h,i} v_{h,j}}{\partial x_i} w_{h,j}\, dx \,, \ n_h^2(\mathbf{u}_h, \mathbf{v}_h, \mathbf{w}_h) \\
&:= -\sum_{T \in \mathbf{T}_h} \int_T \frac{\partial u_{h,i}}{\partial x_i} v_{h,j} w_{h,j}\, dx \,,
\end{aligned}
$$

we replace $n_h^1(\mathbf{u}_h, \mathbf{v}_h, \mathbf{w}_h)$ by $n_h^1(\mathbf{u}_h, \mathbf{v}_h, L_h\mathbf{w}_h)$, and $n_h^2(\mathbf{u}_h, \mathbf{v}_h, \mathbf{w}_h)$ by $n_h^2(\mathbf{u}_h, L_h\mathbf{v}_h, L_h\mathbf{w}_h)$. Then, we modify (cf. [74]) n_h^1 by an upwinded form (replace $\mathbf{v}_h$ by $\mathbf{v}_h^{lk}$), apply the Gauss theorem, and the result is the terms

$$\tilde{n}_h^1(\mathbf{u}_h, \mathbf{v}_h, \mathbf{w}_h) := \sum_l \sum_{k \in \Lambda_l} \oint_{\Gamma_{lk}} (\mathbf{u}_h \cdot \mathbf{n}_{lk}) \mathbf{v}_h^{lk}\, d\gamma\, \mathbf{w}_h(m_l) \,, \tag{3.50}$$

$$\tilde{n}_h^2(\mathbf{u}_h, \mathbf{v}_h, \mathbf{w}_h) := -\sum_l \sum_{k \in \Lambda_l} \oint_{\Gamma_{lk}} \mathbf{u}_h \cdot \mathbf{n}_{lk}\, d\gamma\, \mathbf{v}_h(m_l) \mathbf{w}_h(m_l) \,, \tag{3.51}$$

with

$$\mathbf{v}_h^{lk} := \lambda_{lk} \mathbf{v}_h(m_l) + (1 - \lambda_{lk}) \mathbf{v}_h(m_k) \,, \tag{3.52}$$

in which case the functions λ_{lk} satisfy

$$\lambda_{lk} = 1 - \lambda_{kl} \quad , \quad |\lambda_{lk}| \le c \,. \tag{3.53}$$

Finally, the new form $\tilde{n}_h(\mathbf{u}_h, \mathbf{v}_h, \mathbf{w}_h)$ is defined as

$$
\begin{aligned}
&\tilde{n}_h(\mathbf{u}_h, \mathbf{v}_h, \mathbf{w}_h) \\
&:= \sum_l \sum_{k \in \Lambda_l} \oint_{\Gamma_{lk}} \mathbf{u}_h \cdot \mathbf{n}_{lk}\, d\gamma\, (1 - \lambda_{lk}(\mathbf{u}_h))(\mathbf{v}_h(m_k) - \mathbf{v}_h(m_l)) \mathbf{w}_h(m_l) \,.
\end{aligned}
$$
$$\tag{3.54}$$

We use the following possibilities for λ_{lk} (with $x := \dfrac{1}{\nu}\oint_{\Gamma_{lk}} \mathbf{u}_h \cdot \mathbf{n}_{lk}\, d\gamma$ as measure for the *local Reynolds number*):

1) **Simple** upwind: $\quad \lambda_{lk}(\mathbf{u}_h) := \left\{ \begin{array}{ll} 1 & \text{if } x \geq 0 \\ 0 & \text{otherwise} \end{array} \right\}$.

2) **(weighted) Samarskij** upwind [98]: $\lambda_{lk}(\mathbf{u}_h) := \left\{ \begin{array}{ll} \dfrac{\frac{1}{2} + \alpha x}{1 + \alpha x} & \text{if } x \geq 0 \\[2ex] \dfrac{1}{2(1 - \alpha x)} & \text{otherwise} \end{array} \right\}$.

Here, α is an additional damping parameter which may be chosen by the user. The classical setting is $\alpha = 1$, while $\alpha \to \infty$ leads to the *simple upwind* scheme. However, all values $\alpha \in [0, \infty)$ are possible in general.

For the following convergence results (see [104]) we restrict to the simple upwind scheme. Then, our discrete problem reads:

Find a pair $\{\mathbf{u}_h, p_h\} \in H_h \times L_h$, such that

$$\nu a_h(\mathbf{u}_h, \mathbf{v}_h) + \tilde{n}_h(\mathbf{u}_h, \mathbf{u}_h, \mathbf{v}_h) + b_h(p_h, \mathbf{v}_h) = (\mathbf{f}, \mathbf{v}_h) \quad \forall \mathbf{v}_h \in H_h \quad (3.55)$$
$$b_h(q_h, \mathbf{u}_h) = 0 \quad \forall q_h \in L_h. \quad (3.56)$$

Lemma 6 *There holds for functions in H_h (see [104]):*

1) $\tilde{n}_h(\mathbf{u}_h, \mathbf{v}_h, \mathbf{v}_h) \geq 0 \quad , \forall \mathbf{u}_h \in H_h, \forall \mathbf{v}_h \in H_h$ (Positivity).

2) $|\tilde{n}_h(\mathbf{u}_h^0, \mathbf{v}_h, \mathbf{w}_h) - \tilde{n}_h(\mathbf{u}_h^1, \mathbf{v}_h, \mathbf{w}_h)| \leq c \|\mathbf{u}_h^0 - \mathbf{u}_h^1\|_h \|\mathbf{v}_h\|_h \|\mathbf{w}_h\|_h$ (Continuity).

3) $|n_h(\mathbf{u}_h, \mathbf{v}_h, \mathbf{w}_h) - \tilde{n}_h(\mathbf{u}_h, \mathbf{v}_h, \mathbf{w}_h)| \leq ch |\log h| \|\mathbf{u}_h\|_h \|\mathbf{v}_h\|_h \|\mathbf{w}_h\|_h$ (Approximability).

4) For $\mathbf{f} \in L^2(\Omega)$, there exists at least one solution $\{\mathbf{u}_h, p_h\} \in H_h \times L_h$ (Existence).

Lemma 7 *(Approximation result, see [104])*

For $\nu^{-2}\|\mathbf{f}\|_{-1}$ sufficiently small, the solutions $\{\mathbf{u}, p\}$, resp., $\{\mathbf{u}_h, p_h\}$, are uniquely determined. If additionally $\{\mathbf{u}, p\} \in \mathbf{H}^2(\Omega) \times \mathrm{H}^1(\Omega)$, then there holds

$$\|\mathbf{u} - \mathbf{u}_h\|_h + \|p - p_h\|_0 \leq ch |\log h|. \tag{3.57}$$

For the weighted upwind scheme it can be shown in one dimension (see [98]) that the resulting discretization is of second order accuracy, but the corresponding result in two or three dimensions is still an open problem. For the simple upwind, the corresponding (velocity) stiffness matrix S_h with

$$S_h^{(ij)} := \nu a_h(\mathbf{v}_h^i, \mathbf{v}_h^j) + \tilde{n}_h(\mathbf{u}_h^n, \mathbf{v}_h^i, \mathbf{v}_h^j) \tag{3.58}$$

can be an M–matrix (see [74],[91]), which results in very nice linear algebraic properties, concerning convergence results for Jacobi-, SOR- or ILU–methods. This is essential for our multigrid solution tools which use those as smoothers.

Summarizing the different upwind properties we can state:

- Upwind schemes can be derived for the nonconforming $\tilde{Q}1/Q0$ Stokes elements, for 2D as well as for 3D problems.

- The *simple upwind* scheme leads to M–matrix properties such that the resulting stiffness matrix can be efficiently "inverted" by standard multigrid tools. However, since this approach is only first order accurate, it is suited as preconditioner in a defect correction approach only.

- The *adaptive upwind* scheme (Samarskij) provides higher order approximates which is shown by the following numerical results. However, the exact order and the underlying rigorous analysis are not clear, and the resulting linear algebra properties cannot be predicted by a priori considerations. Nevertheless, the following numerical results will show that this approach is a good candidate for the desired "Black Box" solution tool.

Streamline–diffusion strategies for stabilizing convective terms:

Recently, we have begun to apply streamline–diffusion techniques as an alternative approach for stabilizing convective terms. We were motivated by the following statements which can often be found in literature. For the technical results we refer to the works of Johnson ([59],[61]), Lube [66] and Zhou [123] and the literature cited therein.

- The theoretically derived accuracy for streamline–diffusion techniques in combination with first order finite elements (linear, bilinear) is usually $O(h^{3/2})$ in the L^2-norm if we assume sufficient regularity. In many applications, one can even expect an improved behaviour of order $O(h^2)$ which is optimal in this context. In addition, the streamline–diffusion approach can be easily applied to higher order discretizations (at least theoretically).

- The application of streamline–diffusion techniques results in a *full Galerkin* scheme (see [59],[62]) which allows rigorous a posteriori error control mechanisms even for the Navier–Stokes equations.

We follow the approach of Lube [66] and apply streamline–diffusion stabilization in a *least square* approach for the momentum equation only. All following techniques are explained for the stationary case, not only for reasons of technical simplicity, but also due to some inherent problems for nonstationary equations which are not yet solved. Indeed, all techniques can be applied to the nonstationary case, however some modifications have to be taken into account.

We start again with the standard continuous formulation of the steady Navier–Stokes equations (with $\tilde{\mathbf{u}} = \mathbf{u}$),

$$N(\tilde{\mathbf{u}})\mathbf{u} := -\nu\Delta\mathbf{u} + \tilde{\mathbf{u}}\cdot\nabla\mathbf{u} + \nabla p = \mathbf{f} \quad , \quad \nabla\cdot\mathbf{u} = 0 , \qquad (3.59)$$

as usual with appropriate boundary conditions and a right hand side $\mathbf{f}$. Using

$$n(\mathbf{u}, \mathbf{v}, \mathbf{w}) := \int_\Omega u_i \frac{\partial v_j}{\partial x_i} w_j \, dx \qquad (3.60)$$

as trilinear form, we can introduce again the discrete form $n_h(\cdot, \cdot, \cdot)$, such that

$$n_h(\mathbf{u}_h, \mathbf{v}_h, \mathbf{w}_h) := \sum_{T\in\mathbf{T}_h} \int_T u_{h,i} \frac{\partial v_{h,j}}{\partial x_i} w_{h,j} \, dx . \qquad (3.61)$$

Together with the introduced bilinear forms $a_h(\cdot, \cdot)$ and $b_h(\cdot, \cdot)$, our discrete problem reads:

Find a pair $\{\mathbf{u}_h, p_h\} \in H_h \times L_h$, such that

$$\nu a_h(\mathbf{u}_h, \mathbf{v}_h) + n_h(\mathbf{u}_h, \mathbf{u}_h, \mathbf{v}_h) + b_h(p_h, \mathbf{v}_h) = (\mathbf{f}, \mathbf{v}_h) \quad \forall \mathbf{v}_h \in H_h \quad (3.62)$$
$$b_h(q_h, \mathbf{u}_h) = 0 \qquad \forall q_h \in L_h. \quad (3.63)$$

Then, a possible form of the *least square streamline–diffusion* approach is the following:

Find a pair $\{\mathbf{u}_h, p_h\} \in H_h \times L_h$, such that

$$\nu a_h(\mathbf{u}_h, \mathbf{v}_h) + n_h(\mathbf{u}_h, \mathbf{u}_h, \mathbf{v}_h) + b_h(p_h, \mathbf{v}_h) \qquad\qquad (3.64)$$
$$+ \sum_{T \in \mathbf{T}_h} (N(\mathbf{u}_h)\mathbf{u}_h - \mathbf{f}, \psi(\mathbf{u}_h, \mathbf{v}_h))_{|T} = (\mathbf{f}, \mathbf{v}_h) \quad \forall \mathbf{v}_h \in H_h$$

$$b_h(q_h, \mathbf{u}_h) = 0 \qquad \forall q_h \in L_h. \quad (3.65)$$

The additional term $\sum_{T \in \mathbf{T}_h} (N(\mathbf{u}_h)\mathbf{u}_h - \mathbf{f}, \psi(\mathbf{u}_h, \mathbf{v}_h))_{|T}$ is chosen such that some consistency is guaranteed since this correction term is satisfied by the continuous solution. Further *least square* terms for the continuity equations or *jump terms* due to discontinuous pressure approximations can be added, see for instance [10] and [66], and particularly recent papers by L. Tobiska for more general formulations). Typical settings for the modified test functions $\psi(\mathbf{u}_h, \cdot)$ are

$$\psi(\mathbf{u}_h, \cdot) := \delta_T^1(\mathbf{u}_h \cdot \nabla) \cdot + \delta_T^2(-\nu\Delta)\cdot \quad , \qquad\qquad (3.66)$$

in which case δ_T^i are "free" parameters which have to be chosen accordingly. Employing the nonconforming $\tilde{Q}1/Q0$ Stokes element, some terms are explicitly cancelled. To be precise, for $\mathbf{v}_h \in H_h$ and $q_h \in L_h$, and H_h, resp., L_h denoting the finite element spaces corresponding to the $\tilde{Q}1/Q0$ element pair, we can exploit (see [93])

$$\Delta\mathbf{v}_h = 0 \quad , \quad \nabla q_h = 0, \qquad\qquad (3.67)$$

such that our introduced variational form can be reduced to:

Find a pair $\{\mathbf{u}_h, p_h\} \in H_h \times L_h$, such that

$$\nu a_h(\mathbf{u}_h, \mathbf{v}_h) + n_h(\mathbf{u}_h, \mathbf{u}_h, \mathbf{v}_h)$$
$$+ b_h(p_h, \mathbf{v}_h) + \sum_{T \in \mathbf{T}_h} \delta_T(\mathbf{u}_h \cdot \nabla\mathbf{u}_h, \mathbf{u}_h \cdot \nabla\mathbf{v}_h)_{|T} \qquad (3.68)$$

$$= (\mathbf{f}, \mathbf{v}_h) + \sum_{T \in \mathbf{T}_h} \delta_T (\mathbf{f}, \mathbf{u}_h \cdot \nabla \mathbf{v}_h)_{|T} \quad \forall \mathbf{v}_h \in H_h$$

$$b_h(q_h, \mathbf{u}_h) = 0 \qquad\qquad \forall q_h \in L_h . \quad (3.69)$$

Since in many applications $\mathbf{f}$ can be explicitly set to 0, we cancel the additional right hand side term and end up with a modified variational formulation analogously to the upwind case:

Find a pair $\{\mathbf{u}_h, p_h\} \in H_h \times L_h$, such that

$$\nu a_h(\mathbf{u}_h, \mathbf{v}_h) + \tilde{n}_h(\mathbf{u}_h, \mathbf{u}_h, \mathbf{v}_h) + b_h(p_h, \mathbf{v}_h) \;=\; (\mathbf{f}, \mathbf{v}_h) \quad \forall \mathbf{v}_h \in H_h \quad (3.70)$$

$$b_h(q_h, \mathbf{u}_h) \;=\; 0 \qquad \forall q_h \in L_h . \quad (3.71)$$

In the streamline–diffusion context, $\tilde{n}_h(\cdot, \cdot, \cdot)$ is defined by

$$\tilde{n}_h(\mathbf{u}_h, \mathbf{v}_h, \mathbf{w}_h) := n_h(\mathbf{u}_h, \mathbf{v}_h, \mathbf{w}_h) + \sum_{T \in \mathbf{T}_h} \delta_T (\mathbf{u}_h \cdot \nabla \mathbf{v}_h, \mathbf{u}_h \cdot \nabla \mathbf{w}_h)_{|T} , \quad (3.72)$$

which can be interpreted as elementwise stabilization via the anisotropic diffusion term

$$-\delta_T (u_1^2 \mathbf{u}_{xx} + 2u_1 u_2 \mathbf{u}_{xy} + u_2^2 \mathbf{u}_{yy})_{|T} . \quad (3.73)$$

Beside the modified right hand side term, this "directed" artificial viscosity (which is locally weighted by δ_T) is in contrast to standard *upwind schemes* which can be shown to add "homogeneous" artificial viscosity. This is one of the main reasons why streamline–diffusion approaches have the potential to be more accurate than standard upwind techniques. We refer to the extensive papers by Führer [34], Johnson ([59],[61]), Lube [66] and Zhou [123] which contain many results with respect to approximation and stability properties of the streamline–diffusion techniques. Even the treatment of the (incompressible) Navier–Stokes equations with nonconforming finite elements has been performed and theoretical and numerical results are provided. However, many of the calculated applications are far from "real life" CFD applications such that a fair comparison between upwind (UPW) and streamline–diffusion (SD) approaches is not possible, at least not on the basis of the provided numerical results.

A critical quantity for the efficient computational treatment is the local damping parameter δ_T. A usual setting is the following choice which can be found for instance in [66]. The major idea is similar to the approach performed in the (weighted) *Samarskij upwind*. Again, we introduce the *local Reynolds number* Re_T,

$$Re_T = \frac{\|\mathbf{u}\|_T \cdot h_T}{\nu} .$$

(3.74)

Here, $\|\mathbf{u}\|_T$ means an averaged velocity value over T, and h_T denotes the "local mesh width" (we will discuss this quantity more in detail). Then, we can define

$$\delta_T := \delta^* \cdot \frac{h_T}{\|u\|_\Omega} \cdot \frac{2Re_T}{1 + Re_T} .$$

(3.75)

Obviously, for small local Reynolds numbers, with $Re_T \to 0$, δ_T is decreasing such that we reach in the limit case the standard second order central discretization. Vice versa, for convection dominated flows with $Re_T \gg 1$, we add an anisotropic diffusion term of size $O(h)$ which is aligned to the streamline direction $\mathbf{u}_h$. δ^* is an additional free parameter which can (?) be chosen arbitrarily by the user.

The last remarks indicate already that a convergence behaviour between $O(h^2)$ (for large viscosities/small Reynolds numbers, resp., for fine meshes) and $O(h^{3/2})$ for convection dominated problems may be expected. For rigorous derivations of these approximation properties, for instance for the L^2-error of the velocity, the reader is referred to the papers by Johnson ([59],[61]), Lube [66] and Zhou [123]. The lower bound $O(h^{3/2})$ is in contrast to simple upwind techniques which also stabilize via $O(h)$ terms. Here, the difference is caused by the use of the aligned artificial viscosity term instead of the "homogeneous" artificial viscosity damping in upwind schemes (and by the additional right hand side term which however vanishes for homogeneous data).

Thus, from a theoretical and also numerical point of view, the streamline–diffusion approaches seem to lead to convincing results and the corresponding implementation can be easily performed in standard finite element frameworks. However, from a computational and practical point of view, several problems remain:

- What are the precise definitions for δ^* and h_T depending on the recent application with respect to the underlying mesh? What is h_T on a strongly anisotropic mesh containing large aspect ratios? And what about the sensitivity of the resulting approximation properties with respect to these quantities?

- Can the resulting linear systems efficiently be solved at all? The corresponding stiffness matrices have lost the M-matrix property. What about the performance of typical multigrid solvers which are applied for the numerical solution? Is the range of "good" values δ^* and h_T, in the sense that the performed solvers work in an efficient and robust way, coincident with that range of δ^* and h_T values which lead to satisfying approximation properties?

- The derived convergence results of order $O(h^{3/2})$ are in general proven under the assumption that sufficient regularity is provided. What are the corresponding results for a typical complex CFD application? Are these results asymptotical results only, for mesh widths sufficiently small? And what about the quality on "coarse" meshes due to restrictions by the available hardware components? Moreover, the norms for the measured errors are in most cases global L^2- or H^1-norms. However, in "real" applications one is often interested in local quantities as lift and drag coefficients. What are the corresponding results for these values?

Most of these questions cannot be answered by theoretical considerations, particularly the involved comparisons with upwind techniques. Therefore, we started with numerical experiments and computational comparisons for these two techniques. In the following we will present some of our recent results which we obtained so far. However, it must be stated that this research is not finished and all conclusions have only preliminary character.

We explicitly perform numerical comparisons with stabilization techniques of upwind and streamline–diffusion type, leading to the following variational formulation of the (steady) Navier–Stokes equations:

Find a pair $\{\mathbf{u}_h, p_h\} \in H_h \times L_h$, such that

$$\nu a_h(\mathbf{u}_h, \mathbf{v}_h) + \tilde{n}_h(\mathbf{u}_h, \mathbf{u}_h, \mathbf{v}_h) + b_h(p_h, \mathbf{v}_h) = (\mathbf{f}, \mathbf{v}_h) \quad \forall \mathbf{v}_h \in H_h \quad (3.76)$$

$$b_h(q_h, \mathbf{u}_h) = 0 \qquad \forall q_h \in L_h . \quad (3.77)$$

In the streamline–diffusion context, $\tilde{n}_h(\cdot, \cdot, \cdot)$ is defined as

$$\tilde{n}_h(\mathbf{u}_h, \mathbf{v}_h, \mathbf{w}_h) := n_h(\mathbf{u}_h, \mathbf{v}_h, \mathbf{w}_h) + \sum_{T \in \mathbf{T}_h} \delta_T(\mathbf{u}_h \cdot \nabla \mathbf{v}_h, \mathbf{u}_h \cdot \nabla \mathbf{w}_h)_{|T} , \quad (3.78)$$

while in the upwind approach we set

$$\tilde{n}_h(\mathbf{u}_h, \mathbf{v}_h, \mathbf{w}_h)$$

$$:= \sum_l \sum_{k \in \Lambda_l} \oint_{\Gamma_{lk}} \mathbf{u}_h \cdot \mathbf{n}_{lk} \, d\gamma \, [1 - \lambda_{lk}(\mathbf{u}_h))(\mathbf{v}_h(m_k) - \mathbf{v}_h(m_l)] \mathbf{w}_h(m_l) \,. \tag{3.79}$$

Based on the *local Reynolds number* Re_T,

$$Re_T = \frac{\|\mathbf{u}\|_T \cdot h_T}{\nu} \,, \tag{3.80}$$

we can either define

$$\delta_T := \delta^* \cdot \frac{h_T}{\|\mathbf{u}\|_\Omega} \cdot \frac{2 Re_T}{1 + Re_T} \,, \tag{3.81}$$

respectively,

$$\lambda_{lk}(\mathbf{u}_h) := \left\{ \begin{array}{ll} \dfrac{\frac{1}{2} + \delta^* Re_T}{1 + \delta^* Re_T} & \text{if } Re_T \geq 0 \\[3ex] \dfrac{1}{2(1 - \delta^* Re_T)} & \text{otherwise} \end{array} \right\} \,. \tag{3.82}$$

The quantity h_T is in the following defined as "maximum possible local mesh width" according to a given local velocity $\mathbf{u}_{|T}$ on element T. Hence, the shape of the element containing large *aspect ratios* is respected in view of the flow direction $\mathbf{u}_{|T}$. Then the only remaining free quantity is the parameter δ^*. In the following we denote the actually applied scheme by UPW-δ^*, resp., SD-δ^*. If $\delta^* \to 0$ we reach for both approaches a central discretization scheme; we set C := SD-0. The typical range of practicable values for δ_{SD}^* is $[0.1, \ldots, 2]$ in which case we may expect a convergence behaviour between $O(h)$ and $O(h^2)$. While parameters δ^*, chosen too large, lead in the streamline–diffusion context to *overstabilization* and consequently to decreasing accuracy, we obtain for upwinding in the limit case, $\delta_{UPW}^* \to \infty$, the *simple upwind* scheme which is at least of first order. We denote this approach by UPW-∞, while UPW-1.0 corresponds to the scheme originally proposed by Samarskij.

The following calculations are performed for the same (or slightly modified) configurations as performed in the *1995 DFG Benchmark* (see Section 1.1).

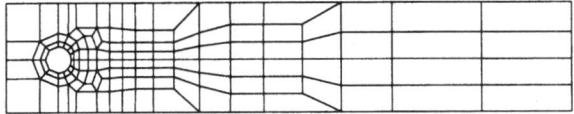

Figure 3.9: Coarse mesh for flow around a cylinder

		$\nu = 1/10$					
		$L_2(\mathbf{u})$	α	$H_1(\mathbf{u})$	α	$L_2(p)$	α
UPW-0.1	3	$2.62 \cdot 10^{-2}$		$1.36 \cdot 10^{-1}$		$4.12 \cdot 10^{-2}$	
	4	$6.77 \cdot 10^{-3}$	1.96	$6.91 \cdot 10^{-2}$	0.98	$1.55 \cdot 10^{-2}$	1.42
	5	$1.70 \cdot 10^{-3}$	2.00	$3.50 \cdot 10^{-2}$	0.98	$6.40 \cdot 10^{-3}$	1.28
SD-1.0	3	$2.35 \cdot 10^{-2}$		$1.25 \cdot 10^{-1}$		$4.03 \cdot 10^{-2}$	
	4	$6.52 \cdot 10^{-3}$	2.07	$6.71 \cdot 10^{-2}$	0.90	$1.56 \cdot 10^{-2}$	1.37
	5	$1.65 \cdot 10^{-3}$	1.99	$3.38 \cdot 10^{-2}$	0.99	$6.38 \cdot 10^{-3}$	1.29
		$\nu = 1/1000$					
		$L_2(\mathbf{u})$	α	$H_1(\mathbf{u})$	α	$L_2(p)$	α
UPW-0.1	3	$6.16 \cdot 10^{-2}$		$3.15 \cdot 10^{-1}$		$3.17 \cdot 10^{-1}$	
	4	$3.43 \cdot 10^{-2}$	0.85	$1.73 \cdot 10^{-1}$	0.87	$1.47 \cdot 10^{-1}$	1.11
	5	$1.51 \cdot 10^{-2}$	1.19	$7.70 \cdot 10^{-2}$	1.17	$6.00 \cdot 10^{-2}$	1.30
SD-1.0	3	$3.04 \cdot 10^{-3}$		$5.55 \cdot 10^{-2}$		$3.45 \cdot 10^{-2}$	
	4	$1.08 \cdot 10^{-3}$	1.50	$3.09 \cdot 10^{-2}$	0.85	$1.89 \cdot 10^{-2}$	0.87
	5	$3.78 \cdot 10^{-4}$	1.52	$1.66 \cdot 10^{-2}$	0.90	$8.24 \cdot 10^{-3}$	1.20

Table 3.8: Approximation results for polynomial solution

We consider flow in a channel around a cylinder and the coarse mesh is shown in Figure 3.9.

This coarse mesh is denoted with *level 1* while all other refinement levels are generated via the usual regular grid refinement algorithm which recursively divides each element into 4 finer elements. We start with prescribing a typical polynomial solution and perform calculations for $\nu = 1/10$ and $\nu = 1/1000$, with a parabolic inflow profile of order 1. The calculations show the resulting errors with respect to the refinement level and the numerically measured value α for the approximation rate $O(h^\alpha)$.

The results are coincident with typical calculations which are presented by other authors for polynomial exact solutions. And indeed, the streamline diffusion technique provides very accurate solutions while in contrast the upwind schemes deteriorate for increasing Reynolds numbers. However, let us perform calculations for a "real" flow. Following the described *1995 DFG Benchmark* configurations, we perform the same stationary computations for $Re = 20$ and additionally for $Re = 50$. Higher Reynolds numbers are impossible with this direct stationary solution approach since then, the corresponding flow is getting nonsteady. In the following tables, we measure the l_2-error (that means in the euclidian norm for the vector of unknowns)

		$l_2(\mathbf{u})$	α	$l_2(p)$	α
UPW-0.1	2	$2.81 \cdot 10^{-2}$		$8.00 \cdot 10^{-3}$	
	3	$9.21 \cdot 10^{-3}$	1.61	$3.10 \cdot 10^{-3}$	1.37
	4	$3.31 \cdot 10^{-3}$	1.48	$9.72 \cdot 10^{-4}$	1.68
	5	$8.83 \cdot 10^{-4}$	1.91	$2.58 \cdot 10^{-4}$	1.92
UPW-1.0	2	$3.82 \cdot 10^{-2}$		$1.67 \cdot 10^{-2}$	
	3	$2.15 \cdot 10^{-2}$	0.83	$8.60 \cdot 10^{-3}$	0.96
	4	$1.03 \cdot 10^{-2}$	1.06	$4.01 \cdot 10^{-3}$	1.10
	5	$4.30 \cdot 10^{-3}$	1.26	$1.59 \cdot 10^{-3}$	1.33
UPW-∞	2	$4.41 \cdot 10^{-2}$		$2.21 \cdot 10^{-2}$	
	3	$2.92 \cdot 10^{-2}$	0.59	$1.29 \cdot 10^{-2}$	0.78
	4	$1.75 \cdot 10^{-2}$	0.74	$7.37 \cdot 10^{-3}$	0.81
	5	$9.74 \cdot 10^{-3}$	0.85	$4.09 \cdot 10^{-3}$	0.85
SD-0.2	2	$8.97 \cdot 10^{-3}$		$5.32 \cdot 10^{-3}$	
	3	$3.64 \cdot 10^{-3}$	1.31	$2.14 \cdot 10^{-3}$	1.32
	4	$1.36 \cdot 10^{-3}$	1.43	$7.46 \cdot 10^{-4}$	1.53
	5	$4.57 \cdot 10^{-4}$	1.58	$2.31 \cdot 10^{-4}$	1.70
SD-1.0	2	$1.53 \cdot 10^{-2}$		$8.08 \cdot 10^{-3}$	
	3	$8.80 \cdot 10^{-3}$	0.80	$4.32 \cdot 10^{-3}$	0.91
	4	$4.59 \cdot 10^{-3}$	0.94	$1.95 \cdot 10^{-3}$	1.15
	5	$2.03 \cdot 10^{-3}$	1.18	$7.62 \cdot 10^{-4}$	1.36
C	2	$1.06 \cdot 10^{-2}$		$7.27 \cdot 10^{-3}$	
	3	$3.62 \cdot 10^{-3}$	1.56	$2.66 \cdot 10^{-3}$	1.46
	4	$9.85 \cdot 10^{-4}$	1.88	$7.91 \cdot 10^{-4}$	1.76
	5	$2.35 \cdot 10^{-4}$	2.07	$2.17 \cdot 10^{-4}$	1.87

Table 3.9: "Real" approximation results for $Re = 20$

with respect to a reference solution which has been calculated on level 7 ($\sim$ 550.000 mesh cells and about 3 millions of unknowns).

The results are somewhat surprising. There are some observations for the global approximation behaviour (l_2 error for velocity and pressure over Ω) which have to be explicitly stated:

1. Both upwind and streamline–diffusion schemes are very sensitive to the remaining free parameter δ^* which has to be precisely fixed by the user (or by another "control mechanism"). It is always worth to try to perform calculations with small values δ^*, going more into direction of a central discretization. However, while we always obtained for the upwinded scheme UPW-0.1 a correspondingly fast convergence behaviour (in fact, there is no big difference in the resulting multigrid rates for the shown upwinded schemes), we have to work harder in the streamline–diffusion case. We explain this "solver topic" more in detail at the end of this section.

2. The central discretization (in combination with the proposed stabilization techniques as solver in a defect correction approach; otherwise the

		$l_2(\mathbf{u})$	α	$l_2(p)$	α
UPW-0.1	2	$1.23 \cdot 10^{-1}$		$9.53 \cdot 10^{-2}$	
	3	$6.01 \cdot 10^{-2}$	1.04	$4.73 \cdot 10^{-2}$	1.00
	4	$2.42 \cdot 10^{-2}$	1.32	$1.90 \cdot 10^{-2}$	1.32
	5	$8.76 \cdot 10^{-3}$	1.47	$6.10 \cdot 10^{-3}$	1.64
UPW-1.0	2	$1.73 \cdot 10^{-1}$		$1.45 \cdot 10^{-1}$	
	3	$1.14 \cdot 10^{-1}$	0.60	$8.55 \cdot 10^{-1}$	0.76
	4	$6.20 \cdot 10^{-2}$	0.88	$4.79 \cdot 10^{-2}$	0.84
	5	$2.80 \cdot 10^{-2}$	1.14	$2.29 \cdot 10^{-2}$	1.06
UPW-∞	2	$1.80 \cdot 10^{-1}$		$1.61 \cdot 10^{-1}$	
	3	$1.31 \cdot 10^{-1}$	0.46	$1.02 \cdot 10^{-1}$	0.66
	4	$8.42 \cdot 10^{-2}$	0.64	$6.38 \cdot 10^{-2}$	0.68
	5	$4.82 \cdot 10^{-2}$	0.81	$3.79 \cdot 10^{-2}$	0.75
SD-0.3	2	$3.75 \cdot 10^{-2}$		$3.11 \cdot 10^{-2}$	
	3	$1.29 \cdot 10^{-2}$	1.54	$1.47 \cdot 10^{-2}$	1.08
	4	$5.56 \cdot 10^{-3}$	1.22	$5.99 \cdot 10^{-3}$	1.30
	5	$2.21 \cdot 10^{-3}$	1.34	$2.08 \cdot 10^{-3}$	1.53
SD-1.0	2	$2.46 \cdot 10^{-2}$		$4.72 \cdot 10^{-2}$	
	3	$2.70 \cdot 10^{-2}$		$2.76 \cdot 10^{-2}$	0.78
	4	$1.48 \cdot 10^{-2}$	0.87	$1.36 \cdot 10^{-2}$	1.02
	5	$7.15 \cdot 10^{-3}$	1.05	$5.90 \cdot 10^{-3}$	1.21
C	2	div.		div.	
	3	$3.72 \cdot 10^{-2}$		$2.65 \cdot 10^{-2}$	
	4	$1.11 \cdot 10^{-2}$	1.80	$1.22 \cdot 10^{-2}$	1.12
	5	$2.67 \cdot 10^{-3}$	2.06	$4.04 \cdot 10^{-3}$	1.60

Table 3.10: "Real" approximation results for $Re = 50$

linear problems arising from the central discretization cannot be solved directly by multigrid) lead to the best approximation results and show – if the mesh is sufficiently fine – an asymptotically quadratic convergence behaviour.

3. The streamline–diffusion approach leads in fact to better results than the corresponding upwind schemes. However, the differences are smaller than expected and the predicted asymptotical convergence behaviour seems to be inbetween of order $O(h^{3/2})$ and $O(h^2)$.

4. The well-tuned upwind scheme UPW-0.1 is clearly superior in comparison to the classical upwind scheme UPW-1.0 and especially to the first order scheme UPW-∞. However, it is also preferrable with regard to some streamline–diffusion schemes which do not perform the "optimal" setting for δ^*.

5. Surprisingly, the pressure error also behaves like $O(h^{3/2})$ up to $O(h^2)$. Here, we have to keep in mind that a piecewise constant ansatz is utilized, and no other postprocessing instead of interpolating to the nodes has been performed.

	δ^*	level 3	level 4	level 5	level 6	level 7	reference
		Drag for $Re = 20$					
	0.1	5.5810	5.5608	5.5657	5.5718	5.5755	**5.576**
UPW	1.0	6.0199	5.7054	5.6011	5.5793	5.5771	**5.576**
	∞	6.4863	6.0203	5.7802	5.6717	5.6230	**5.576**
SD	0.2	5.6770	5.6299	5.5953	5.5821	5.5787	**5.576**
	1.0	6.3836	5.9674	5.7269	5.6282	5.5931	**5.576**
C		5.5549	5.5638	5.5677	5.5724	5.5756	**5.576**
		Drag for $Re = 50$					
	0.1	3.9213	3.7387	3.6958	3.6915	3.6924	**3.69**
UPW	1.0	4.3140	3.8904	3.7308	3.6939	3.6908	**3.69**
	∞	4.5370	4.0750	3.8503	3.7585	3.7221	**3.69**
SD	0.3	3.9764	3.8409	3.7521	3.7137	3.7000	**3.69**
	1.0	4.4380	4.0945	3.8705	3.7617	3.7168	**3.69**
C		3.9356	3.7700	3.7111	3.6966	3.6939	**3.69**
		Lift for $Re = 20$					
	0.1	0.00487	0.00913	0.01014	0.01043	0.01054	**0.0106**
UPW	1.0	-0.00280	0.00647	0.00993	0.01058	0.01063	**0.0106**
	∞	-0.00235	0.00673	0.01040	0.01128	0.01119	**0.0106**
SD	0.2	0.00613	0.00796	0.00949	0.01014	0.01045	**0.0106**
	1.0	-0.00117	0.00434	0.00779	0.00944	0.01016	**0.0106**
C		0.00538	0.00803	0.00973	0.01031	0.01052	**0.0106**
		Lift for $Re = 50$					
	0.1	-0.01353	-0.01013	-0.01018	-0.01060	-0.01072	**-0.0108**
UPW	1.0	-0.02223	-0.01367	-0.01051	-0.01014	-0.01043	**-0.0108**
	∞	-0.02414	-0.01606	-0.01245	-0.01113	-0.01074	**-0.0108**
SD	0.3	-0.01098	-0.01154	-0.01122	-0.01099	-0.01085	**-0.0108**
	1.0	-0.01652	-0.01362	-0.01194	-0.01127	-0.01098	**-0.0108**
C		-0.00844	-0.01109	-0.01109	-0.01090	-0.01083	**-0.0108**

Table 3.11: Drag and lift coefficients for $Re = 20$ and $Re = 50$

If we additionally take into account the local error behaviour with respect to the provided drag and lift coefficients which are defined on the contour of the cylinder only, we obtain the following results. Further conclusions can be drawn.

6. Almost the same conclusions are valid, especially with respect to the sensitivity of the parameter δ^*. However, the *weighted upwind* scheme UPW-0.1 leads in general to the best results! Only on very fine meshes, the streamline–diffusion and also the central schemes are becoming superior.

7. There seems to be a large difference between global and local error behaviour. Up to now the theoretical framework is still missing since most approximation results are typically derived in global norms.

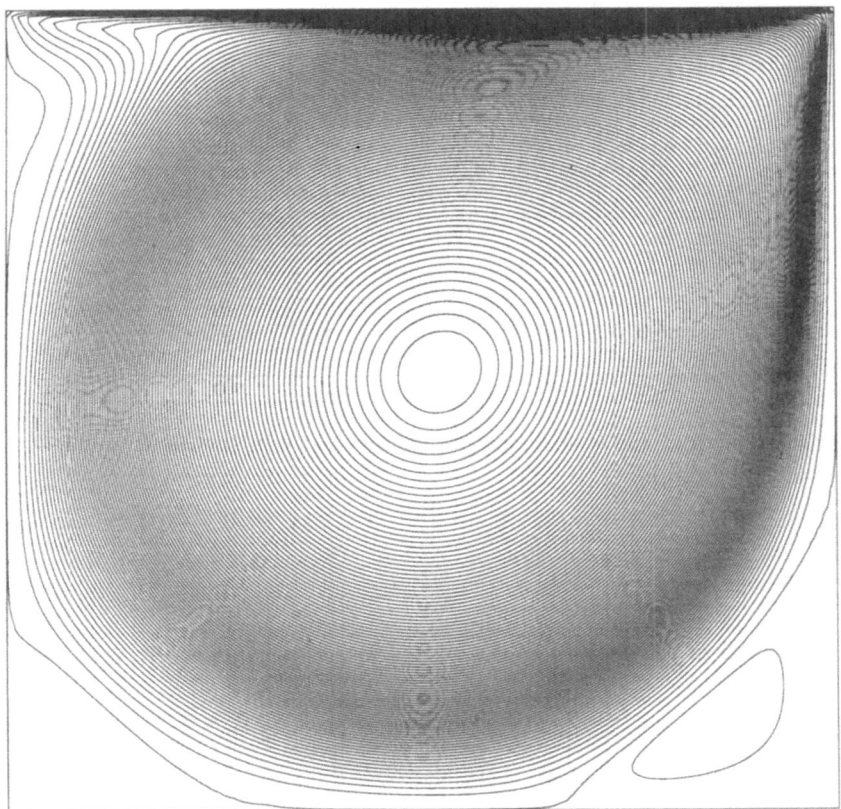

Figure 3.10: Typical flow pattern for *lid driven cavity* at $Re = 2,000$

Before we finish this section concerning stabilization techniques for convective terms, we provide some additional computational results for a typical *lid driven cavity* calculation for $Re = 1/\nu = 2,000$. For this configuration, the typical flow pattern looks like in the following Figure 3.10.

The main vortex in the interior of the domain contains most of the flow energy and initiates for the chosen viscosity parameter both secondary vortices (at the lower left and right corner) and even a smaller third separation zone at the upper left side. However, we are here more interested in the quality of the approximated main flow and particularly the main vortex since this is the source for all other derived small scale physical phenomenons.

We perform calculations on a uniformly refined mesh covering the unit square such that refinement level i corresponds to an equidistant mesh width of $h = 2^{i-1}$. In the following tables we show the results for the flux, resp., the

	δ^*	level 6	level 7	level 8	level 9	level 10
		Flux for $1/\nu = 2,000$				
UPW	0.1	-0.1200	-0.1228	-0.1267	-0.1251	-0.1221
	1.0	-0.1230	-0.1389	-0.1424	-0.1401	-0.1330
SD	0.25	-0.0630	-0.0815	-0.0985	-0.1101	-0.1162
	1.0	-0.0607	-0.0775	-0.0899	-0.1024	-0.1115
	4.0	-0.0432	-0.0575	-0.0711	-0.0856	-0.0994
C		-0.0520	-0.0780	-0.1005	-0.1132	-0.1181
		Energy for $1/\nu = 2,000$				
UPW	0.1	0.345	0.345	0.337	0.322	0.311
	1.0	0.355	0.365	0.367	0.356	0.336
SD	0.25	0.210	0.239	0.268	0.288	0.298
	1.0	0.202	0.225	0.250	0.273	0.289
	4.0	0.161	0.183	0.210	0.239	0.266
C		0.179	0.229	0.271	0.293	0.301

Table 3.12: Flux values and energy of the velocity in the *lid driven cavity* at $Re = 2,000$

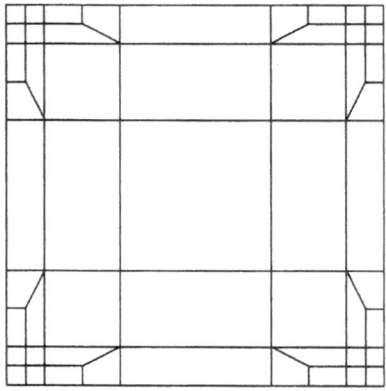

Figure 3.11: Semi-adapted coarse mesh for *lid driven cavity* at $Re = 10,000$

normal velocity, passing between the center point $(0.5, 0.5)$ and the boundary point $(0.5, 0.0)$ at the bottom (in fact this is the streamfunction value at $(0.5, 0.5)$). Additionally, we provide results for the total *energy* of the flow, that means the measured l_2-norm of the overall velocity vector.

The typical phenomenon which occurs in this configuration and also for other problems involving higher Reynolds numbers is that on "coarse meshes", $h = 1/32$ or $h = 1/64$, the central scheme has problems with approximating the **main flow**. Large errors of more than $30 - 50\%$ may occur and the resulting flow pattern may look "unphysically".

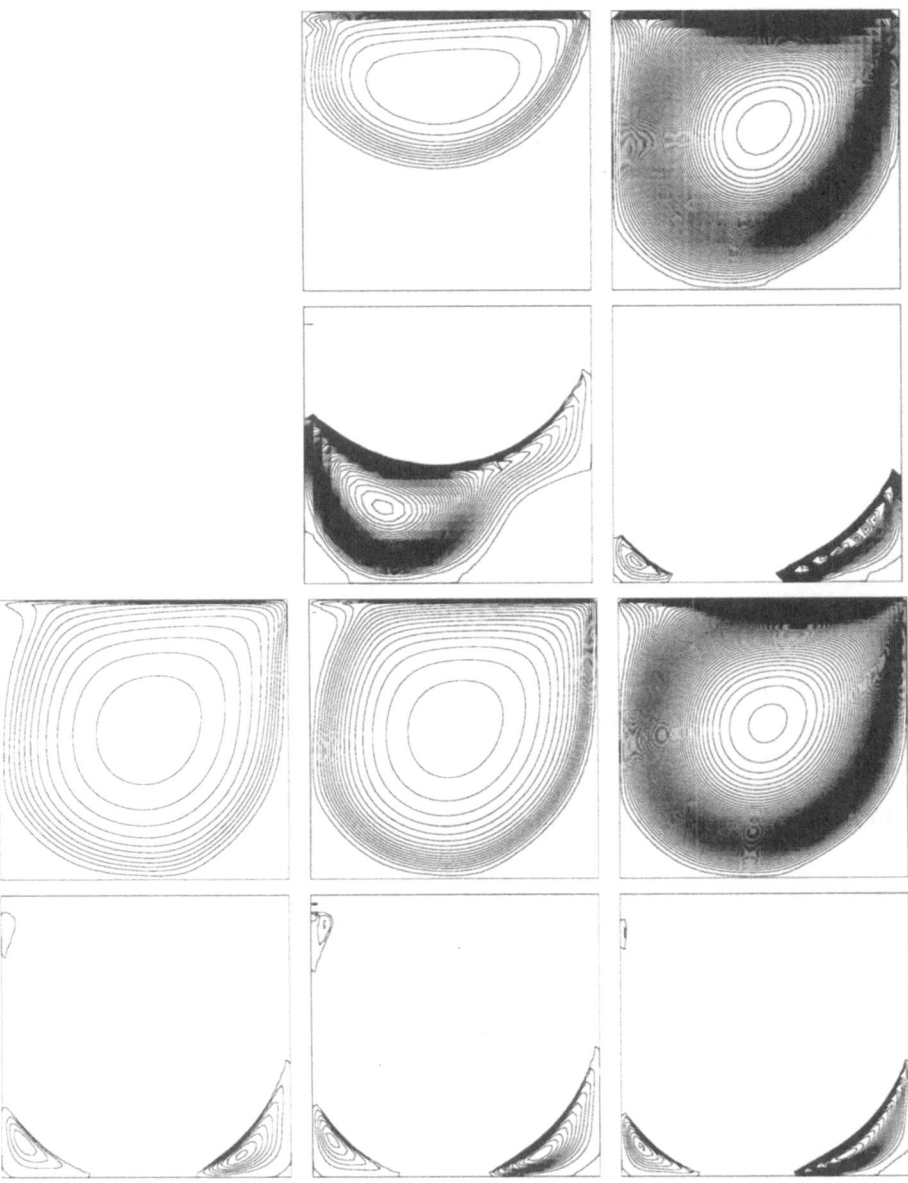

Figure 3.12: Negative streamlines (in the range [-0.125:0], 51 isolines, equidistantly distributed) and positive streamlines (37 isolines, for [0:0.0065]) for *lid driven cavity* at $Re = 10,000$, with central (left), streamline–diffusion (middle) and upwind discretization (right). The first two rows are for $h = 1/32$, the third and fourth for $h = 1/64$. For $h = 1/32$, the *central* scheme did not converge.

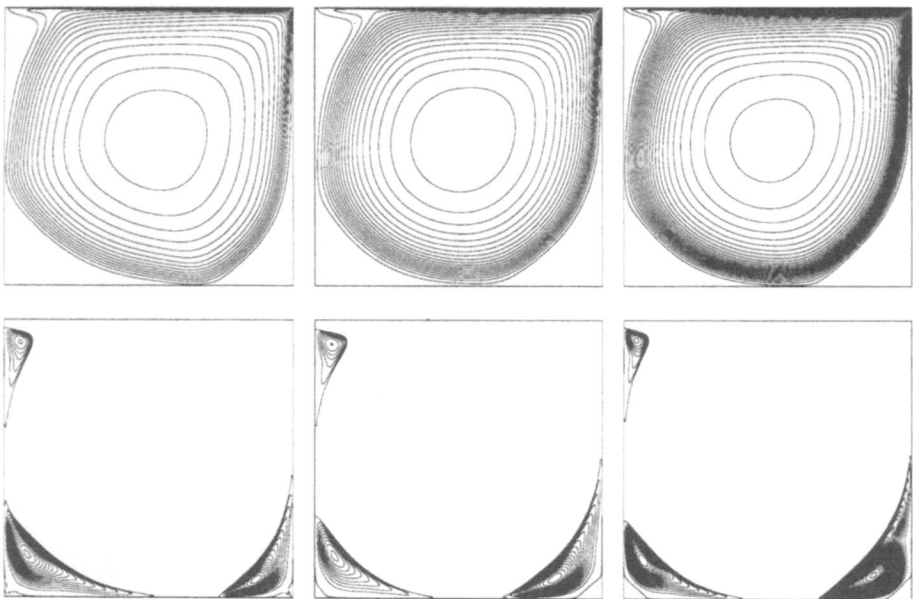

Figure 3.13: Negative streamlines (in the range [-0.125:0], 51 isolines, equidistantly distributed) and positive streamlines (37 isolines, for [0:0.0065]) for *lid driven cavity* at $Re = 10,000$, with central (left), streamline–diffusion (middle) and upwind discretization (right). The results are obtained for the mesh in Figure 3.11 which has been four times refined.

While for $Re = 2,000$ these effects are not so spectacular, this observation is more visible for the case $Re = 10,000$. The figures in 3.12 show results for $h = 1/32$ (here, the central discretized solution for $h = 1/32$ could not be obtained at all) and for $h = 1/64$. The pictures in Figure 3.13 show analogous results on the semi-adapted mesh (see 3.11). While the mesh width in the interior is only $h = 1/40$, the grid resolution near the walls is much finer.

As can be easily seen by counting the streamlines in Figure 3.13, the absolute values of the streamfunction for the streamline–diffusion ansatz are smaller than compared with the upwind technique, while the central scheme leads to even smaller values. Additionally, the *total energy* in the central case is 0.156 while for streamline-diffusion the value is 0.181 and for upwinding 0.254. In the central case, the main structure of the flow is wrong which results in unphysically secondary vortices in the lower part: the maximum value of the left vortex is larger than for the right one! Nevertheless, the more "spectacular" third separation zone near the left upper corner is surprisingly well approximated.

We can conclude that the central scheme is potentially most accurate but only if the mesh is fine enough (?). This well-know fact is repeated by these calculations. However, also the streamline-diffusion stabilization technique does not seem to lead to significant improvement.

Or probably: up to now we have not found the "right" parameters involved. Both schemes, the central and the streamline–diffusion methods, provide a higher convergence behaviour in view of asymptotical mesh widths, but they seem to fail on certain meshes which are too coarse for a given Reynolds number, at least in our computations. The background for these mesh requirements comes from 3D calculations. One always has to keep in mind that a 3D configuration for $h = 1/64$ leads to already 262.144 mesh cells while in 2D only 4.096 elements are the result. Thus, (finite element) calculations for $h = 1/64$ up to $h = 1/128$ are for many computers today the maximum performable configurations in 3D. And if we cannot guarantee to obtain quantitatively precise results with these coarse meshes, we want to simulate at least a qualitatively correct physical behaviour in 3D. This qualitative behaviour seems to be much better obtained by the performed upwind schemes. Indeed, the corresponding asymptotical convergence may not be monotone and of lower order only, but for coarse meshes the results always represent the "overall physical" behaviour in a quite good manner.

Up to now, we have completely neglected the question of efficient linear solvers of multigrid type for the stabilization techniques proposed. As already seen in the section concerning finite element upwinding techniques, some M-matrix properties for the simple, first order upwind scheme UPW-∞ in combination with the nonconforming rotated multilinear finite elements can be guaranteed. Consequently, standard iterative schemes as SOR or ILU can be applied as smoothers in a multigrid approach for the discretized momentum equations such that the resulting nonsymmetric stiffness matrices can be efficiently solved. In practise, the multigrid rates for solving these nonsymmetric problems can be even better than for symmetric diffusive problems. The reason for this behaviour is that upwinding in the limit case of dominating convection leads to lower triangular matrices if appropriate renumbering strategies are applied. Hence, both SOR or ILU schemes are in the limit case exact solvers and imply excellent convergence rates if applied as smoothers. These desired M-matrix properties cannot be guaranteed for the weighted versions, but our numerical experience shows a similarly excellent behaviour, even down to the damping parameter $\delta^* \geq 0.1$. Below that value, numerical instabilities usually occur and the numerical solution efficiency is diminished.

In contrast, the streamline–diffusion approaches show a different (multigrid) convergence behaviour. In the limit case of pure convection the stabilization term is an anisotropic diffusion operator according to the streamlines such that the resulting matrices in the momentum equation may tend to be tridiagonal. However it is well-known from the numerical treatment of such scalar equations that only ILU–methods (or line solvers) can provide a robust solution behaviour. Hence, in combination with appropriate renumbering techniques, probably of *streamline renumbering* type as proposed by Hackbusch [50], the corresponding multigrid tools may be as robust and efficient as for upwind discretizations. Up to now, we did not apply this approach and as far as we know the corresponding techniques for 3D flows are not sufficiently verified at all. Moreover, one can easily see that overstabilization, that means the choice of too large values δ^*, may even lead to less efficient solution behaviour since we approach the "tridiagonal matrix" limit. This is to see in comparison to the upwind approach in which case larger values δ^* tend to matrices with correspondingly better properties. So, it is not clear at all if the range of "optimal" values δ^* for obtaining good multigrid rates is the same as for providing accurate approximations. All these remarks concerning the solution behaviour of streamline–diffusion approaches require further intensified research activities.

We have discussed so far the stationary case only. However, also for nonstationary calculations corresponding stabilization techniques are absolutely necessary. The reason is not so much the solution efficiency (small time steps help to improve convergence rates for nonsteady problems), but one still has to take care that robust discretization and hence accurate approximative solutions are provided. While the application of the upwind techniques can be directly performed without any modification, the streamline–diffusion case is somewhat delicate. While the proposed *least square* ansatz applied to generalized Navier–Stokes equations leads to unnecessary stabilization of terms involving zero order terms, we simply perform a stabilization by modifying the convective terms only. Another approach might be obtained via *least square* stabilization for space–time finite elements which then leads to *full Galerkin schemes* in space and time ($\sim DG(n)$ schemes, include the Backward Euler). These schemes provide a posteriori error control in space and time and may be an essential component of future algorithmic approaches, but so far these techniques are theoretically verified only, and they may be tested for simpler test cases, but not often for complex CFD applications.

We can summarize our considerations concerning stabilization techniques for convective terms by some concluding remarks. The next issues are not only based on the results presented in this section, but include also our complete numerical experience with these schemes during the last years. Obviously they are only personal and subjective, and particularly the results obtained for the streamline–diffusion techniques are not complete. But nevertheless we suggest they are of importance for other CFD specialists, too, and should be published in this context as starting point for further discussion and research.

- Streamline–diffusion techniques for stabilizing the convective term in the incompressible Navier–Stokes equations offer interesting features for mathematicians. They can be derived and rigorously analyzed in the context of *Galerkin least square* schemes and seem to be appropriate to work in the recently favourized a posteriori error control context. From theoretical point of view, their application to higher order elements and inside of fully reliable error control mechanisms seems to be under control, even for the Navier–Stokes equations. The crucial points are hidden in the practical application of these techniques:

  - The user is advised to fix some remaining free parameters which may also include local mesh size parameters. Our experience shows that the complete approach is very sensitive with respect to the specific setting of these quantities. Hence, in the context of "Black Box" software packages, our recent version of the streamline–diffusion method is not fully reliable and efficient as long as these free parameters cannot be appropriatly fixed by a self-adaptive control during the calculation.

  - The range of these admissible free parameters due to high approximation properties in comparison to values which allow fast iterative solvers is not guaranteed to be coincident. From a computational point of view, the consequence may be that two different streamline–diffusion methods have to be applied, one determining the accuracy and the other one as preconditioner in a defect–correction approach to iterate towards the solution.

  - Another aspect is that streamline–diffusion stiffness matrices involve large numerical cost during the assembling process. For instance, in 3D the trilinear form consists of 13 terms which all have to be generated by numerical quadrature and reference element transformations in the finite element context. This explains the huge CPU cost for the nonstationary projection schemes PP3D-SD in the *FEATFLOW Benchmark* in comparison to the upwind approach UPW-0.1 (see Section 1.2). The remedy might be a more clever implementation, but nevertheless the numerical amount of work for the assembling process remains large and is very difficult to improve on modern hardware platforms.

  - As far as we know, only approximation properties with respect to global norms have been derived. From theoretical point of view (and additionally also due to our numerical results) there is no indication concerning the expected accuracy of streamline–diffusion and also upwind techniques if local quantities as drag and lift coefficients are required. At least our recent numerical results show no improved behaviour in comparison to the upwind schemes.

– Additionally, our numerical results show that on a <u>fixed</u> mesh for a given Reynolds number, the applied streamline–diffusion ansatz may not lead to satisfying results: The results are somewhat disappointing up to now, as soon as the Reynolds numbers is "high". The remedy may be again the adaptive fixing of the involved free parameters, but so far the upwind schemes often provide significantly better results (in a qualitative <u>and</u> also quantitative sense) on relatively coarse meshes.

• The upwind schemes provide comparative approximation properties for global and especially for local quantities (sometimes even better!). We prefer to set the free parameter δ^* to $\delta^*_{UPW} = 0.1$ since the resulting scheme UPW-0.1 is a good compromise between high accuracy and excellent solution performance. Moreover, upwind stiffness matrices are easy to implement, in 2D and 3D, and lead to excellently balanced behaviour of the assembling and the solution part with respect to modern processors. The disadvantage for all that is the insufficient underlying mathematical analysis. From theoretical point of view, the approximation order is still not proven, and the techniques for applying a fully reliable a posteriori error control are not yet completely solved (for a recent overview, see [34] and [93]).

As final conclusion we can state:

Streamline–diffusion techniques may have a great potential for future numerical tools in CFD software since they provide high accuracy and are necessary components for a fully reliable and efficient a posteriori error control. However, much more research is needed to optimize the numerical and computational aspects such that "Black Box" components for CFD software packages can be provided.

Since this optimization process will take some time, our favourized stabilization technique is the upwind scheme UPW-0.1 which is recently the best compromise with respect to high accuracy and excellent solution performance, in 2D as well as in 3D. Both schemes should be considered in future approaches since so far it is not clear which approach has the greater potential for efficient "real life" CFD simulation tools.

3.1.5   A posteriori error control mechanisms for finite element approaches

For early works on a posteriori error control for finite element methods, we refer to the pioneering papers of Babuška and Rheinboldt [4] and Bank and Weiser [5]. The developed approaches have been recently surveyed in Verfürth [117]. The concept of *a posteriori error control by duality arguments* has been introduced by Eriksson, Johnson and their co-workers (see [30] and the literature cited therein). Heuristically based error indicators for elliptic problems have been devised, e.g., by Zienkiewicz and Zhu [124]. All of them may lack of the two problems:

1. It is not clear how to apply them to realistic CFD problems. The control is mostly performed for global norms while in practical applications local quantities as drag and lift coefficients are often more interesting.

2. The approaches lead mostly to equilibrated errors over the mesh, but a quantitatively precise prediction of the true error is not provided.

One may state that these approaches are better called *error indicators* since a sharp estimation for the error can be given only in some specific configurations. In the last years, some new approaches have been developed which seem to overcome these failures and which can be formulated in a very general framework such that also the Navier–Stokes equations and other CFD relevant models can be treated. For a short but motivating presentation of the recent state of the art we follow papers by Rannacher and Becker (see [8],[9]).

The strategies for error control and mesh refinement used in the finite element context are mostly based on a posteriori estimates in global norms, e.g., the L^2- or the energy norm, which involve local residuals of the computed solution. The resulting mesh refinement process aims at equilibrating these local *error indicators*. However, meshes generated on the basis of such *global* error estimates may not be appropriate in cases of strongly varying coefficients and for controlling the accuracy in approximating local quantities as, e.g., lift and drag coefficients or pressure distribution on contours. For this task one needs more detailed information on the mechanism of error propagation which can be obtained by employing suitable duality arguments known from the a priori error analysis as the *Aubin-Nitsche trick* (see, e.g., [16]). The corresponding *dual solutions* then yield the appropriate weight factors to be used in the a posteriori error estimates.

To give an illustrative example, we consider as starting point the Poisson equation

$$-\Delta u = f \quad \text{in } \Omega, \qquad u = 0 \quad \text{on } \partial\Omega, \tag{3.83}$$

in a bounded domain $\Omega \subset \mathbf{R}^2$ which, for the moment, is assumed to be convex polygonal. The restriction to two dimensions is only for simplicity as the extension to arbitrary dimensions is straightforward.

By $(\cdot, \cdot)$ we denote the L^2 inner product and by $\| \cdot \|$ the corresponding norm on Ω. Let (3.83) be discretized by a standard finite element Galerkin method using piecewise polynomials, for instance linear or bilinear shape functions on meshes $\mathbf{T}_h = \bigcup\{T\}$ satisfying the usual regularity conditions (see, e.g., [16]). For each $\mathbf{T}_h$, let $h_{max} = \max_{T \in \mathbf{T}_h} h_T$, where $h_T = \text{diam } (T)$. The corresponding finite element subspaces are $V_h \subset V := H_0^1(\Omega)$, but in principle also nonconforming approximations are allowed (see Führer [34]). The approximation properties can usually be characterized in terms of local approximation estimates (see [16]),

$$\max\left\{\|v - I_h v\|_T, h_T^{1/2}\|v - I_h v\|_{\partial T}\right\} \le C_{i,T} h_T^{1+r} \|\nabla^{1+r} v\|_{\tilde{T}}, \quad r \in \{0, 1\}, \tag{3.84}$$

for $v \in V \cap H^{1+r}(\Omega)$, where $I_h v \in V_h$ is some locally defined approximation to v, and $\| \cdot \|_B$ denotes the L^2-norm over a set B. For $r = 0$, $\tilde{T}$ is the union of all neighbouring elements of T, while for $r = 1$, one simply has $\tilde{T} = T$.

Starting from the variational formulation of (3.83), the finite element method seeks to determine approximations $u_h \in V_h$, such that

$$(\nabla u_h, \nabla \varphi_h) = (f, \varphi_h) \quad \forall \varphi_h \in V_h. \tag{3.85}$$

Subtracting (3.85) from the variational formulation of (3.83) results in the following orthogonality relation for the error $e := u - u_h$,

$$(\nabla e, \nabla \varphi_h) = 0 \quad \forall \varphi_h \in V_h. \tag{3.86}$$

This discretization allows for optimal-order a priori estimates in the energy and L^2-norm the following approximation result,

$$\|e\| + h_{max}\|\nabla e\| \leq C\, h_{max}^2 \|\nabla^2 u\|, \tag{3.87}$$

provided that the solution u is sufficiently regular. Corresponding estimates also hold with respect to the maximum norm (for references see, e.g., [16]). For error control in the energy or L^2-norm, one may proceed as follows. Using the Galerkin orthogonality (3.86) and integration by parts on each element T yields

$$(\nabla e, \nabla z) = \sum_{T \in \mathbf{T}_h} \left\{ (f + \Delta u_h, z - z_h)_T - \frac{1}{2}(n \cdot [\nabla u_h], z - z_h)_{\partial T} \right\}, \tag{3.88}$$

for any $z \in V$, where $[\nabla u_h]$ denotes the jump of ∇u_h across the element boundary, and $z_h \in V_h$ is a suitable approximation of z. On edges along the boundary, we set $[\nabla u_h] = \nabla u_h$.

Then, using Hölder's inequality on each element, we obtain

$$|(\nabla e, \nabla z)| \leq \sum_{T \in \mathbf{T}_h} \rho_T\, \omega_T, \tag{3.89}$$

with local *residuals* ρ_T and *weights* ω_T. These are defined via

$$\rho_T \quad := \quad h_T\|f + \Delta u_h\|_T + \frac{1}{2}h_T^{1/2}\|n \cdot [\nabla u_h]\|_{\partial T} \tag{3.90}$$

$$\omega_T \quad := \quad \max\left\{ h_T^{-1}\|z - z_h\|_T, h_T^{-1/2}\|z - z_h\|_{\partial T} \right\}. \tag{3.91}$$

By virtue of (3.84), the approximation z_h may be chosen such that for $r \in \{0, 1\}$ holds

$$\omega_T \leq C_{i,T} h_T^r \|\nabla^{1+r} z\|_{\tilde{T}}. \tag{3.92}$$

Consequently, it follows immediately that

$$|(\nabla e, \nabla z)| \leq C_i \left(\sum_{T \in \mathbf{T}_h} h_T^{2r} \rho_T^2 \right)^{1/2} \|\nabla^{1+r} z\|, \tag{3.93}$$

for $z \in H^{1+r}(\Omega)$. The *interpolation constant* C_i is usually of moderate size, $C_i = O(1)$, depending on the shape of the elements T, and can be explicitly calculated in general. Taking the supremum over $z \in V \cap H^{1+r}(\Omega)$, one obtains the following a posteriori error bound in the energy norm (for $r = 0$) or in the L^2-norm (for $r = 1$),

$$\|\nabla^{1-r} e\| \le C_s C_i \Big(\sum_{T \in \mathbf{T}_h} h_T^{2r} \rho_T^2 \Big)^{1/2} =: \eta(u_h) . \tag{3.94}$$

The *stability constant* C_s measures the stability properties of the *dual problem*

$$(\nabla \varphi, \nabla z) = (\nabla^{1-r} e, \nabla^{1-r} \varphi) \quad \forall \varphi \in V , \tag{3.95}$$

in terms of the global a priori estimate

$$\|\nabla^{1+r} z\| \le C_s \|\nabla^{1-r} e\| , \tag{3.96}$$

which is trivial for $r = 0$. By analogous arguments, a posteriori error estimates may also be derived with respect to the maximum norm.

Based on an a posteriori error estimate of type (3.94) the corresponding mesh refinement process may be organized according to the *error per cell strategy* (see [9]). For some prescribed error tolerance TOL, the goal is to reach a most economical mesh $\mathbf{T}_h$ on which $\eta(u_h) \approx TOL$. Accordingly, the mesh refinement process aims at equilibrating the *local error indicators* $\eta_T := h_T^r \rho_T$ by refining (or coarsening) the elements $T \in \mathbf{T}_h$ according to the criterion

$$\eta_T \approx \frac{TOL}{\sqrt{NEL} \, C_i C_s} , \qquad NEL = \#\{T \in \mathbf{T}_h\} . \tag{3.97}$$

The refinement strategies based on the conventional a posteriori error estimate (3.94) rely on the assumption that the local error indicators

$$\eta_T := h_T^r \rho_T \tag{3.98}$$

properly describe the dependence of the global error on the local mesh size h_T. However, this may not be true in certain situations since the a posteriori error estimate (3.94) contains information about the mechanism of error propagation only through the *global* stability constant C_s.

To overcome this deficiency, it has been proposed in [9] to use the quantities ω_T in the estimate (3.89) as weight-factors multiplied by the local residuals ρ_T and to compute them numerically. These weights contain all information about the local approximation properties of the spaces V_h, as well as the local stability properties of the underlying continuous problem. This correspondence can be used for the mesh refinement algorithm through a *feed-back process*: In the course of the refinement the dual solution is calculated on the current mesh yielding approximate weights. On the basis of the resulting a posteriori error estimate the mesh is refined according to one of the criteria described below. This process is repeated yielding more and more accurate weights, i.e., a posteriori error bounds, until the prescribed stopping criterion is fulfilled. This approach allows one to construct almost optimal meshes for various kinds of error quantities, where "optimal" can mean "most economical" for achieving a prescribed accuracy TOL or "most accurate" for a given maximum number NEL_{max} of mesh cells.

The described concept for deriving a posteriori error estimates via a duality argument directly generalizes to the case that the quantity to be computed is given in terms of an arbitrary linear functional $J(\cdot)$ defined on the space V (or on a suitable subspace containing the finite element space V_h and the exact solution u). Relevant cases are, for instance, the mean value of u over Ω (*torsion moment*), the value of its gradient $\nabla u(a)$ at some point $a \in \Omega$ (*stress values*) or the line integral of some directional derivative $\beta \cdot \nabla u$ over the boundary $\partial \Omega$ (*total surface tension*),

$$J(\varphi) = \int_\Omega \varphi(x)\, dx \quad , \quad J(\varphi) = \nabla \varphi(a) \quad , \quad J(\varphi) = \int_{\partial \Omega} \beta \cdot \nabla \varphi\, ds. \quad (3.99)$$

Weighted a posteriori error estimate for the energy or L^2-norm can be obtained within this framework by taking the error functional

$$J(\varphi) = \|\nabla^r e\|^{-1} \int_\Omega \nabla^r \varphi \nabla^r e\, dx, \quad (3.100)$$

for $r = 0$ or $r = 1$, respectively. In this case, the evaluation of the weight factors ω_T, i.e., the computation of the dual solution z, requires at first an approximation of the functional $J(\cdot)$. This may be achieved by replacing the unknown error e by some approximation $\tilde{e} = \tilde{u}_h - u_h$ obtained by extrapolation from two consecutively refined meshes.

A general approach for residual-based error control:

For a general approach to residual-based error control for abstract variational problems, let V be a Hilbert space with inner product $(\cdot, \cdot)$ and corresponding norm $\| \cdot \|$, and $a(\cdot\,;\cdot)$ a semi-linear form. We seek a solution to the abstract variational problem

$$\text{Find } u \in V, \text{ such that:} \quad a(u\,;\varphi) = 0 \quad \forall \varphi \in V. \tag{3.101}$$

This problem is approximated by a Galerkin method using a sequence of finite dimensional subspaces "$V_h \subset V$", parameterized by a discretization parameter h, such that:

$$\inf_{\varphi \in V_h} \|u - \varphi\| \to 0 \quad (h \to 0). \tag{3.102}$$

Then, the discrete problems read:

$$\text{Find } u_h \in V_h, \text{ such that:} \quad a(u_h\,;\varphi) = 0 \quad \forall \varphi \in V_h. \tag{3.103}$$

With the Fréchet derivative a' of a, we have the following orthogonality relation for the error $e := u - u_h$:

$$\int_0^1 a'(u_h + te\,;e, \varphi)\, dt = a(u\,;\varphi) - a(u_h\,;\varphi) = 0 \quad \forall \varphi \in V_h. \tag{3.104}$$

This suggests to use the bilinear form

$$L(u, u_h\,;\varphi, z) = \int_0^1 a'(u_h + te\,;\varphi, z)\, dt \tag{3.105}$$

in the duality arguments. Suppose that the quantity $J(u)$ has to be computed, where J is a linear functional defined on V, or on a suitable subspace containing the ansatz spaces V_h and the solution u. For representing the error $J(e)$, we use the duality argument:

$$\text{Find } z \in V, \text{ such that:} \quad L(u, u_h\,;\varphi, z) = J(\varphi) \quad \forall \varphi \in V. \tag{3.106}$$

Assuming that this problem has a unique solution, and using the Galerkin orthogonality (3.104), we easily obtain the error representation

$$J(e) = L(u, u_h; e, z - \bar{z}),\qquad(3.107)$$

with some approximation $\bar{z} \in V_h$. The goal is to evaluate the right hand side numerically, in order to get a sharp a posteriori estimate for the quantity $J(e)$ and thus a criterion for the optimal local adjustment of the discretization.

For the further discussion, we become more specific about the setting of the above problem. Let the variational problem (3.101) originate from a semi-linear second-order partial differential equation of the form

$$A(u)u = -\sum_{i,j=1}^{d} \partial_i\{a_{ij}(\cdot,u)\partial_j u\} + \sum_{j=1}^{d} a_{0j}(\cdot,u)\partial_j u + a_{00}(\cdot,u)u = f,\quad(3.108)$$

on a bounded domain $\Omega \subset \mathbf{R}^d$ with, for simplicity, homogeneous Dirichlet boundary conditions, $u|_{\partial\Omega} = 0$. Hence, the natural solution space is $V = H_0^1(\Omega)$. The following discussion assumes (3.108) to be a scalar equation, but everything directly carries over to systems. Accordingly, the semi-linear form a and its linearization a' have their natural meaning. In this setting, the error representation (3.107) has the following concrete form:

$$J(e) = \sum_{T \in \mathbf{T}_h} \left\{ (f - A(u_h)u_h, z - \bar{z})_T - \frac{1}{2}([\partial_n^A u_h], z - \bar{z})_{\partial T} \right\},\qquad(3.109)$$

where $\partial_n^A = \sum_{i,j=1}^{d} n_i a_{ij}(\cdot,u_h)\partial_j$. To evaluate this formula, one may replace the unknown solution u in the bilinear form $L(u, u_h; \cdot, \cdot)$ by the computed approximation u_h, and solve the corresponding perturbed dual problem by the same method as used in computing u_h, yielding an approximation $z_h \in V_h$ to the exact dual solution z,

$$\text{Find } z_h \in V_h, \text{ such that: } \quad L(u_h, u_h; \varphi, z_h) = J(\varphi) \quad \forall \varphi \in V_h.\qquad(3.110)$$

Controlling the effect of this perturbation on the accuracy of the resulting error estimator

$$J(e) \approx \tilde{\eta}(u_h) := L(u_h, u_h; e, z_h - \bar{z}_h) \tag{3.111}$$

may be a delicate task and depends strongly on the particular problem considered. A rigorous analysis can be given in the case of ordinary differential equations. Numerical experiences of Becker/Rannacher indicate that this problem seems to be less critical in stable situations. The more crucial problem is the numerical computation of the perturbed dual solution z_h as described above.

Then, one possibility is to evaluate (3.109) directly by replacing the local errors $z - \bar{z}$ by $\tilde{z}_h - \bar{z}_h$, where $\tilde{z}_h$ is some local higher-order interpolant of z_h and $\bar{z}_h$ its interpolation in V_h. Another approach is to convert (3.109) into an estimate,

$$|J(e)| \leq \sum_{T \in \mathbf{T}_h} \omega_T \left\{ h_T^2 \|f - A(u_h)u_h\|_T + \frac{1}{2} h_T^{3/2} \|[\partial_n^A u_h]\|_{\partial T} \right\}, \tag{3.112}$$

with the weights

$$\omega_T := \max \left\{ h_T^{-2} \|z - \bar{z}\|_T, h_T^{-3/2} \|z - \bar{z}\|_{\partial T} \right\}. \tag{3.113}$$

By the usual approximation properties of finite elements there holds

$$\omega_T \leq C_{i,T} \|\nabla^2 z\|_T, \tag{3.114}$$

where the interpolation constant $C_{i,T}$ may be assumed to be of size $C_{i,T} \approx O(1)$ or may be directly calculated. Estimates for the weights ω_T may again be obtained from the approximate dual solution z_h simply by taking appropriate second-order difference quotients,

$$\omega_T \approx |T|^{1/2} |D_h^2 z_h(x_T)|, \quad x_T \text{ midpoint of } T. \tag{3.115}$$

In [9] it is shown that the a posteriori error estimate (3.112) asymptotically, for $h \to 0$, takes the form

$$|J(e)| \approx \sum_{T \in \mathbf{T}_h} h_T^4 |D^2 z(x_T)| |D_A^2 u(x_T)|, \qquad (3.116)$$

which shows that the variation of the error quantity $J(e)$ with respect to changes of the local mesh width h_T is essentially determined by the dual solution.

Application to the Navier-Stokes equations:

As an example we consider the computation of R. Becker [8] for the drag (and the resulting drag coefficient) and the *pressure difference* for the incompressible flow around a cylinder. The underlying configuration is identical with the 2D stationary test example in the previously described *1995 DFG Benchmark*. The quantities to be computed are the *drag force*

$$J_{\mathrm{drag}}(\mathbf{u}, p) = \int_S (\nu \frac{\partial v_t}{\partial \mathbf{n}} n_y - p n_x) \, dS \qquad (3.117)$$

with the following notations: circle S, normal vector $\mathbf{n}$ on S with x-component n_x and y-component n_y, tangential velocity v_t on S and tangent vector $\mathbf{t} = (n_y, -n_x)$.

As a further reference value the *pressure difference* $\Delta p = p(x_a, y_a) - p(x_e, y_e)$ is defined, with the front and end point of the cylinder $(x_a, y_a) = (0.15, 0.2)$ and $(x_e, y_e) = (0.25, 0.2)$, respectively,

$$J_{\Delta p}(\mathbf{u}, p) = p(x_a, y_a) - p(x_e, y_e, t). \qquad (3.118)$$

For discretizing this problem, a finite element method based on the conforming quadrilateral $Q1/Q1$-Stokes element is applied (all calculations were performed by R. Becker and can be found in [8] and more recent papers). For treating the velocity–pressure coupling as well as the convective term, consistent least–squares–stabilization á la Hughes, Franca and Balestra [58] is employed and a "hanging" node approach with corresponding optimal multi-grid tools is applied.

The *weighted* a posteriori error estimates can be obtained following the argument described above. In computing the drag force, the (approximate) dual problem reads

Find $z \in V$, such that:

$$\nu(\nabla\varphi, \nabla\mathbf{z}_h) + (\mathbf{u}_h \cdot \nabla\varphi, \mathbf{z}_h) - (q_h, \nabla\cdot\varphi) - (\nabla\cdot\mathbf{z}_h, \xi) = J_{\mathrm{drag}}(\varphi, \xi), \quad (3.119)$$

for all $\{\varphi, \xi\} \in H_h \times L_h$, where

$$J_{\mathrm{drag}}(\varphi, \xi) = \int_S \left\{ \nu \frac{\partial\varphi_t}{\partial\mathbf{n}} n_y - \xi n_x \right\} dS. \quad (3.120)$$

Denoting its solution by $\{\mathbf{z}_h, q_h\}$, the a posteriori estimate for the drag error becomes

$$\begin{aligned} &|J_{\mathrm{drag}}(\mathbf{u}, p) - J_{\mathrm{drag}}(\mathbf{u}_h, p_h)| \\ &\le C_I \sum_{T\in\mathbf{T}_h} \left\{ \omega_T^{(1)}\left(\eta_T^{(1)} + \eta_T^{(2)}\right) + \omega_T^{(2)}\eta_T^{(3)} + \omega_T^{(3)}\eta_T^{(4)} \right\} \end{aligned} \quad (3.121)$$

with the *local error indicators*

$$\eta_T^{(1)} = h_T^2 \|f + \nu\Delta\mathbf{u}_h - \mathbf{u}_h \cdot \nabla\mathbf{u}_h - \nabla p_h\|_T, \quad (3.122)$$
$$\eta_T^{(2)} = h_T^{3/2}\nu\|[\partial_n\mathbf{u}_h]\|_{\partial T}, \quad (3.123)$$
$$\eta_T^{(3)} = h_T\|\nabla\cdot\mathbf{u}_h\|_T, \quad (3.124)$$
$$\eta_T^{(4)} = \delta_T\|f + \nu\Delta\mathbf{u}_h - \mathbf{u}_h \cdot \nabla\mathbf{u}_h - \nabla p_h\|_T. \quad (3.125)$$

and the weights

$$\omega_T^{(1)} = \|\nabla^2\mathbf{z}_h\|_T, \quad \omega_T^{(2)} = \|\nabla q_h\|_T, \quad \omega_T^{(3)} = \|U \cdot \nabla\mathbf{z}_h + \nabla q_h\|_T. \quad (3.126)$$

The bounds for the dual solution $\{\mathbf{z}_h, q_h\}$ are obtained computationally by replacing the unknown solution $\mathbf{u}$ in the convection term by its approximation $\mathbf{u}_h$ and solving the resulting linearized problem on the same mesh. From the obtained approximate dual solution $\{\mathbf{Z}_h, \mathbf{Q}_h\}$ suitable difference quotients are taken to approximate the weights $\omega_T^{(i)}$. The interpolation constant may again be determined analytically or simply taken as $C_I = 1$.

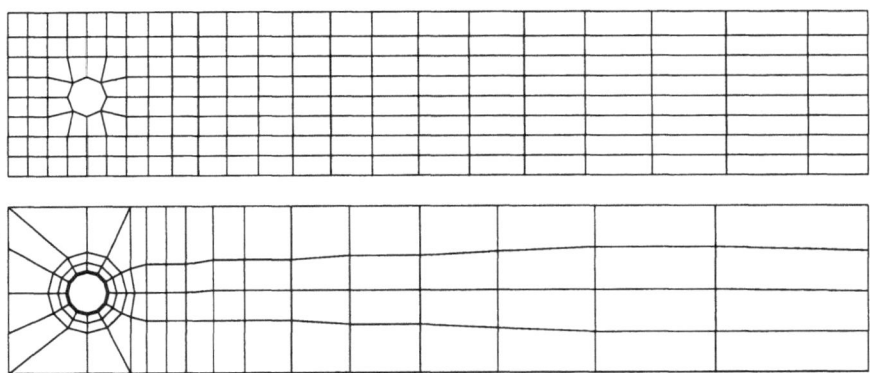

Figure 3.14: Almost uniform and pre-adapted coarse meshes

The complete implementation of the above strategy for error control via weighted a posteriori error estimates for the benchmark problem is just under work by R. Becker (see recent papers). In a preliminary version of the code a more heuristic strategy for the mesh refinement based on the global error bound has been used. Since the weights corresponding to the error in the drag force as well as those for the pressure difference are expected to become large in the neighborhood of the cylinder contour S, in each refinement step additionally all elements adjacent to S are refined. Table 3.13 shows the corresponding results for the pressure difference computed on

1. hierarchically refined meshes, starting from a coarse mesh with almost uniform width.

2. hierarchically refined meshes, starting from a coarse mesh which is refined towards the cylinder contour.

3. adaptively generated meshes starting from the coarse mesh in 1.

These results demonstrate the superiority of the adaptive algorithm, as it produces an error of less than 1% already after 6 refinement steps on a mesh with about 20,000 vertices in about 50 seconds while the other algorithms need at least 75,000 vertices and 200 seconds to achieve the same accuracy. Further, it is obvious that during the refinement process the complexity of the adaptive algorithm scales linearly with the number of unknowns.

Finally, we briefly discuss how a mesh refinement process may be organized on the basis of an a posteriori error estimate of the proposed type. All following algorithms and remarks concerning numerical experience are taken from the paper [9] by Becker/Rannacher.

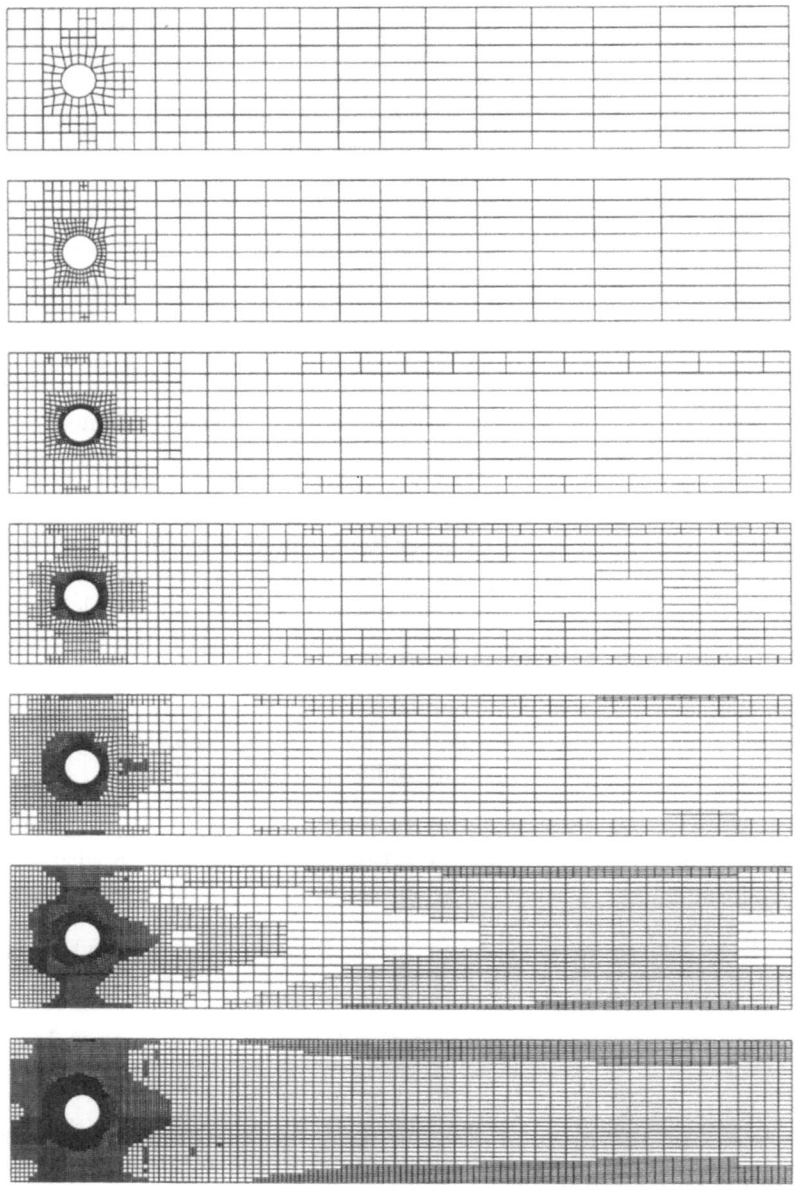

Figure 3.15: A sequence of successively refined meshes for the cylinder flow problem

Uniform Refinement, Grid1		
i	NVT	Δp	sec.
1	2268	0.109389	4.0
2	8664	0.110513	15.2
3	33840	0.113617	61.7
4	133728	0.115488	254.2
5	531648	0.116486	1001.1
Uniform Refinement, Grid2		
1	1296	0.106318	3.3
2	4896	0.112428	11.7
3	19008	0.115484	46.9
4	74880	0.116651	195.1
5	297216	0.117098	796.2

Adaptive Refinement, Grid1		
i	NVT	Δp	sec.
1	825	0.105085	2.2
2	1362	0.105990	2.9
3	2616	0.111482	5.4
4	5334	0.113978	11.3
5	10281	0.115794	23.4
6	21546	0.116915	51.6
7	40584	0.117253	108.2
8	86259	0.117379	233.0
9	164046	0.117499	406.9
10	330930	0.117530	830.4

Table 3.13: Results of the cylinder flow computations by R. Becker for various types of mesh refinements, (reference value $\Delta p = 0.1175\ldots$)

Suppose that some error tolerance TOL and maximum number NEL_{max} of mesh cells are given. The goal is to find a most economical mesh $\mathbf{T}_h$ on which

$$|J(e)| \approx \eta(u_h) = \sum_{T \in \mathbf{T}_h} \eta_T \approx TOL, \qquad (3.127)$$

with the *local error indicators*

$$\eta_T := \omega_T \rho_T. \qquad (3.128)$$

Usually, one starts from an initial coarse mesh which is then successively refined. There are essentially three alternative strategies:

1. **Error per cell strategy**

   The mesh generation aims at equilibrating the local error indicators η_T, by refining (or coarsening) the elements $T \in \mathbf{T}_h$ according to the criterion

   $$\eta_T \approx \frac{TOL}{NEL}, \qquad NEL = \#\{T \in \mathbf{T}_h\}. \qquad (3.129)$$

   Since NEL depends on the result of the refinement decision, this strategy is implicit and usually needs iteration. It is common practice to work with a varying value NEL on each refinement level which is permanently updated according to the refinement process. The result is a mesh on which $\eta(u_h) \approx TOL$, provided that NEL_{max} is not exceeded.

2. **Fixed fraction strategy**

   In each refinement cycle, the elements are ordered according to the size of η_T and either a fixed portion (say 30%) of the elements with largest η_T or the portion of elements which make up for a certain part of the estimator, $\kappa\eta(u_h)$, is refined. The appropriate choice of the parameter κ is crucial and depends very much on the particular situation. For "regular" functionals, one may choose $\kappa = 0.6-0.8$, while for "singular" functionals a smaller choice $\kappa = 0.1 - 0.2$ is advisable, in order to enhance local refinement. This process is repeated until the stopping criterion $\eta(u_h) \approx TOL$ is fulfilled, or NEL_{max} is exceeded.

3. **Tolerance reduction strategy**

   One works with a varying tolerance TOL_{var}. If on a mesh $\mathbf{T}_h$ a discrete solution u_h^{old} has been obtained with corresponding error estimator $\eta(u_h^{old})$, the tolerance is set to

   $$TOL_{var} := \sigma\eta(u_h^{old}), \qquad (3.130)$$

   with some reduction factor $\sigma \in (0,1)$ (usually $\sigma = 0.5$). In the next step, one (or more) cycles of the *error per cell strategy* are applied with tolerance TOL_{var} yielding a refined mesh $\mathbf{T}_h^{new}$ and the new solution u_h^{new} with corresponding error bound $\eta(u_h^{new})$. Then, the tolerance is reduced again and a new refinement cycle begins. This process is repeated until $TOL_{var} \leq TOL$, or NEL_{max} is exceeded.

The *error per cell strategy* is very delicate as its performance strongly depends on the parameters in the refinement decision. In certain cases it may happen that an inappropriate choice leads to very slow mesh refinement and turns into an inefficient overall solution. On the other hand, the *fixed fraction strategy* guarantees that in each refinement cycle a sufficiently large number of elements is refined. However, in its pure form, it does not allow for mesh coarsening and in certain cases it may tend to over-refine the mesh. Therefore, for practical computations, Becker/Rannacher recommend the *tolerance reduction strategy* which seems to be the most robust and accurate strategy among the three. Its use is particularly advisable if the weight-function $\omega(x)$ has nonintegrable singularities which would otherwise dominate the refinement process.

The proposed approach is rather universal and has, on a heuristic basis, already been successfully applied to rather complex problems like the Navier-Stokes equations in fluid mechanics, the Navier-Lamé equations in linear elasticity and the Prandtl-Reuss model in elasto-plasticity. It has also proven to be efficient in solving large dimensional radiative transfer problems in astrophysics. The same concept can be used in the context of ordinary differential equations where residual-based error estimation was introduced in [60], and then further developed in the spirit of the recent approach. Finally, weighted a posteriori error analysis may also be useful in deriving stopping criteria for iteration processes in finite element methods as studied in [10].

These considerations show that error control for partial differential equations is possible in general, even for the incompressible Navier–Stokes equations. Although our own experience with these error control mechanisms is still very small, we decided to present these recent results in this book. Thus, let us thank again Roland Becker and Rolf Rannacher for providing these research results which are very promising for the future. The next generations of CFD software will and <u>must</u> contain error control concepts with appropriate fully adaptive solution mechanisms. Even if in complex applications the sharp bounds for the error estimators might not (and will not!) be guaranteed, we will obtain at least meshes with essentially reduced number of grid points which are correspondingly chosen due to user-specified flow quantities as drag and lift.

These error control, resp., discretization aspects and the main topic of this book, the efficient numerical "pure" solution approaches, will be part of optimized numerical tools. Together with the indicated software techniques, further accelerations for typical CFD simulations by a factor of 100 or even more should be possible. However, up to now this is still a dream for which we have to work hard in the next years. We do not expect to obtain – before the year 2000 – the corresponding software packages which have realized all these features and which are able to treat "real life" problems, but we mean the proposed concepts are convincing enough to try hard in future.

3.2 Time stepping techniques

In this section we examine some specific aspects which arise if (standard) time discretization techniques are applied to the nonstationary incompressible Navier–Stokes equations,

$$\mathbf{u}_t - \nu\Delta\mathbf{u} + \mathbf{u}\cdot\nabla\mathbf{u} + \nabla p = \mathbf{f} \quad , \quad \nabla\cdot\mathbf{u} = 0\,, \qquad (3.131)$$

for given force $\mathbf{f}$ and viscosity ν, with prescribed boundary values and initial conditions.

We will not examine numerical schemes like the *discontinuous space–time Galerkin methods* (see Johnson [61]) and *characteristic methods* (see Pironneau [78]). These both classes are (probably the only) schemes which we did not perform during the last years such that own numerical experience regarding their efficiency and accuracy cannot be provided. Our implementation of these approaches is just under process (see [48]), and we hope to compare them later with our recently applied schemes. In this book, we mainly concentrate on the traditional solution approach which is a separate standard discretization in space and time.

3.2.1 The Fractional–step–θ scheme and other One–step–θ schemes

We first (semi-) discretize the nonstationary Navier–Stokes equations in time by one of the usual methods known from the treatment of ordinary differential equations (ODE), such as the Forward or Backward Euler-, the Crank–Nicolson-, the Fractional–step–θ–scheme or others, and obtain – for each time step – a sequence of generalized stationary Navier–Stokes problems:

Given $\mathbf{u}^n$ *and the time step* $k = t_{n+1} - t_n$, *then solve for* $\mathbf{u} = \mathbf{u}^{n+1}$ *and* $p = p^{n+1}$

$$\frac{\mathbf{u} - \mathbf{u}^n}{k} + \theta[-\nu\Delta\mathbf{u} + \mathbf{u} \cdot \nabla\mathbf{u}] + \nabla p = \mathbf{g}^{n+1} \quad , \quad \nabla\cdot\mathbf{u} = 0 , \qquad (3.132)$$

with right hand side

$$\mathbf{g}^{n+1} := \theta\mathbf{f}^{n+1} + (1 - \theta)\mathbf{f}^n - (1 - \theta)[-\nu\Delta\mathbf{u}^n + \mathbf{u}^n \cdot \nabla\mathbf{u}^n] . \qquad (3.133)$$

The parameter θ has to be chosen depending on the time–stepping scheme, e.g., $\theta = 1$ for the Backward Euler-, or $\theta = 1/2$ for the Crank–Nicolson-scheme. Also *multi–step* schemes are allowed, as the BDF(2)-scheme, which lead in general to modifications of the right hand side terms. In all following cases, we end up with the task of solving, in each time step, a nonlinear coupled problem which has then to be discretized in space.

Given $\mathbf{u}^n$, *parameters* $k = k(t_{n+1})$, $\theta = \theta(t_{n+1})$ *and* $\theta_i = \theta_i(t_{n+1}), i = 1,\ldots,3$, *then solve for* $\mathbf{u} = \mathbf{u}^{n+1}$ *and* $p = p^{n+1}$

$$[I + \theta k N(\mathbf{u})]\mathbf{u} + k\nabla p = [I - \theta_1 k N(\mathbf{u}^n)]\mathbf{u}^n + \theta_2 k\mathbf{f}^{n+1} + \theta_3 k\mathbf{f}^n , \quad \nabla\cdot\mathbf{u} = 0 . \qquad (3.134)$$

Here and in the following, we use the compact form for the diffusive and convective part

$$N(\tilde{\mathbf{u}})\mathbf{u} := -\nu\Delta\mathbf{u} + \tilde{\mathbf{u}} \cdot \nabla\mathbf{u} . \qquad (3.135)$$

The terms θ_i, $i = 1, \ldots, 3$, are quantitites according to the implicitly suggested *One–step–θ–schemes*. If *Multi–step schemes* are applied, additional terms involving "older" solutions $\mathbf{u}^{n-k}$ and right hand sides $\mathbf{f}^{n-k}$ have to be added. The pressure term $\nabla p = \nabla p^{n+1}$ may be replaced by

$$\theta \nabla p^{n+1} + (1 - \theta) \nabla p^n , \tag{3.136}$$

but, with appropriate post–processing, both strategies lead to solutions of the same accuracy. However, the fully implicit treatment of the incompressibility constraint, $\nabla \cdot \mathbf{u}(t_{n+1}) = 0$, is very important since the appearance of explicit terms $(1 - \theta) \nabla \cdot \mathbf{u}(t_n)$ in the right hand side requires initial values which already satisfy the continuity equation.

The resulting differential algebraic equation system is highly stiff, with a stiffness ratio of order $O(\nu h^{-2})$ or even $O(\nu h^{-4})$ in the case of the *discretely divergence–free* formulation. In the past, explicit time–stepping schemes have been commonly used in nonstationary flow calculations, but because of the severe stability problems inherent in this approach, the required small time steps often prohibit the accurate long time solution of really time–dependent flows. Due to the high stiffness, implicit schemes in the choice of time–stepping methods for solving this problem should be preferred. Since implicit methods have become feasible, thanks to more efficient linear solvers, the schemes most frequently used are either the simple first–order Backward Euler–scheme (BE), with $\theta = 1$, or the second–order Crank–Nicolson–scheme (CN), with $\theta = 1/2$. These two methods belong to the group of *One–step–θ–schemes*. The CN–scheme occasionally suffers (at least for simple scalar problems) from unexpected instabilities because of its only weak damping property (not strongly *A–stable*), while the BE–scheme is of first order accuracy only. We will discuss the corresponding properties and problems with regard to CFD problems in a few moments.

Alternative schemes of higher order accuracy are based on the (diagonally) implicit Runge–Kutta formulas or the backward differencing multi–step methods, being well-known from the ODE literature. We omit these schemes by mainly two reasons: It is not clear at all if higher than second order accuracy for the nonlinear Navier–Stokes equations can be numerically obtained. And for second order accuracy the proposed One–step schemes seem to be superior with respect to their computational complexity, particularly if the construction and performance of fully implicit time step control mechanisms is taken into account.

Another method which seems to have the potential to excel in this competition is the *Fractional–step–θ–scheme* (FS). It was first proposed by Glowinski et al. (see [40]), in the form of an operator splitting scheme separating the two problems "nonlinearity" and "incompressibility". However, this scheme deserves also attention as a mere time stepping method as will become clear by the following discussion.

Comparing with the One–step–θ–schemes the FS scheme uses three different values for θ and for the time step k at each time level. For a realistic comparison with respect to accuracy but also computational efficiency we define a macro time step with $K = t_{n+1} - t_n$ as a sequence of 3 time (sub-) steps of (variable) size k_i,

$$K = t_{n+1} - t_n \quad , \quad K = \sum_{i=1}^{3} k_i . \tag{3.137}$$

Then, in the case of the Backward Euler- or the Crank–Nicolson–scheme, we perform 3 substeps with the same θ as above, $\theta = 0.5$ or 1, and equidistant time steps $k = K/3$.

Backward Euler–scheme:

$$[I + \tfrac{K}{3} N(\mathbf{u}^{n+\frac{1}{3}})]\mathbf{u}^{n+\frac{1}{3}} + \tfrac{K}{3}\nabla p^{n+\frac{1}{3}} = [I - 0 \cdot N(\mathbf{u}^n)]\mathbf{u}^n + \tfrac{K}{3}\mathbf{f}^{n+\frac{1}{3}}$$
$$\nabla \cdot \mathbf{u}^{n+\frac{1}{3}} = 0,$$
$$[I + \tfrac{K}{3} N(\mathbf{u}^{n+\frac{2}{3}})]\mathbf{u}^{n+\frac{2}{3}} + \tfrac{K}{3}\nabla p^{n+\frac{2}{3}} = \mathbf{u}^{n+\frac{1}{3}} + \tfrac{K}{3}\mathbf{f}^{n+\frac{2}{3}}$$
$$\nabla \cdot \mathbf{u}^{n+\frac{2}{3}} = 0,$$
$$[I + \tfrac{K}{3} N(\mathbf{u}^{n+1})]\mathbf{u}^{n+1} + \tfrac{K}{3}\nabla p^{n+1} = \mathbf{u}^{n+\frac{2}{3}} + \tfrac{K}{3}\mathbf{f}^{n+1}$$
$$\nabla \cdot \mathbf{u}^{n+1} = 0.$$

Crank–Nicolson–scheme:

$$[I + \tfrac{K}{6} N(\mathbf{u}^{n+\frac{1}{3}})]\mathbf{u}^{n+\frac{1}{3}} + \tfrac{K}{6}\nabla p^{n+\frac{1}{3}} = [I - \tfrac{K}{6} \cdot N(\mathbf{u}^n)]\mathbf{u}^n + \tfrac{K}{6}\mathbf{f}^{n+\frac{1}{3}} + \tfrac{K}{6}\mathbf{f}^n$$
$$\nabla \cdot \mathbf{u}^{n+\frac{1}{3}} = 0,$$
$$[I + \tfrac{K}{6} N(\mathbf{u}^{n+\frac{2}{3}})]\mathbf{u}^{n+\frac{2}{3}} + \tfrac{K}{6}\nabla p^{n+\frac{2}{3}} = [I - \tfrac{K}{6} \cdot N(\mathbf{u}^{n+\frac{1}{3}})]\mathbf{u}^{n+\frac{1}{3}} + \tfrac{K}{6}\mathbf{f}^{n+\frac{2}{3}}$$
$$+ \tfrac{K}{6}\mathbf{f}^{n+\frac{1}{3}}$$
$$\nabla \cdot \mathbf{u}^{n+\frac{2}{3}} = 0,$$
$$[I + \tfrac{K}{6} N(\mathbf{u}^{n+1})]\mathbf{u}^{n+1} + \tfrac{K}{6}\nabla p^{n+1} = [I - \tfrac{K}{6} \cdot N(\mathbf{u}^{n+\frac{2}{3}})]\mathbf{u}^{n+\frac{2}{3}} + \tfrac{K}{6}\mathbf{f}^{n+1}$$
$$+ \tfrac{K}{6}\mathbf{f}^{n+\frac{2}{3}}$$
$$\nabla \cdot \mathbf{u}^{n+1} = 0.$$

For the Fractional–step–θ–scheme we proceed as follows. Choosing $\theta = 1 - \frac{\sqrt{2}}{2}$, $\theta' = 1 - 2\theta$, and $\alpha = \frac{1-2\theta}{1-\theta}$, $\beta = 1 - \alpha$, the macro time step $t_n \rightarrow t_{n+1} = t_n + K$ is split into three consecutive substeps (with $\tilde{\theta} := \alpha\theta K = \beta\theta'K$):

$$[I + \tilde{\theta}N(\mathbf{u}^{n+\theta})]\mathbf{u}^{n+\theta} + \theta K\nabla p^{n+\theta} = [I - \beta\theta KN(\mathbf{u}^n)]\mathbf{u}^n + \theta K\mathbf{f}^n$$
$$\nabla\cdot\mathbf{u}^{n+\theta} = 0,$$
$$[I + \tilde{\theta}N(\mathbf{u}^{n+1-\theta})]\mathbf{u}^{n+1-\theta} + \theta'K\nabla p^{n+1-\theta} = [I - \alpha\theta'KN(\mathbf{u}^{n+\theta})]\mathbf{u}^{n+\theta}$$
$$+\theta'K\mathbf{f}^{n+1-\theta}$$
$$\nabla\cdot\mathbf{u}^{n+1-\theta} = 0,$$
$$[I + \tilde{\theta}N(\mathbf{u}^{n+1})]\mathbf{u}^{n+1} + \theta K\nabla p^{n+1} = [I - \beta\theta KN(\mathbf{u}^{n+1-\theta})]\mathbf{u}^{n+1-\theta}$$
$$+\theta K\mathbf{f}^{n+1-\theta}$$
$$\nabla\cdot\mathbf{u}^{n+1} = 0.$$

The FS–scheme is based on the settings

$$\theta \approx 0.293 \quad , \quad \theta' \approx 0.414 \quad , \quad \alpha \approx 0.585 \quad , \quad \beta \approx 0.414,$$

and can be interpreted as a variable time stepping scheme which is 3–cyclic for each time step. That means that in the first and third substep other local time steps are chosen than in the second substep (compare with $k_i = K/3$ for the One–step–θ–schemes!),

$$k_1 = k_3 \approx 0.293 \cdot K \quad , \quad k_2 \approx 0.414 \cdot K. \tag{3.138}$$

The local weight $\bar{\theta}$ in comparison to the Crank–Nicolson–scheme can be approximatly defined in each substep as $\bar{\theta} \approx 0.514$ which is slightly larger than the corresponding value for the Crank–Nicolson–scheme ($\bar{\theta} = 0.5$). However, this slight modification is sufficient to provide strong A–stability while preserving the second order accuracy (see Müller–Urbaniak [72]).

Following this notation, it is obvious that Backward Euler-, Crank–Nicolson- and the FS–scheme lead to about the same numerical complexity in each (macro) time step. There are some differences in generating right hand sides, but in implicit approaches the numerical cost for their assembling can be (almost) neglected. The open question is the resulting accuracy and stability. For the analysis of the employed schemes it is most instructive to look first at the scalar linear test equation

$$\dot{x}(t) + \lambda x(t) = 0 \quad , \quad t \geq 0, \tag{3.139}$$

where $\lambda \in \mathbf{C}$, $Re\lambda \geq 0$. A time stepping scheme applied to this equation, with constant time step size Δt, generates a sequence of values $x_n \sim x(t_n)$, with $t_n := n\Delta t$. The behaviour of the scheme as $t \to \infty$, depending on the parameter λ, is usually characterized by the *amplification factor* $\omega = \omega(\lambda \Delta t)$. In particular, for the One–step schemes which we use, there holds $x_n = \omega^n x_0$. In terms of ω, we can formulate the following desirable properties of time stepping schemes:

1) $|\omega(\lambda \Delta t)| \leq 1$ *(local stability)*

2) $\lim\limits_{Re\lambda \to \infty} |\omega(\lambda \Delta t)| \leq 1 - O(\Delta t)$ *(global regularity)*

3) $\lim\limits_{Re\lambda \to \infty} |\omega(\lambda \Delta t)| \leq 1 - \delta < 1$ *(smoothing property: strong A–stability)*

4) $|\omega(\lambda \Delta t)| \sim 1$ for $Re\lambda = 0$ *(non dissipative)*.

The discussion of the proposed One–step schemes is given in a form, which applies to the general linear evolution equation

$$u_t + A(t)u = f(t) \quad , \quad A_n = A(t_n) \quad , \quad f_n = f(t_n) \,. \tag{3.140}$$

The approximation properties can be expressed in terms of the amplification factor $\omega(\lambda \Delta t)$ with λ eigenvalue of $A(t)$. Then, the hope is to carry these conclusions also to the nonlinear Navier–Stokes equations.

Classical One–step θ–schemes:

$$[I + \theta \Delta t A_{n+1}]u_{n+1} = [I - (1-\theta)\Delta t A_n]u_n + \theta \Delta t f_{n+1} + (1-\theta)\Delta t f_n , \tag{3.141}$$

with amplification factors

$$\omega(z) = \frac{1 - (1-\theta)z}{1 + \theta z} \,. \tag{3.142}$$

We will also perform the Forward Euler scheme ($\theta = 0$) in our numerical tests, even it is only conditionally stable ($\Delta t \leq 1/\lambda$) and of first order accuracy. However, some modifications of the right hand side terms may lead to explicit second order schemes.
Its implicit counterpart, the Backward Euler scheme ($\theta = 1$), is strongly A–stable ($|\omega(z)| \to 0$ for $Re\lambda \to \infty$), but it tends to damp out free oscillations

$(|\omega(i\Delta t)| < 1, |\omega| = 0.995$ for $\Delta t = 0.1)$, and is also of first order accuracy. The Crank–Nicolson scheme $(\theta = 1/2)$ is of second order, but has only a low damping property $(|\omega(z)| \to 1$ for $Re\lambda \to \infty)$ and is only A–stable. However, free oscillations are well preserved $(|\omega(i\Delta t)| = 1)$.

Summing up, all schemes have some advantages and some disadvantages: The Backward Euler is very robust, but inaccurate and strongly damping. So, it should only be used for nonstationary calculations which aim to iterate towards the stationary limit. In contrast, the Crank–Nicolson is more accurate, but seems to tend to become unstable. A scheme, having the advantages of both of those is the FS–scheme:

The Fractional–step θ–scheme:

Choosing $\theta \in (0, 1)$, $\theta' = 1 - 2\theta$, and $\alpha \in [0, 1]$, $\beta = 1 - \alpha$, the macro time step $t_n \to t_{n+1}$ is split into the three following substeps:

$$
\begin{aligned}
[I + \alpha\theta\Delta t A_{n+\theta}] \quad u_{n+\theta} &= [I - \beta\theta\Delta t A_n]u_n + \theta\Delta t f_n \\[2mm]
[I + \beta\theta'\Delta t A_{n+1-\theta}] \quad u_{n+1-\theta} &= [I - \alpha\theta'\Delta t A_{n+\theta}]u_{n+\theta} + \theta'\Delta t f_{n+1-\theta} \\[2mm]
[I + \alpha\theta\Delta t A_{n+1}] \quad u_{n+1} &= [I - \beta\theta\Delta t\, A_{n+1-\theta}]u_{n+1-\theta} + \theta\Delta t f_{n+1-\theta}
\end{aligned}
$$
$$\tag{3.143}$$

with

$$
\omega(z) = \frac{(1 - \beta\theta z)^2(1 - \alpha\theta' z)}{(1 + \alpha\theta z)^2(1 + \beta\theta' z)}. \tag{3.144}
$$

For comparisons with the *One–step θ–schemes* we always have to take into account that both schemes are performed for macro time steps consisting of three substeps. For the special choice $\theta = 1 - \frac{\sqrt{2}}{2}$, this scheme is of second order. Taking $\alpha = \frac{1-2\theta}{1-\theta}$, then the coefficient matrices are the same in all substeps. Further, the scheme is strongly A–stable $(\lim_{Re\lambda\to\infty} |\omega(z)| = \frac{\beta}{\alpha} \sim 0.7)$, and free oscillations are well preserved $(|\omega(i\Delta t)| \sim 0.9998$ for $\Delta t = 0.8)$.

Since the FS–scheme is a strongly A–stable time stepping approach, it possesses the full smoothing property which is important in the case of rough initial or boundary data. Further, it contains only very little numerical dissipation which is crucial in the computation of non–enforced temporal oscillations. For a more detailed discussion of these aspects see the papers of

Rannacher [81], Rannacher et al. [73], and Müller–Urbaniak [72]. In particular Müller–Urbaniak [72] has given a rigorous theoretical analysis of the FS–scheme applied to the Navier–Stokes problem. The main result is a second order error estimate of the form

$$\|\mathbf{u}_h^n - \mathbf{u}(\cdot, t_n)\| \leq C(t_n) \left(h^2 + \tau_n^{-1} \Delta t^2 \right) \quad , \quad \tau_n := \min\{1, t_n\}. \qquad (3.145)$$

This result has several aspects:

- It shows the stability of the scheme for sufficiently small time steps Δt (independent of the mesh size h!) and additionally the actual second order convergence on finite time intervals.

- It guarantees second order accuracy at times $t \geq t_0 > 0$, even for generic initial data.

- The error constant $C(t)$ remains bounded for $t \to \infty$, if the boundary data and the body force are bounded and if the exact solution $\mathbf{u}$ is *exponentially stable*. The dependence of $C(t)$ on the viscosity ν, resp., the Reynolds number, is considered in [62].

The numerical and computational counterpart to these theoretical works of Rannacher and Müller–Urbaniak concerning the applicability of these methods including the FS–scheme on CFD problems can be found in [73] and especially in [102]. In this paper, we have performed comparative studies of time stepping schemes, including also differences between the described Backward Euler- (BE), the Crank–Nicolson- (CN) and the Fractional–step θ–scheme (FS). These tests are applied to several configurations with the aim to figure out corresponding accuracy and robustness results. We repeat some of them in the following.

3.2.2   Numerical comparisons
of some time discretization schemes

The two test problems we consider (see [56], [101] and [102] for more details) are:

1.) Von Kármán vortex shedding behind an inclined plate in a channel.

2.) Flow in a Venturi pipe (a dynamic water pump device in a sail boat).

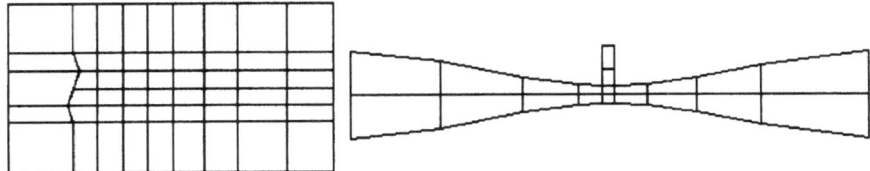

Figure 3.16: Coarse grids

All results have been obtained by performing the fully coupled solver CC2D ($\sim$ local MPSC) or PP2D ($\sim$ global MPSC/*discrete projection*). Their mathematical part is described in all details in the previous Chapter 2 and in [102], and both codes can be downloaded at:

> http://www.iwr.uni-heidelberg.de/~featflow

Figure 3.16 shows the coarsest grids used which are refined systematically by connecting opposite midpoints. The resulting meshes in our multigrid calculations for the channel flow range from 13,500 elements (coarse level) to 54,000 elements (fine level), and for the Venturi pipe from 20,500 (coarse level) to 82,000 elements (fine level).

1) Plate problem:

The total length of the channel is $L_t = 10$, the height is $H_t = 5$, and the length of the plate is $L_p = 1$. At the inlet we prescribe a parabolic velocity profile with $U_{max} = 1$, while at the outlet a Neumann–type outflow condition is used (see [56]). The prescribed viscosity is $\nu = 1/500$. Figure 3.17 shows a typical snapshot of the streamlines for this problem and the oscillating time behaviour (up to $T = 60$, starting with Stokes flow) for the mean pressure difference across the plate defined by

$$P_{diff} := \oint_{plate_{left}} p\,ds - \oint_{plate_{right}} p\,ds . \qquad (3.146)$$

The reason for choosing this quantity in our tests is the possible appearance of pressure boundary layers. Since P_{diff} is an important physical quantity, a "good" method should be able to give accurate results for this quantity.

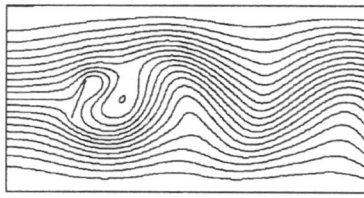 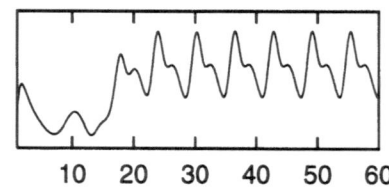

Figure 3.17: Streamline snapshot and mean pressure difference for the plate problem

2) Venturi pipe problem:

The total length of the Venturi pipe is $L_t = 32$, the height at the inlet is $H_t = 5$, the height in the interior is $H_i = 1$, and the width of the small upper channel is $W_i = 0.8$. At the upper small "inlet" and the right "outlet" we prescribe the zero mean pressure condition (see [56]), while at the left inlet a parabolic velocity profile with $U_{max} = 1$ is prescribed, leading to a maximum velocity of about 7 in the interior. At the narrowing a lower pressure is generated ("Bernoulli principle") which enforces an incoming flux from the upper inlet, at least for the viscosity parameter $\nu = 1/1,000$ used. Figure 3.18 shows a typical snapshot of the streamfunction and pressure for this problem, and the (long) time behaviour (up to $T = 30$) of the pressure and first velocity component at a certain mesh point in the right upper half of the domain.

Differences between first and second order schemes

We first concentrate on the results for the first order Backward Euler–scheme (BE) for the momentum equations. We show that this method usually leads to results comparable to those obtained by the schemes of second order, if the time step is drastically reduced (in our tests by at least a factor 10). Further, no essential speedup in efficiency can be found: The evaluation of the right hand side in the BE–scheme can be slightly accelerated, but this is not essential in implicit methods. In Figure 3.19 we show relative streamlines (i.e., after subtracting the Stokes flow) for the channel flow example at $T = 10$, computed with the fully nonlinear coupled versions (CC2D-n) of the CN- and FS–scheme (with $k = 0.33$), and with the BE–method for $k = 0.33$ and $k = 0.033$. Here, and in all other calculations, the corresponding time step for the BE–scheme has to be chosen smaller by at least a factor 10.

Figure 3.20 shows the time behaviour of the flux through the upper inlet in the Venturi pipe. We show corresponding results for CC2D-n via the FS- and BE–scheme, compared with the reference solution. Again, the BE–method forces us to choose at least $k = 0.011$ to obtain similar results.

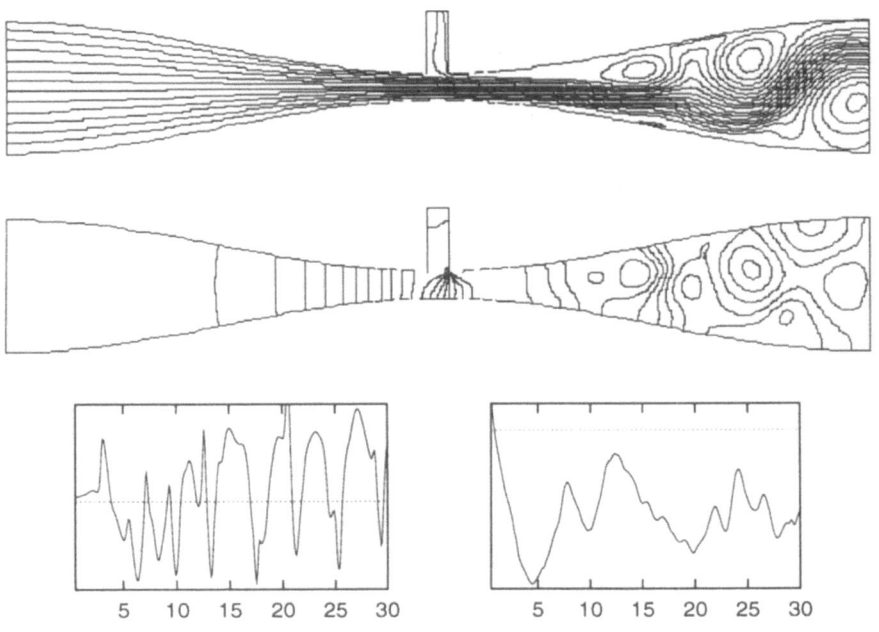

Figure 3.18: Streamline/pressure plot and velocity/pressure behaviour for the Venturi pipe

Differences between Crank–Nicolson- and Fractional–step–θ–scheme

As mentioned before, the essential theoretical results are: Both are of second order accuracy and lead to comparable numerical cost. However, the CN–scheme is only A–stable while the FS–method is strongly A–stable. That means, that for rough initial values or boundary conditions the classical CN–scheme may lead to numerical oscillations which are damped for smaller time steps only. We confirm these theoretical considerations numerically. The result is that for time steps k sufficiently small both schemes lead to approximate solutions with no significant differences, even for long time calculations and high Reynolds numbers.

However, if the time step is too coarse, the CN–scheme tends to produce unphysical oscillations, which means that even for moderate time steps, the CN–scheme may be less robust and accurate compared to the FS–scheme. By applying the adaptive time step control, these oscillations are damped, but generally a smaller time step has to be chosen to ensure sufficient robustness.

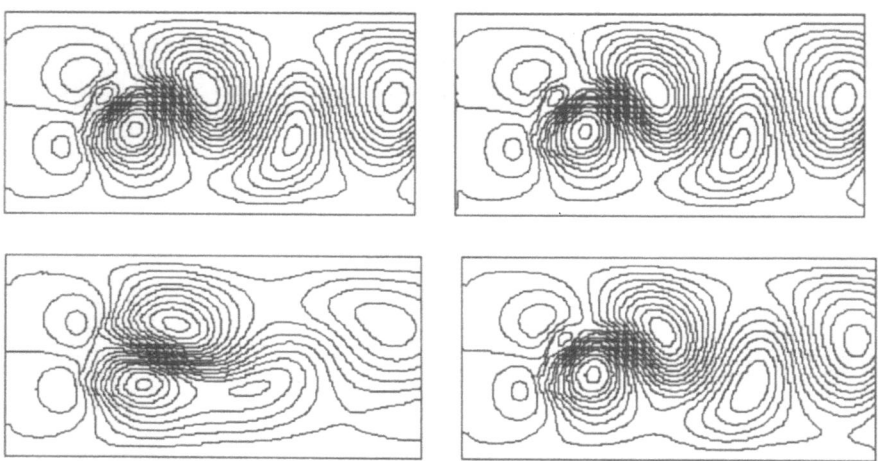

Figure 3.19: Relative streamlines of vortex shedding for the channel flow. First line CN($k = 0.33$, left) and FS($k = 0.33$, right), second line BE($k = 0.33$, left) and BE($k = 0.033$, right)

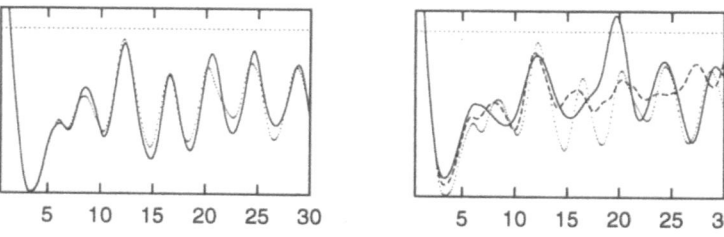

Figure 3.20: Total flux through the upper inlet, with FS (left) for $k = 0.11$ (line), and BE (right) for $k = 0.11$ (line) and for $k = 0.05$ (dashed), compared to a reference solution (dotted)

Figures 3.21 and 3.22 show the time behaviour of the mean pressure drop across the plate, resp., that of the total flux through the upper inlet, compared to reference solutions on the same mesh. Both calculations are performed with the scheme CC2D-n.

These results confirm that both schemes have about the same accuracy for realistic step sizes. We now consider their stability properties. As mentioned before we expect problems for the CN–scheme in coping with high frequency perturbations caused by rough data. First, Figure 3.23 shows an instability effect of the CN–scheme in the computation of the flow around a plate. Here, the mean pressure drop shows non–physical fluctuations for $k = 0.11$ which

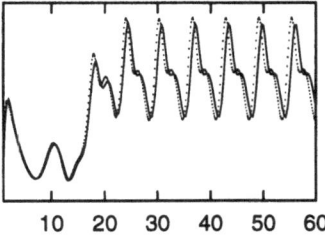 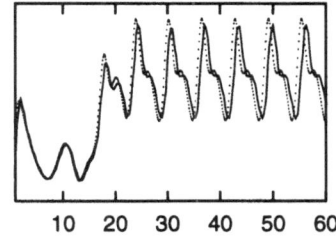

Figure 3.21: Mean pressure drop for the channel flow, calculated by CC2D-n-CN (left) and CC2D-n-FS (right) with $k = 0.33$, compared to a reference solution (dotted)

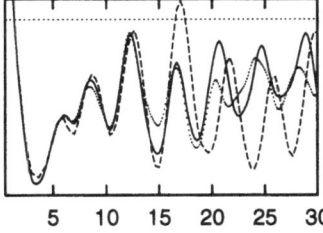 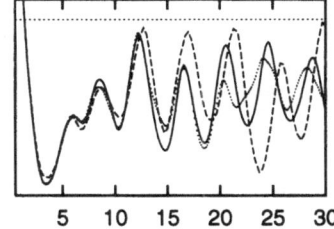

Figure 3.22: Flux for the Venturi pipe, calculated by CC2D-n-CN (left) and CC2D-n-FS (right), with $k = 0.33$ (dashed) and $k = 0.11$ (line), compared to a reference solution (dotted)

disappear for $k = 0.05$. For the same step size the FS–scheme is stable. For even coarser step sizes k, both schemes exhibit unstable behaviour. These results are obtained by the fully coupled solver CC2D-1 with linear extrapolation (of 2nd order) of the advection direction, but the same behaviour can be obtained with the pure projection schemes PP2D. Additionally, our results are not limited to the case of pressure values on the boundary, the same holds also for points in the interior, i.e., all numerical oscillations are global effects.

Figure 3.24 demonstrates the last assertion: Even the flux (measured through the upper inlet in the Venturi pipe) exhibits an unstable behaviour if calculated by the CN–scheme.

These results show that the classical CN–scheme and the FS–scheme are essentially of the same accuracy and efficiency. Nevertheless, there may occur problems for the CN–scheme, due to the loss of strong A–stability, which can produce unphysical fluctuations of the solution.

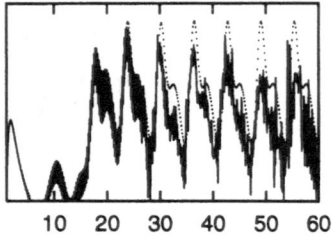 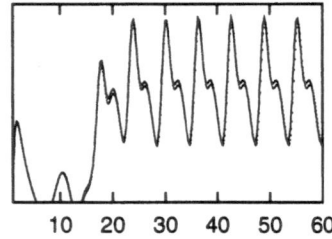

Figure 3.23: Mean pressure drop for the channel flow, calculated by CC2D-l-CN (left) and CC2D-l-FS (right) with $k = 0.11$, compared to a reference solution (dotted)

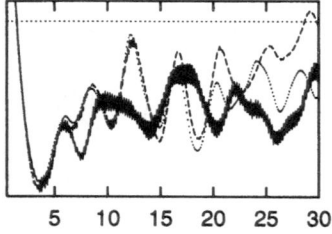 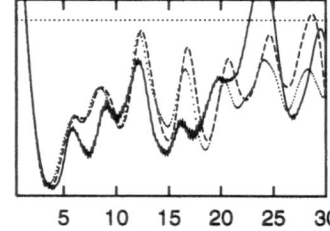

Figure 3.24: Flux for Venturi pipe, calculated by CC2D-l-CN (left) and CC2D-l-FS (right), with $k = 0.11$ (line) and $k = 0.033$ (dashed), compared to a reference solution (dotted)

3.2.3 Adaptive time stepping for incompressible flow problems

The remaining part of this section concerning time stepping techniques for the Navier–Stokes equations is related to the problem of how to perform an adaptive time step control for highly implicit approaches. We emphasize the key word *implicit* since this restriction eliminates many "simple" error control mechanisms of *predictor* (for velocity only) – *corrector* type from the list of possible approaches. Assuming that no a priori information for complex flows (in space and time) is available, then "control by hand", which is often useful for testing accuracy and robustness but not for practical applications, is easy to explain: We perform tests with a sequence of fixed time steps until no significant differences can be detected in comparison to a reference solution. This "exact" solution is calculated with a very small time step, but on the same spatial mesh.

A practicable approach is the following heuristic technique which is based on the estimation of the *local truncation error*. Even if the mathematical motivation looks like being rather crude, the result is in fact a working error indicator. For that, we assume that the applied time stepping scheme is second order accurate in time (an analogous procedure can be derived for first or higher order schemes). The former time level is t_n and we want to calculate a solution at $t_{n+1} = t_n + k$. We denote by $v_k = v_k(t_{n+1})$ the discrete solution pair $v_k = \{\mathbf{u}_k, p_k\} = \{\mathbf{u}_k(t_{n+1}), p_k(t_{n+1})\}$ which is obtained by using time step size k, starting from the previous approximate $\{\mathbf{u}(t_n), p(t_n)\}$ at t_n. Let $v = v(t_{n+1})$ denote the exact solution at t_{n+1}.

Our aim is to find an appropriate value for k such that the following relation holds for the functional $J(\cdot)$, applied to the exact and the discrete solution at the new time level t_{n+1},

$$|||J(v) - J(v_k)||| \sim TOL. \tag{3.147}$$

The functional $J(\cdot)$ may represent directly the velocity and/or pressure values, measured in a properly chosen norm, $||| \cdot ||| := \| \cdot \|$, for instance the euclidian vector norm,

$$|||J(v) - J(v_k)||| = \|v - v_k\|, \tag{3.148}$$

or $J(\cdot)$ denotes local quantities as drag or lift coefficients, and we set $||| \cdot ||| := | \cdot |$,

$$|||J(v) - J(v_k)||| = |\, \text{lift}(\mathbf{u}(t_{n+1}), p(t_{n+1})) - \text{lift}(\mathbf{u}_k(t_{n+1}), p_k(t_{n+1}))|\,. \tag{3.149}$$

However, also quantities as certain point values, pressure differences on contours and even fluxes can be controlled by appropriate definitions of $J(\cdot)$. One essential background for this approach is that we implicitly make the (heuristic) assumption that the error at starting level t_n is zero. Further, we assume the asymptotic error expansion

$$J(v) - J(v_k) \sim k^2\, e(v) + O(k^4), \tag{3.150}$$

with an error term $e(v)$ which is independent of the time step. This relation is assumed both for velocity and pressure components and for certain linear functionals containing derived quantities as drag and lift terms.

We perform two calculations for step sizes $\tilde{k}$ and $3\tilde{k}$, that means we apply three substeps with $\tilde{k}$ and one step with $3\tilde{k}$ (due to the FS–scheme, otherwise the same procedure can be practized with 2 substeps). In the case of the Fractional–step–θ–scheme we compare the three substeps with one Crank–Nicolson step with step size $3\tilde{k}$ (which has a slightly different constant in the error expansion). This approach cannot be verified by rigorous mathematical arguments, but all performed calculations have leaded to convincing results. Moreover, it should not be forgotten that all these techniques lead to *error indicators* in contrast to rigorous *error estimators* such that somewhat heuristic arguments may be allowed.

So, we calculate the *local differences* $REL_{\tilde{k}}$,

$$REL_{\tilde{k}} := J(v_{3\tilde{k}}) - J(v_{\tilde{k}}) . \tag{3.151}$$

By a linear combination of relation (3.150) for $\tilde{k}$ and $3\tilde{k}$, we can obtain for $e(v)$ that

$$e(v) \sim \frac{J(v_{3\tilde{k}}) - J(v_{\tilde{k}})}{8\tilde{k}^2} , \tag{3.152}$$

and hence

$$J(v) - J(v_k) \sim \frac{k^2}{8\tilde{k}^2} \left[J(v_{3\tilde{k}}) - J(v_{\tilde{k}}) \right] = \frac{k^2}{8\tilde{k}^2} REL_{\tilde{k}} . \tag{3.153}$$

This last relation leads to the following estimate for k if the relative error shall be related to the given tolerance TOL as demanded in (3.147),

$$k^2 \sim TOL \frac{8\,\tilde{k}^2}{||| REL_{\tilde{k}} |||} . \tag{3.154}$$

The corresponding time step control is rather easy to implement since no additional tools have to be added. However, the additional numerical cost may be quite large since our *predictor* step consists of solving a complete (linear or nonlinear) stationary Navier–Stokes problem, and this with the larger time step $3\tilde{k}$ even. The resulting CPU time for solving this single step may be as large as the cost for the other three substeps. Additionally, the solution of this subproblem is even (almost) worthless since it is less accurate (three times larger time steps!) and it is needed for the control of the time step size only. However, this approach is one of the few which work efficiently and particularly robust for fully implicit schemes with corresponding time steps which are not at all restricted due to any stability constraint.

Simple adaptive time step control:

1. Given a step size $\tilde{k}$, we perform three (sub-) steps with parameter $\tilde{k}$, and one step with $3\tilde{k}$. We calculate the relative changes $|||REL_{\tilde{k}}|||$ and use this estimator together with (3.154) to compute the necessary time step size k which is required such that the error in (3.147), $|||J(v) - J(v_k)|||$, is related to TOL.

2. If the resulting value k is much smaller than the actually used time step $\tilde{k}$ for its prediction, we go back to 1. and repeat the last calculation with $\tilde{k} = k$.

3. If the value for k is larger than the used $\tilde{k}$, or only slightly smaller (say less than 50%) we accept the result and perform the next macro time step, now with k and $3k$.

4. Finally, we can obtain a even higher order accuracy for $v_k(t_{n+1})$ if we perform an additional linear extrapolation step with the partial solutions $v_{\tilde{k}}$ and $v_{3\tilde{k}}$.

There is another important reason why such complicated – since nonlinear – subproblems have to be treated. Often the standard control of any global norm for the velocity or the pressure is not sufficient. Comparing with the results from the *1995 DFG Benchmark* (see Section 1.1 and [87]), the interesting quantities are lift and drag coefficients, depending on time. The derived framework allows special definitions of the control function J and is able to handle these specific coefficients. However, typical calculations show that in many cases the control of exactly one specific quantity is not sufficient. If we consider the typical *von Kármán* vortex shedding behind a cylinder, the lift is the most sensitive (and seemingly the most interesting) quantity as typical representant for other technical issues. The lift value is highly oscillating in time, between $+1$ and -1 for instance, hereby passing the x–axis. If this occurs, both exact and discrete values are very small such that a time step control based exclusively on this quantity increases the time step which may lead to catastrophical results. The remedy is to control several quantities at the same time, for instance lift and/or drag coefficients, together with global norms of velocity and pressure errors. Nevertheless, one practical problem remains as usual in such adaptive error control concepts:

how to fix the tolerance parameters TOL for these different quantities in a given application?

The proposed time step control is well known in the field of ordinary differential equations and works well in many calculations. The basic assumption is that we are already in the asymptotical range such that the error representation in (3.150) is true which, however, cannot be guaranteed in general. A second conceptual problem is that we can only give estimates for the local discretization error, hereby assuming "exact" starting values for every time step. Thus, we have performed so far an *adaptive error indicator* only: we are planning in the future to incorporate a global residual based error control (see [62]) which follows the concepts of the proposed error control from the previous section.

3.3   Nonlinear iteration techniques

We discuss some specific numerical techniques which have shown to be well-suited for the treatment of the nonlinearity in the stationary and nonstationary incompressible Navier–Stokes equations. These procedures are explained first in detail for the following abstract nonlinear problem which is assumed to be stationary,

$$T(u)u = f. \tag{3.155}$$

$T(.)$ may be viewed to represent the continuous or discretized Navier–Stokes or Burgers operator. Analogously to the linear case we start with a *nonlinear basic iteration* which can be formulated as

---

Nonlinear basic iteration for the abstract problem:

Given: Iterate u^{n-1}
Perform: One (nonlinear) relaxation step to obtain u^n

$$u^n = u^{n-1} - \omega^{n-1}[\tilde{T}(u^{n-1})]^{-1}\left(T(u^{n-1})u^{n-1} - f\right) \tag{3.156}$$

The quantity ω^{n-1} is an additional relaxation parameter which has to be chosen appropriately. We may stop this iteration scheme as usual if a maximum iteration counter is exhausted, $n \geq N_{max}$, or if the nonlinear residual is "sufficiently small" in a certain norm,

$$\|T(u^n)u^n - f\| \leq TOL. \tag{3.157}$$

The operator $\tilde{T}(u^{n-1})$ represents an approximation of the Frechét–derivative of T with respect to the last iterate u^{n-1} and allows a wide variety of resulting schemes. In algorithmic notation, we can reformulate our nonlinear basic iteration as in the following diagram.

Reformulated nonlinear basic iteration:

Given: Iterate u^{n-1}
Perform: Three substeps to obtain u^n

1. Calculate the nonlinear residual d^{n-1},

$$d^{n-1} = T(u^{n-1})u^{n-1} - f. \tag{3.158}$$

2. Solve an auxiliary subproblem for y^{n-1} with right hand side d^{n-1},

$$\tilde{T}(u^{n-1})y^{n-1} = d^{n-1}. \tag{3.159}$$

3. Update u^{n-1} via the auxiliary solution y^{n-1} and the relaxation parameter ω^{n-1}, and obtain u^n,

$$u^n = u^{n-1} - \omega^{n-1}y^{n-1}. \tag{3.160}$$

If we choose $\tilde{T}(u^{n-1})$ to be the exact Frechét–derivative, we obtain the standard Newton iteration scheme. Typical (numerical) properties of this classical approach are:

- Newton's method is characterized by the fact that it is usually a quadratically converging process, if at all. Therefore, once it converges, it requires only a few iterations. However, particularly in nonsteady processes, similar iteration rates can be reached by simpler methods.

- In general, "good" initial iterates u^0 are required to obtain a convergent scheme.

- The construction of the Newton matrix $\tilde{T} = \tilde{T}^N(\cdot)$ in each iteration is often expensive. In general, one has two choices to perform if partial differential equations are treated. One can calculate first the Frechét–derivative on continuous level and then discretize the resulting operator. Since continuous techniques are applied to derive discrete preconditioners, this approach may lead to non–optimal iteration schemes. Alternatively, one can discretize first, and then try to calculate the Frechét–derivative of $T = T_h$. However, especially for CFD problems this approach often results in very expensive matrix calculations via numerical differentiation and some sparsity of the matrices may be lost. For the treatment of partial differential equations one usually prefers the first approach.

- For nonlinearities which typically arise in Navier–Stokes or Burgers equations, the discretized Newton matrix consists of "good" convective terms which can be stabilized numerically, and "bad" reactive terms which may cause instabilities with regard to discretization and solution aspects (see later).

Other possibilities for $\tilde{T}$ lead to the so-called *fixed point* schemes. The approximate Frechét–derivative may be chosen as the operator itself,

$$\tilde{T}(u^{n-1}) = T(u^{n-1}), \tag{3.161}$$

or $\tilde{T}$ contains only linear parts of T even. The corresponding schemes are less accurate than the Newton scheme (with linear or super-linear convergence only) but the convergence properties for the iterative solution of auxiliary linear subproblems due to $[\tilde{T}(u^{n-1})]^{-1}$ may be more efficient. If further optimization strategies for selecting the damping parameters ω^n are applied (see the *adaptive fixed point defect correction methods*), the resulting overall convergence behaviour of these quasi–Newton schemes may be even superior to the classical Newton scheme.

In the nonstationary case, we can also stop with $n = N_{max} = 1$ and $\omega^n = 1$, and the choice $\tilde{T} = T(\bar{u}(t_{old}))$. Here, $\bar{u}(t_{old})$ may be the solution from a previous time level or is obtained by a linear combination of already calculated solutions. This approach can be interpreted as a simple linearization via extrapolation backwards in time. The corresponding time–stepping scheme is potentially of arbitrary order accurate in time, but the full implicitness may be lost. As an advantage, we have to build up the corresponding system matrices and right hand sides only once in each time step. Then, the resulting linear systems are nonsymmetric or even symmetric in the case of explicit approaches.

It is obvious that the fully nonlinear iteration schemes cause more numerical cost in each time step than the linearization techniques (with only one iteration). Further, a fully explicit treatment consumes the least effort. However, the question not included in these considerations is that for the necessary size of the time step. The fully nonlinear iteration is expected to be more stable and accurate while the price to be paid for the more explicit schemes is a smaller time step size. So, asking for the total CPU time, which schemes are the more efficient ones for nonstationary nonlinear problems? The same question arises for the stationary problems: Summing up the total CPU cost, what is the best compromise between the complexity of the performed approximate Frechét–derivative and the corresponding number of nonlinear iterations?

We cannot answer these questions for problems which arise from general partial differential equations. However, we will discuss in more detail some experiences with these techniques for the incompressible Navier–Stokes equations. The following considerations will be – if possible – supplemented by numerical examples.

3.3.1   The "adaptive fixed point defect correction" method

We consider the following continuous formulation for generalized incompressible Navier–Stokes equations which arise in the treatment of stationary or nonstationary problems,

$$\alpha \mathbf{u} - \nu \Delta \mathbf{u} + \mathbf{u} \cdot \nabla \mathbf{u} + \nabla p = \mathbf{f} \quad , \quad \nabla \cdot \mathbf{u} = g, \tag{3.162}$$

and introduce the following approximate Frechét operator $\tilde{T}(\cdot)$ which corresponds to linearized Navier–Stokes problems,

$$\tilde{T}(\mathbf{v}) := \begin{bmatrix} \tilde{S}(\mathbf{v}) & \nabla \\ \nabla \cdot & 0 \end{bmatrix}. \tag{3.163}$$

Then, the continuous formulation of the nonlinear basic iteration reads:

Continuous nonlinear basic iteration:

Given: Iterates $\mathbf{u}^{n-1}$, p^{n-1}
Perform: One relaxation step to obtain $\mathbf{u}^n$, p^n

$$\begin{bmatrix} \mathbf{resu}^{n-1} \\ resp^{n-1} \end{bmatrix} = \begin{bmatrix} \alpha \mathbf{u}^{n-1} - \nu\Delta\mathbf{u}^{n-1} + \mathbf{u}^{n-1}\cdot\nabla\mathbf{u}^{n-1} + \nabla p^{n-1} - \mathbf{f} \\ \nabla\cdot\mathbf{u}^{n-1} - g \end{bmatrix}$$

$$\begin{bmatrix} \mathbf{u}^n \\ p^n \end{bmatrix} = \begin{bmatrix} \mathbf{u}^{n-1} \\ p^{n-1} \end{bmatrix} - \omega^{n-1} \begin{bmatrix} \tilde{S}(\mathbf{u}^{n-1}) & \nabla \\ \nabla\cdot & 0 \end{bmatrix}^{-1} \begin{bmatrix} \mathbf{resu}^{n-1} \\ resp^{n-1} \end{bmatrix}$$

Based on this algorithm, the <u>efficient</u> numerical treatment of the nonlinear Navier–Stokes equations is reduced to two subproblems:

1. How to select the damping parameters ω^{n-1}?

2. How to choose the preconditioning operator $\tilde{S}(\mathbf{u}^{n-1})$?

The classical Newton approach:

The preconditioning operator $\tilde{S} = \tilde{S}^N(\cdot)$ in the exact Frechét–derivative of the continuous Navier–Stokes operator is derived (for prescribed $\tilde{\mathbf{U}}$) as

$$\tilde{S}(\tilde{\mathbf{U}})\mathbf{u} := \alpha\mathbf{u} - \nu\Delta\mathbf{u} + \tilde{\mathbf{U}}\cdot\nabla\mathbf{u} + \mathbf{u}\cdot\nabla\tilde{\mathbf{U}}. \qquad (3.164)$$

Consequently, for given $\mathbf{u}^{n-1}$, we have to solve the following auxiliary convection–diffusion–reaction problem (with incompressibility constraint) in the n–th nonlinear sweep, with resulting solution $\{\mathbf{v}, q\}$,

$$\alpha\mathbf{v} - \nu\Delta\mathbf{v} + \mathbf{u}^{n-1}\cdot\nabla\mathbf{v} + \mathbf{v}\cdot\nabla\mathbf{u}^{n-1} + \nabla q = \mathbf{rhs}(\mathbf{u}^{n-1}) \quad (3.165)$$

$$\nabla\cdot\mathbf{v} = rhs(p^{n-1}). \quad (3.166)$$

If we denote again by M the mass matrix, by L the discretized Laplacian, by $N(\mathbf{u}^{n-1})$ the convective matrix depending on $\mathbf{u}^{n-1}$, and by B, resp., B^T the gradient and divergence matrix, we obtain the following discretized coupled system of equations:

$$\alpha M v_1 + \nu L v_1 + N(\mathbf{u}^{n-1})v_1 + M(\nabla u_1^{n-1})v_1$$
$$+ M(\nabla u_1^{n-1})v_2 + B_1 q = rhs_1(\mathbf{u}^{n-1}) \qquad (3.167)$$

$$\alpha M v_2 + \nu L v_2 + N(\mathbf{u}^{n-1})v_2 + M(\nabla u_2^{n-1})v_1$$
$$+ M(\nabla u_2^{n-1})v_2 + B_2 q = rhs_2(\mathbf{u}^{n-1}) \qquad (3.168)$$

$$B_1^T v_1 + B_2^T v_2 = rhs(p^{n-1}) \qquad (3.169)$$

The indices $(\cdot)_1$, resp., $(\cdot)_2$, mean the first, resp., the second component of the corresponding vector–valued function. $M(\nabla w_i)$ denotes a special mass matrix with coefficients which depend on the gradient of the i–th component of the vector $\mathbf{w}$. These together form the full Newton-matrix which has to be "inverted" in each nonlinear step. Our numerical experience with this approach can be summarized as follows:

- If the Newton iteration is converging, the observed convergence rates show mostly a quadratic behaviour. However, it is not clear if this behaviour is visible from the first iteration step on, or if these are only "asymptotical" results after one has already approached the solution. Since in many CFD applications it is enough to gain 1-5 digits of accuracy (due to the discretization error which is typically of this rough order, or in nonstationary calculations with good starting values) it might appear that the asymptotical quadratic behaviour is not essential in many practical applications.

- While the convective terms $N(\mathbf{u}^{n-1})$ can be stabilized by the previously explained techniques of upwind or streamline–diffusion type, the zero–order reactive terms $M(\nabla u_i^{n-1})$ lead to matrices with diagonal elements related to ∇u_i^{n-1}. Their sign and size cannot be controlled (only by very rough techniques of suppressing negative values) such that no M-matrix properties can be obtained. In fact, the robust and efficient multigrid solution of problems involving terms like $M(\nabla u_i^{n-1})$ is – at least to our knowledge – still under research. If direct solvers are performed these problems do not occur, but how to apply Gaussian elimination for realistic 3D problems?

- Both components of the momentum equations are additionally coupled via

$$M(\nabla u_1^{n-1})v_1 + M(\nabla u_1^{n-1})v_2 \quad \text{resp.,} \quad M(\nabla u_2^{n-1})v_1 + M(\nabla u_2^{n-1})v_2 .$$

This may lead to further severe problems for the iterative solution process.

The Newton scheme leads to only few nonlinear iteration steps, but the corresponding numerical expense in solving the discrete linear systems is enormous such that we omit this approach. An alternative is the following technique:

The adaptive fixed point defect correction approach:

- Employ an approximate Frechét–derivative $\tilde{S} = \tilde{S}^F(\mathbf{v})$ only. Doing that, one has to find a compromise between the requirements that $\tilde{S}^F$ is near to the exact operator $\tilde{S}^N$ which arises in the classical Newton approach (this determines the number of nonlinear iterations) and that $[\tilde{S}^F]^{-1}$ can be efficiently applied (this determines the cost per nonlinear iteration).

- Accelerate the speed of the nonlinear convergence via adaptively calculated "optimal" damping parameters ω^n.

The easiest choice is to take the linear part $\tilde{S} = \tilde{S}^L$ only (*Picard iteration*),

$$\tilde{S}^L \mathbf{u} := \alpha \mathbf{u} - \nu \Delta \mathbf{u}. \qquad (3.170)$$

However, this leads in general to very poor convergence results only. Indeed, the strong restriction to this symmetric case is completely unnecessary, since fast multigrid tools have become feasible which allow to solve nonsymmetric Oseen problems almost as efficient as pure Stokes equations.

Our actually favourized scheme is the following *adaptive fixed point defect correction* approach which is performed on discrete level. Assume that we have already discretized the stationary Navier–Stokes equations

$$\alpha \mathbf{u} - \nu \Delta \mathbf{u} + \mathbf{u} \cdot \nabla \mathbf{u} + \nabla p = \mathbf{f} \quad , \quad \nabla \cdot \mathbf{u} = g. \qquad (3.171)$$

with the following discrete variational approach:

Find a pair $\{\mathbf{u}_h, p_h\} \in H_h \times L_h$, such that

$$\tilde{a}_h^L(\mathbf{u}_h, \mathbf{v}_h) + n_h^1(\mathbf{u}_h, \mathbf{u}_h, \mathbf{v}_h) + b_h(p_h, \mathbf{v}_h) = (\mathbf{f}, \mathbf{v}_h) \; \forall \mathbf{v}_h \in H_h \quad (3.172)$$
$$b_h(q_h, \mathbf{u}_h) = (g, q_h) \; \forall q_h \in L_h. \quad (3.173)$$

$n_h^1(\cdot, \cdot, \cdot)$ corresponds to the discretized convective term, either by a streamline–diffusion or an upwind ansatz. Also central discretization is allowed. Let $\tilde{a}_h^L(\cdot, \cdot)$ include the linear parts, corresponding to $\alpha \mathbf{u} - \nu \Delta \mathbf{u}$. Further, we employ the same matrix notations as before and define the matrix $\tilde{S} = \tilde{S}^F$ as

$$\tilde{S}^F(\mathbf{v}) := \alpha M + \nu L + N^2(\mathbf{v}). \tag{3.174}$$

$n_h^2(\cdot, \cdot, \cdot)$, resp., N^2 may be another discretization of the convective term. Then, in corresponding matrix–vector notation our approach reads:

Discrete nonlinear basic iteration:

Given: Iterates $\mathbf{u}^{n-1}$, p^{n-1}

Perform: Solve an auxiliary Oseen problem to obtain $\mathbf{v}$, q

$$\left[\begin{array}{c} \mathbf{resu}^{n-1} \\ resp^{n-1} \end{array} \right] = \left[\begin{array}{c} \alpha M \mathbf{u}^{n-1} + \nu L \mathbf{u}^{n-1} + N^1(\mathbf{u}^{n-1})\mathbf{u}^{n-1} + B p^{n-1} - \mathbf{f} \\ B^T \mathbf{u}^{n-1} - g \end{array} \right]$$

$$\left[\begin{array}{cc} \tilde{S}^F(\mathbf{u}^{n-1}) & B \\ B^T & 0 \end{array} \right] \left[\begin{array}{c} \mathbf{v} \\ q \end{array} \right] = \left[\begin{array}{c} \mathbf{resu}^{n-1} \\ resp^{n-1} \end{array} \right]$$

Perform: Choose an appropriate ω^{n-1} and obtain $\mathbf{u}^n$, p^n

$$\left[\begin{array}{c} \mathbf{u}^n \\ p^n \end{array} \right] = \left[\begin{array}{c} \mathbf{u}^{n-1} \\ p^{n-1} \end{array} \right] - \omega^{n-1} \left[\begin{array}{c} \mathbf{v} \\ q \end{array} \right]$$

The essential motivation for this splitting ansatz is the following:

- Choose a discretization $\tilde{n}_h^1$ for the convective term which is **stable** and **accurate**. If the nonlinear solver – which is a fixed point defect correction approach – converges at all, then towards the solution according to n_h^1, resp., N^1.

- Choose a discretization n_h^2 for the convective term which leads to linear systems with **"nice" matrix–properties** for N^2. N^2, resp., $\tilde{S}^F(\mathbf{u}^{n-1})$, is used as preconditioner only which has to be efficiently "inverted" in each nonlinear step.

We modify the exact Frechét–derivative not only by omitting the "bad" reaction terms from the Newton scheme (which might be still included in

$n_h^1(\cdot,\cdot,\cdot))$, but also by applying different discretizations n_h^1 and n_h^2 for the convective term; in fact we allow to use completely different discretizations for all terms and even different values for the viscosity parameter. If this iteration scheme converges at all, it achieves the solution due to the discretization $n_h^1(\cdot,\cdot,\cdot)$. To obtain high accuracy, we may choose damped upwind or streamline–diffusion discretizations. Even the pure central scheme is allowed.

The choice of the discretization involved in $n_h^2(\cdot,\cdot,\cdot)$ is very delicate. It is obvious that the number of nonlinear steps is minimal if n_h^2 is "near" to n_h^1, in some sense. On the other hand, n_h^2 must be a stable discretization which allows the fast solution of linear Oseen–problems (or convection–diffusion problems if applied to Burgers equations) involving the matrix N^2. Typical examples are accurate discretizations with "highly damped" upwind or streamline–diffusion approaches for n_h^1 on the right hand side, while the preconditioning scheme may be chosen due to the simple upwind UPW-∞, the weighted scheme UPW-α ($\alpha \geq 0.1$, see the previous section) or the typical streamline–diffusion ansatz SD-1.0.

In contrast to the classical Newton scheme, the expected convergence rates are only linear, but the range of convergence in dependence of the initial values is very large. Nevertheless, due to the different discretizations n_h^1 and n_h^2 the convergence behaviour may slow down if the difference between n_h^1 and n_h^2 is too large (for instance if a central scheme is preconditioned by the simple first order upwind).

The adaptive choice of the damping parameters:

Further essential stabilization and acceleration can be obtained by the following *step length control* technique which tries to control adaptively the size of the update vector $\{\mathbf{v}, q\}$. We explain this technique first for the abstract nonlinear problem $T(u)u = f$ in which case $T(.)$ may be viewed to represent the continuous or discretized Navier–Stokes or Burgers operator.

The typical idea is to determine the relaxation parameter ω^{n-1} such that the error between the new iterate $u^n = u^{n-1} - \omega^{n-1}y^{n-1}$ and the exact solution u is minimized in an appropriate norm. A good candidate is the (discrete) defect norm, resp., euclidian norm $\|\cdot\|_E$ of the defect vector (for other norms, see the following section). That means that ω^{n-1} may be chosen to satisfy:

$$\omega^{n-1} = \min_{\omega} \|T(u^{n-1} - \omega y^{n-1})(u^{n-1} - \omega y^{n-1}) - f\|_E \qquad (3.175)$$

This optimization process leads in general to an additional one-dimensional nonlinear problem which requires the multiple evaluation of terms $T_\omega :=$

$T(u^{n-1} - \omega y^{n-1})$ in each of the resulting minimization steps. Since particularly for the Navier–Stokes equations the corresponding numerical cost are rather large we simplify the functional in the optimization step:

$$\omega^{n-1} = \omega^{n-1}(\tilde{\omega}) = \min_{\omega} \|T(u^{n-1} - \tilde{\omega} y^{n-1})(u^{n-1} - \omega y^{n-1}) - f\|_E \quad (3.176)$$

In fact, we linearize the fully nonlinear operator $T(u^{n-1} - \omega y^{n-1})$ by $T(u^{n-1} - \tilde{\omega} y^{n-1})$ with $\tilde{\omega}$ given. Two possible approaches are:

1. **fixed choice:** $\tilde{\omega} = \omega^{fixed}$ (for instance $\omega^{fixed} = 0.8$ for damping)

2. **adaptive choice:** $\tilde{\omega} = \omega^{n-2}$

We prefer the second approach which utilizes for the linearization the last damping parameter ω^{n-2} due to the idea that no "large differences" may occur in the step from ω^{n-2} to ω^{n-1}. Then, the optimal ω^{n-1} according to this linear minimization problem is easily obtained via the following formula (with $\langle \cdot, \cdot \rangle_E$ denoting the euclidian scalar product)

$$\omega^{n-1} = \frac{\langle T(u^{n-1} - \tilde{\omega} y^{n-1}) y^{n-1}, f - T(u^{n-1} - \tilde{\omega} y^{n-1}) u^{n-1} \rangle_E}{\langle T(u^{n-1} - \tilde{\omega} y^{n-1}) y^{n-1}, T(u^{n-1} - \tilde{\omega} y^{n-1}) y^{n-1} \rangle_E}. \quad (3.177)$$

The corresponding cost are 1 evaluation of the operator $T(u^{n-1} - \tilde{\omega} y^{n-1})$ and 2 operator applications to y^{n-1}, resp., to u^{n-1}. Formulating this technique for the incompressible Navier–Stokes equations, our complete discrete nonlinear basic iteration reads as follows:

The adaptive fixed point defect correction method:

Given: Iterates $\mathbf{u}^{n-1}$, p^{n-1}

Perform: 5 substeps to obtain new iterates $\mathbf{u}^n$, p^n

1. Calculate defect vectors $\mathbf{resu}^{n-1}$, $resp^{n-1}$ containing the nonlinear residual:

$$\begin{bmatrix} \mathbf{resu}^{n-1} \\ resp^{n-1} \end{bmatrix} = \begin{bmatrix} \alpha M \mathbf{u}^{n-1} + \nu L \mathbf{u}^{n-1} + N^1(\mathbf{u}^{n-1})\mathbf{u}^{n-1} + B p^{n-1} - \mathbf{f} \\ B^T \mathbf{u}^{n-1} - g \end{bmatrix}$$

---

2. Solve an auxiliary problem in $\mathbf{v}$, q with $\mathbf{resu}^{n-1}$, $resp^{n-1}$ as right hand side:

$$\begin{bmatrix} \alpha M + \nu L + N^2(\mathbf{u}^{n-1}) & B \\ B^T & 0 \end{bmatrix} \begin{bmatrix} \mathbf{v} \\ q \end{bmatrix} = \begin{bmatrix} \mathbf{resu}^{n-1} \\ resp^{n-1} \end{bmatrix}$$

3. Build the matrix S^{n-1} involving $\mathbf{u}^{n-1}$, $\mathbf{v}$ and $\tilde{\omega}$ (with $\tilde{\omega} = \omega^{n-2}$ or fixed):

$$S^{n-1} := \alpha M + \nu L + N^1(\mathbf{u}^{n-1} - \tilde{\omega}\mathbf{v})$$

4. Calculate the "optimal" ω^{n-1} via:

$$\omega^{n-1} = \frac{\left\langle \begin{bmatrix} S^{n-1} & B \\ B^T & 0 \end{bmatrix} \begin{bmatrix} \mathbf{v} \\ q \end{bmatrix}, \begin{bmatrix} \mathbf{f} \\ g \end{bmatrix} - \begin{bmatrix} S^{n-1} & B \\ B^T & 0 \end{bmatrix} \begin{bmatrix} \mathbf{u}^{n-1} \\ p^{n-1} \end{bmatrix} \right\rangle_E}{\left\langle \begin{bmatrix} S^{n-1} & B \\ B^T & 0 \end{bmatrix} \begin{bmatrix} \mathbf{v} \\ q \end{bmatrix}, \begin{bmatrix} S^{n-1} & B \\ B^T & 0 \end{bmatrix} \begin{bmatrix} \mathbf{v} \\ q \end{bmatrix} \right\rangle_E}$$

5. Update $\mathbf{u}^n$, p^n via ω^{n-1} and $\mathbf{v}$, q:

$$\begin{bmatrix} \mathbf{u}^n \\ p^n \end{bmatrix} = \begin{bmatrix} \mathbf{u}^{n-1} \\ p^{n-1} \end{bmatrix} - \omega^{n-1} \begin{bmatrix} \mathbf{v} \\ q \end{bmatrix}$$

The proposed nonlinear iteration may not be optimal with respect to the convergence speed but the resulting efficiency and especially the robustness behaviour for solving incompressible Navier–Stokes problems is mostly sufficient. Particularly the adaptive minimization step may lead to a more robust convergence behaviour which is needed to provide algorithmic tools for "Black Box" software packages.

One essential background idea for the design of our nonlinear solution schemes is the strong separation between **nonlinear treatment** and **linear solution tools**. Our approach approximates nonlinear effects on the finest mesh only, while multigrid is exclusively applied to solve linear subproblems in a preconditioning step. Consequently, we have to take care that at least the finest mesh is sufficiently accurate to describe all important physical effects. However, the coarser meshes are – roughly spoken – only involved to "provide some spectral approximation for the acceleration of matrix problems" for the performed linear multigrid. In this context, multigrid is only applied as very efficient linear **preconditioning tool**, separated from the discretization aspects regarding the nonlinearity. This is a very essential difference to *direct nonlinear multigrid* approaches which require the approximation of analogous nonlinear problems on the coarse meshes, too. In contrast, due to the separation of both tasks, we can apply standard multigrid tools without any modification and can guarantee at least the fast treatment of the subproblems in each nonlinear iteration step.

This approach may not be the best for some other problems in which case nonlinear multigrid approaches can lead to better results. However, our explicit experience in the case of the incompressible Navier–Stokes equations shows explicitly that this "separation approach" seems to be superior, at least if we compare the available codes.

3.3.2 Numerical aspects of nonlinear (and linear) iteration schemes

We complement this section with some numerical examples which demonstrate the considered theoretical behaviour. Further, we address the aspect of stopping criterions for discrete linear and nonlinear problems. Most iterative schemes are stopped if the euclidian norm of the defect vector $D_h := A_h U_h - F_h$ is sufficiently small,

$$\|D_h\|_E \leq TOL. \tag{3.178}$$

We will not discuss the delicate question of how to choose TOL. Instead we concentrate on the choice of more appropriate norms $\| \cdot \|$ instead of $\| \cdot \|_E$. If we take into account that our numerical schemes are based on finite element formulations, we might be interested in finding a relation between this euclidian norm estimate and the L^2-norm of a corresponding finite element residual. The essential key is the knowledge that the defect vector D_h can be interpreted as the coefficient vector of the residual involving the corresponding discrete differential operator, projected onto the finite element space.

Assuming that the continuous problem is given in operator notation, $Lu = f$, we can usually define a corresponding discrete differential operator L_h which is determined with respect to the linear forms $a(\cdot, \cdot)$, resp., $a_h(\cdot, \cdot)$, and the finite element subspace V_h. We denote by $u_h \in V_h$ the finite element solution, with corresponding coefficient vector U_h (and vector F_h with respect to the right hand side f), and by A_h and M_h, resp., $M_{h,l}$, the resulting stiffness and mass matrices, resp., the diagonal *lumped* mass matrix. Let $\varphi_h^{(j)} \in V_h, j =$

$1, \ldots, NEQ_h$, be the basic functions, then there holds for the components of the defect vector D_h:

$$D_h^{(i)} = a_h(u_h, \varphi_h^{(i)}) - (f, \varphi_h^{(i)}) = (L_h u_h - f, \varphi_h^{(i)}) \qquad (3.179)$$

Hence, if we define the discrete finite element residual $res_h \in V_h$ with coefficient vector RES_h via L^2-projection,

$$(res_h, \varphi_h) = (L_h u_h - f, \varphi_h) \quad , \quad \forall \varphi_h \in V_h , \qquad (3.180)$$

we can derive the following approximate relation between defect vector D_h and finite element residual res_h,

$$D_h = M_h RES_h \approx M_{h,l} RES_h . \qquad (3.181)$$

As a consequence, we have the following estimates for the discrete defect vector D_h and the finite element residual res_h, measured in the L^2-norm $\| \cdot \|$:

$$\|res_h\|^2 \approx \sum_i (M_{h,l}^{(i)})^{-1}(D_h^{(i)})^2 \quad , \quad \|D_h\|_E^2 = \frac{1}{NEQ_h} \sum_i (D_h^{(i)})^2 . \quad (3.182)$$

In general the terms $(M_{h,l}^{(i)})^{-1}$ and NEQ_h can be approximated by $O(h^{-2})$, and both expressions are of the same order if the underlying mesh contains elements only which all are of about the same size. However, for meshes which are locally refined or which contain large aspect ratios, both approaches differ significantly.

In the given representation in (3.182), the second term is weighted with $\frac{1}{NEQ_h}$ instead of NEQ_h. This scaling is often performed to preserve the discrete norm of the constant unity vector, but the finite element background of D_h being a *finite element defect* is hereby neglected. Consequently, both norms in (3.182) differ by h^2, and in fact, the resulting stopping criterion in the second euclidian norm is weakened with decreasing mesh width.

From a mathematical point of view we prefer the variant which measures approximate weighted residuals via the relation,

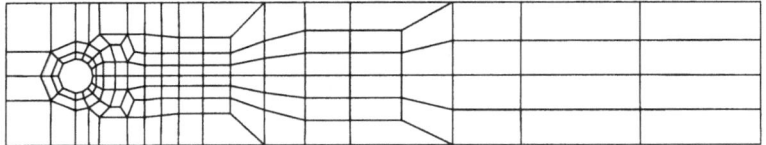

Figure 3.25: Coarse mesh for flow around a cylinder

$$\|res_h\|^2 \approx \sum_i (M_{h,l}^{(i)})^{-1}(D_h^{(i)})^2 , \qquad (3.183)$$

since hereby the underlying mesh topology as weighting factors is much better incorporated into the control of the iterative process. Further, the strength of the corresponding stopping criterions is related to finite element functions and, with respect to the absolute value, independent of the number of unknowns.

The aim of the subsequent computations is to demonstrate the behaviour of the proposed nonlinear iterative schemes as direct nonlinear solvers for the stationary Navier–Stokes equations. We examine the following aspects:

1. The dependence of the nonlinear convergence from the Reynolds number, the mesh width and shape of the mesh and the required accuracy for solving the auxiliary linear Oseen equations.

2. The dependence of the nonlinear defect correction approach from the chosen discretization for the preconditioning and for the defect calculation step.

3. The influence of the adaptive step length control and the choice of the norm to be controlled: weighted finite element residuals or standard l_2-norm of the defect vector.

The computations are strongly related to the tests in Section 4 which contains most of our numerical results, and consequently the flow configurations are more or less identical. We start again with the same configuration as performed in the *1995 DFG Benchmark* (see Sections 1.1). We consider 'flow in a channel around a cylinder', and the coarse mesh is shown in Figure 3.25.

This coarse grid is denoted with *level 1* while all other refinement levels are generated via the usual regular mesh refinement algorithm which recursively divides each element into 4 finer elements.

We perform the same stationary computations as in Section 3.1.4 for $Re = 20$ and additionally for $Re = 50$: Higher Reynolds numbers are impossible with this direct stationary solution approach since then, the corresponding flow is getting nonsteady. More precisely, we apply the following settings:

- We perform the control via the weighted finite element residuals.

- The nonlinear stopping criterion is $\epsilon_{non} = 10^{-3}$.

- The linear stopping criterion for the Oseen problems in each nonlinear iteration is to "gain one digit", that means $\epsilon_{lin} = 10^{-1}$; we apply a total number of 8 smoothing steps and 9 linear multigrid sweeps in the maximum.

- We apply the adaptive step length control and perform additionally calculations with prescribed values for ω.

The following Tables give the total number of nonlinear iterations (**NL**) needed to reach the given accuracy ϵ_{non} and – even more decisive for the understanding of the total work – the total number of performed linear multigrid sweeps (**MG**). We start with calculations which iterate towards an upwinded solution in the defect correction process. Then we show the results for calculations which iterate towards an solution which is discretized via streamline–diffusion or central discretization (see Section 3.1.4 for a detailed description).

First of all, the following conclusions can be drawn:

1. The dependence on the mesh width is rather small: if both discretizations for the solution and the preconditioner are the same, then almost no differences with respect to the mesh width are visisble. Only in that situation that both are different, e.g., particularly in those cases when we try to obtain the central solution C, the results improve essentially with finer mesh widths. The reason may be that with decreasing mesh width h, the diffusive part gets dominant! Finally, if h is sufficiently small, the nonlinear convergence rates in the defect correction approaches seem to reach the same behaviour as for those configurations with both discretizations being the same.

2. As we have shown in the previous Section 3.1.4, the use of smaller damping factors in the upwinding (UPW-0.1) or in the streamline–diffusion terms (SD-0.25) leads in general to improved accuracy results. However, the results here indicate that at the same time the numerical efficiency with respect to the application of standard multigrid deteriorates. The Tables show that the use of more accurate discretizations as

LEV	Re = 20 ω=adaptive NL / MG	Re = 20 ω=1.0 NL / MG	Re = 20 ω=0.8 NL / MG	Re = 50 ω=adaptive NL / MG	Re = 50 ω=1.0 NL / MG	Re = 50 ω=0.8 NL / MG
			Discretization UPW-∞ – solver UPW-∞			
3	5 / 15	5 / 15	7 / 21	7 / 22	8 / 28	11 / 41
4	5 / 15	5 / 15	7 / 21	9 / 26	9 / 26	12 / 35
5	5 / 14	6 / 17	7 / 19	10 / 29	10 / 29	12 / 35
			Discretization UPW-1.0 – solver UPW-∞			
3	5 / 15	6 / 18	7 / 21	8 / 25	9 / 32	12 / 45
4	5 / 15	6 / 18	8 / 24	10 / 29	10 / 29	13 / 38
5	5 / 14	6 / 17	8 / 22	12 / 35	11 / 32	14 / 41
			Discretization UPW-1.0 – solver UPW-1.0			
3	5 / 15	5 / 15	7 / 22	8 / 25	8 / 30	11 / 41
4	5 / 15	6 / 18	7 / 21	10 / 29	10 / 29	12 / 35
5	5 / 14	6 / 17	8 / 23	11 / 32	10 / 29	13 / 38
			Discretization UPW-0.1 – solver UPW-∞			
3	6 / 18	8 / 24	10 / 30	11 / 33	12 / 42	15 / 57
4	6 / 18	7 / 21	9 / 27	13 / 39	12 / 36	16 / 48
5	6 / 17	7 / 20	9 / 25	13 / 38	12 / 35	15 / 44
			Discretization UPW-0.1 – solver UPW-1.0			
3	5 / 15	6 / 18	8 / 24	11 / 36	11 / 42	14 / 54
4	6 / 18	6 / 18	8 / 24	12 / 36	11 / 33	14 / 42
5	6 / 18	7 / 21	8 / 24	13 / 38	11 / 32	14 / 41
			Discretization UPW-0.1 – solver UPW-0.1			
3	5 / 17	6 / 20	7 / 24	10 / 34	9 / 32	12 / 47
4	5 / 15	6 / 18	8 / 24	11 / 33	10 / 30	13 / 39
5	5 / 15	6 / 18	8 / 24	12 / 35	11 / 32	13 / 38

Table 3.14: Nonlinear convergence results (I) for the 'Flow around a cylinder' configuration

preconditioners (UPW-0.1, SD-0.25) may lead to less nonlinear sweeps, but the total number of required linear multigrid steps increases! Thus, the results indicate that the use of different discretizations, one being accurate and another efficient one which allows fast Numerical Linear Algebra tools, may be preferrable. However, as the results for the central discretizations show, the use of discretizations in the preconditioning step which are "not too far away" from the discretization performed for the solution itself, might be recommendable, at least in the case of the case Re = 50.

3. The adaptive step length control is preferrable! The setting of fixed damping parameters may lead to better results in special cases, but the results obtained via this optimization step are at least of the same order, and the additional numerical amount is negligible. Moreover,

LEV	$Re=20$ ω=adaptive NL / MG	$Re=20$ ω=1.0 NL / MG	$Re=20$ ω=0.8 NL / MG	$Re=50$ ω=adaptive NL / MG	$Re=50$ ω=1.0 NL / MG	$Re=50$ ω=0.8 NL / MG
Discretization SD-1.0 – solver SD-1.0						
3	6 / 18	6 / 18	8 / 24	10 / 38	10 / 38	12 / 44
4	6 / 18	6 / 18	8 / 24	12 / 39	10 / 37	13 / 48
5	6 / 17	7 / 20	8 / 23	12 / 36	12 / 36	15 / 49
Discretization SD-0.25 – solver SD-1.0						
3	10 / 30	15 / 46	19 / 57	20 / 74	23 / 70	30 / 91
4	9 / 27	12 / 35	16 / 48	20 / 64	21 / 61	27 / 78
5	7 / 21	9 / 25	12 / 26	15 / 46	17 / 42	22 / 53
Discretization SD-0.25 – solver SD-0.25						
3	5 / 17	6 / 22	8 / 28	11 / 38	10 / 35	13 / 48
4	5 / 16	6 / 19	8 / 24	14 / 41	12 / 37	14 / 44
5	5 / 14	6 / 17	8 / 23	12 / 35	11 / 32	14 / 41
Discretization C – solver UPW-∞						
3	10 / 30	13 / 34	16 / 41	72 / 216	– / –	50 / 147
4	7 / 21	9 / 27	11 / 31	19 / 57	22 / 57	27 / 67
5	7 / 20	8 / 23	10 / 28	14 / 41	15 / 43	18 / 51
Discretization C – solver UPW-1.0						
3	9 / 27	11 / 33	14 / 42	100 / 302	– / –	– / –
4	7 / 21	8 / 24	10 / 30	17 / 51	20 / 59	24 / 69
5	6 / 18	7 / 21	9 / 27	13 / 38	13 / 38	16 / 47
Discretization C – solver UPW-0.1						
3	7 / 24	7 / 23	9 / 29	– / –	– / –	– / –
4	6 / 18	7 / 21	8 / 24	– / –	– / –	– / –
5	6 / 18	6 / 18	8 / 24	13 / 38	11 / 32	14 / 41
Discretization C – solver SD-1.0						
3	22 / 66	36 / 109	46 / 139	85 / 266	100 / 237	>100 / >234
4	13 / 40	20 / 72	28 / 101	53 / 160	66 / 246	94 / 324
5	8 / 24	11 / 31	16 / 41	30 / 91	37 / 129	53 / 176
Discretization C – solver SD-0.25						
3	8 / 25	11 / 34	14 / 43	25 / 85	30 / 94	37 / 116
4	6 / 19	7 / 22	10 / 30	17 / 51	19 / 54	26 / 78
5	5 / 15	6 / 18	8 / 24	13 / 39	12 / 35	17 / 47

Table 3.15: Nonlinear convergence results (II) for the 'Flow around a cylinder configuration'

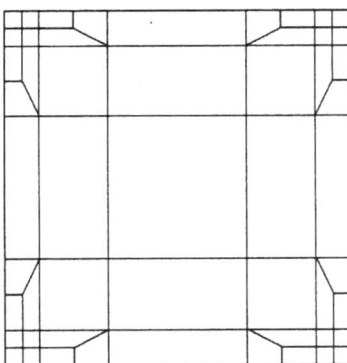

Figure 3.26: Semi-adapted coarse mesh for *lid driven cavity*

this adaptive process introduces a higher degree of robustness such that for (almost) all configurations the corresponding solutions could be calculated.

Next, we examine a configuration which allows larger Reynolds numbers, hereby still leading to stationary solutions: the standard *lid driven cavity*, for $Re = 2,000$ and $Re = 10,000$. As in Section 3.1.4 we employ as coarsest mesh the semi-adapted mesh in Figure 3.26.

For the next calculations, we apply the following setting:

- We perform the control via the weighted finite element residuals and the standard l_2 euclidian vector norm of the defect.

- The nonlinear stopping criterion is $\epsilon_{non} = 10^{-3}$ (weighted), resp., $\epsilon_{non} = 10^{-8}$ (l_2).

- The linear stopping criterion for the Oseen problems in each nonlinear iteration is again $\epsilon_{lin} = 10^{-1}$.

- We apply the adaptive step length control and perform also calculations with prescribed values for ω.

- If not explicitly denoted, we utilize the (nonparametric) midpoint–oriented finite element space $H_h^{(b)}$ with the Midpoint rule for evaluating the convective term.

The following Tables show again the total number of nonlinear iterations (**NL**) needed to reach the given accuracy ϵ_{non} and the corresponding total

Discret./Solver	LEV	$Re = 2,000$ ω=adaptive NL / MG	$Re = 2,000$ ω=1.0 NL / MG	$Re = 2,000$ ω=0.7 NL / MG	$Re = 2,000$ ω=adapt.(l_2) NL / MG
UPW-∞/UPW-∞	5	12 / 29	12 / 27	15 / 36	12 / 34
	6	13 / 33	13 / 33	17 / 41	13 / 33
	7	14 / 34	13 / 32	13 / 29	10 / 26
UPW-1.0/UPW-∞	5	12 / 30	11 / 26	15 / 34	12 / 32
	6	14 / 35	13 / 32	17 / 37	13 / 34
	7	12 / 28	12 / 25	15 / 28	10 / 24
UPW-1.0/UPW-1.0	5	12 / 30	14 / 33	16 / 40	12 / 35
	6	14 / 36	14 / 36	17 / 43	14 / 40
	7	12 / 28	12 / 25	14 / 32	9 / 23
UPW-0.1/UPW-∞	5	14 / 30	15 / 26	23 / 35	16 / 44
	6	15 / 29	16 / 26	25 / 35	15 / 35
	7	15 / 28	16 / 26	24 / 27	12 / 23
UPW-0.1/UPW-1.0	5	13 / 34	13 / 31	19 / 39	14 / 41
	6	15 / 39	13 / 31	19 / 39	12 / 31
	7	12 / 25	12 / 28	16 / 28	9 / 22
UPW-0.1/UPW-0.1	5	10 / 27	17 / 43	12 / 32	15 / 55
	6	13 / 36	12 / 33	13 / 34	11 / 45
	7	11 / 35	11 / 36	14 / 42	9 / 36

Table 3.16: Nonlinear convergence results (I) for 'Lid Driven Cavity' at $Re = 2,000$

number of performed linear multigrid sweeps (**MG**) which is a good measure for the total numerical complexity.

Similar conclusions as before seem to be valid:

1. The dependence of the mesh width is small. Only in that case that both discretizations are different, and particularly in those cases that we try to obtain the central solution C, the results can essentially improve with finer mesh widths.

2. The use of more accurate discretizations as preconditioners (UPW-0.1, SD-0.35) may lead to less nonlinear sweeps, but the total number of linear multigrid steps increases.

3. Surprisingly, the use of simple upwind schemes as preconditioners leads to very efficient nonlinear solvers; only on very coarse meshes, the use of streamline–diffusion for preconditioning seems to be superior.

4. The adaptive step length control is preferrable, for robustness and efficiency reasons.

5. As motivated in the previous theoretical considerations, the standard use of the l_2 euclidian vector norm for controlling the defect leads to mesh width dependent results since the norm of the measured values is already mesh dependent.

Discret./Solver	LEV	$Re = 2,000$ ω=adaptive NL / MG	$Re = 2,000$ ω=1.0 NL / MG	$Re = 2,000$ ω=0.7 NL / MG	$Re = 2,000$ ω=adapt.(l_2) NL / MG
SD-1.0/SD-1.0	5	11 / 28	21 / 69	14 / 31	13 / 39
	6	13 / 31	17 / 32	15 / 32	16 / 41
	7	14 / 31	14 / 30	17 / 39	10 / 26
SD-0.35/SD-1.0	5	12 / 25	13 / 27	29 / 30	17 / 52
	6	15 / 29	15 / 29	30 / 30	15 / 36
	7	15 / 29	15 / 29	30 / 30	12 / 26
SD-0.35/SD-0.35	5	10 / 26	21 / 58	12 / 29	15 / 53
	6	13 / 34	14 / 36	15 / 36	11 / 38
	7	13 / 34	12 / 30	16 / 47	11 / 34
C/UPW-∞	5	– / –	– / –	163 / 163	– / –
	6	30 / 48	32 / 33	47 / 47	33 / 55
	7	19 / 36	21 / 23	31 / 31	15 / 30
C/UPW-1.0	5	– / –	– / –	– / –	– / –
	6	28 / 43	25 / 34	36 / 37	35 / 54
	7	14 / 27	14 / 27	21 / 36	11 / 24
C/UPW-0.1	5	– / –	– / –	– / –	– / –
	6	28 / 64	– / –	17 / 41	26 / 68
	7	11 / 35	11 / 38	14 / 45	9 / 36
C/SD-1.0	5	128 / 189	177 / 182	507 / 512	251 / 480
	6	107 / 119	141 / 148	407 / 407	118 / 178
	7	72 / 80	93 / 99	258 / 258	52 / 67
C/SD-0.35	5	50 / 89	54 / 70	182 / 187	99 / 247
	6	41 / 73	48 / 61	146 / 146	43 / 102
	7	33 / 62	34 / 47	94 / 94	25 / 57

Table 3.17: Nonlinear convergence results (II) for 'Lid Driven Cavity' at $Re = 2,000$

Next, we examine the results in dependence of ϵ_{non} for controlling the nonlinear iteration process. Here, in the case of the weighted residuals, ϵ_{non} varies from $\epsilon_{non} = 10^{-1}$ to $\epsilon_{non} = 10^{-3}$. Table 3.18 shows the results for some selected discretization/solver configurations.

As can be seen, the main work has to be done to decrease the nonlinear residual from $\epsilon_{non} = 10^{-2}$ down to $\epsilon_{non} = 10^{-3}$ while the limit $\epsilon_{non} = 10^{-2}$, resp., $\epsilon_{non} = 10^{-1}$, is achieved quite fast. This result might indicate that the performed nonlinear iteration of fixed point-like type is very efficient for satisfying "weaker" stopping limits or for obtaining starting values, while for the process of getting higher accuracy for the measured nonlinear residuals, an iteration scheme of Newton-type might be more efficient. So, we should invest more research in applying "real" Newton-like methods to the Navier-Stokes equations, probably using the demonstrated fixed–point techniques as preconditioners or for getting "good" initial values.

Discret./Solver	LEV	$Re = 2,000$ $\epsilon_{non} = 10^{-3}$ NL / MG	$Re = 2,000$ $\epsilon_{non} = 10^{-2}$ NL / MG	$Re = 2,000$ $\epsilon_{non} = 10^{-1}$ NL / MG
UPW-1.0/UPW-1.0	5	12 / 30	8 / 18	4 / 7
	6	14 / 36	9 / 21	4 / 7
	7	12 / 28	7 / 15	4 / 7
UPW-0.1/UPW-0.1	5	10 / 27	6 / 15	3 / 6
	6	13 / 36	7 / 18	4 / 9
	7	11 / 35	7 / 20	5 / 13
SD-1.0/SD-1.0	5	11 / 28	6 / 13	4 / 8
	6	13 / 31	6 / 11	4 / 7
	7	14 / 31	7 / 14	4 / 7
SD-0.35/SD-0.35	5	10 / 26	6 / 14	4 / 8
	6	13 / 34	7 / 15	4 / 8
	7	13 / 34	8 / 19	4 / 8
C/UPW-∞	5	– / –	– / –	8 / 14
	6	30 / 48	16 / 30	7 / 12
	7	19 / 36	12 / 22	7 / 12
C/UPW-1.0	5	– / –	– / –	7 / 13
	6	28 / 43	13 / 25	6 / 11
	7	14 / 27	9 / 17	5 / 9
C/UPW-0.1	5	– / –	– / –	4 / 11
	6	28 / 64	8 / 21	4 / 9
	7	11 / 35	7 / 20	5 / 13
C/SD-1.0	5	128 / 180	36 / 62	6 / 12
	6	107 / 119	37 / 48	6 / 11
	7	72 / 80	33 / 41	7 / 13
C/SD-0.35	5	50 / 89	16 / 32	4 / 8
	6	41 / 73	15 / 29	4 / 7
	7	33 / 62	14 / 28	5 / 10

Table 3.18: Nonlinear convergence results (III) for 'Lid Driven Cavity' at $Re = 2,000$: Different parameters ϵ_{non} for the control of the nonlinear iteration; adaptive step length control

The next tests concern the question of how to evaluate the convective term if the streamline–diffusion, resp., the central discretization is performed. We compare:

- The (nonparametric) midpoint–oriented finite element space $H_h^{(b)}$ with the Midpoint rule (M) for evaluating the convective term.

- The (nonparametric) meanvalue–oriented (on edges) finite element space $H_h^{(a)}$ with the Midpoint rule (M) for evaluating the convective term.

- The (nonparametric) midpoint–oriented finite element space $H_h^{(b)}$ with the 2×2 Gaussian formula (G) for evaluating the convective term.

Discret./Solver	$Re = 2,000$ $H_h^{(b)}$ / M NL / MG	$Re = 2,000$ $H_h^{(b)}$ / G NL / MG	$Re = 2,000$ $H_h^{(a)}$ / M NL / MG	$Re = 2,000$ $H_h^{(a)}$ / G NL / MG
SD-1.0/SD-1.0	13 / 31	10 / 20	10 / 22	13 / 25
SD-0.35/SD-1.0	15 / 29	17 / 30	15 / 29	19 / 35
SD-0.35/SD-0.35	13 / 34	– / –	11 / 24	12 / 26
C/UPW-∞	30 / 48	25 / 48	97 /103	40 / 47
C/UPW-1.0	28 / 43	21 / 41	89 / 96	39 / 47
C/UPW-0.1	28 / 64	12 / 32	– / –	57 / 86
C/SD-1.0	107 /119	179 /197	129 /152	206 / 223
C/SD-0.35	41 / 73	– / –	45 / 69	81 / 114

Table 3.19: Nonlinear convergence results (IV) for 'Lid Driven Cavity' at $Re = 2,000$: Different elements ($H_h^{(a)}$ and $H_h^{(b)}$) and varying quadrature formulas ($M \sim$ Midpoint rule and $G \sim 2 \times 2$ Gaussian formula); all on level 6, adaptive, weighted, $\epsilon_{non} = 10^{-3}$

- The (nonparametric) meanvalue–oriented (on edges) finite element space $H_h^{(a)}$ with the 2×2 Gaussian formula (G) for evaluating the convective term.

Obviously, the different choice of both nonconforming finite element variants and the kind of quadrature formula can lead to completely different numerical behaviour of the linear and nonlinear solvers. They even lead to divergent behaviour in some configurations if we always apply the same parameter settings for both linear and nonlinear solvers. In some cases, the divergence was caused by a failure of the linear solvers which did not converge; or: the linear solvers even converged, but nevertheless the nonlinear iteration diverged.

For the following computations, we further increase the Reynolds number up to $Re = 10,000$ (still the 2D flow seems to remain steady!). The following Table 3.20 shows the results on level 6 and 7: again the increase of unknowns, resp., a finer mesh width, leads to better results for the central cases. As before, the use of upwinding as preconditioner leads to satisfying results on level 7 (level 6 seems to be too coarse to get the central solution!), while the application of streamline–diffusion leads to worse results, at least in the given examples. Additionally, it must be stated that the linear multigrid in combination with upwinding worked fine with 8 smoothing steps, while the use of streamline–diffusion required 16 smoothing steps totally.

The final examples address the question of the required accuracy for solving the auxiliary linear Oseen problems. While in the previous examples we always performed the strategy "gain one digit", that means $\epsilon_{lin} = 10^{-1}$, we additionally show in Table 3.21 some exemplary results if we spend more

Discret./Solver	LEV	$Re = 10,000$ $\epsilon_{non} = 10^{-3}$ NL / MG	$Re = 10,000$ $\epsilon_{non} = 10^{-2}$ NL / MG	$Re = 10,000$ $\epsilon_{non} = 10^{-1}$ NL / MG
UPW-∞/UPW-∞	6	20 / 53	11 / 26	4 / 7
	7	23 / 66	12 / 31	4 / 7
UPW-1.0/UPW-∞	6	21 / 54	12 / 27	4 / 7
	7	24 / 67	12 / 30	4 / 6
UPW-1.0/UPW-1.0	6	20 / 56	12 / 29	4 / 7
	7	23 / 70	12 / 32	4 / 8
UPW-0.1/UPW-∞	6	24 / 66	14 / 36	6 / 12
	7	20 / 52	12 / 28	7 / 14
UPW-0.1/UPW-1.0	6	25 / 71	13 / 35	5 / 10
	7	24 / 67	12 / 31	5 / 9
UPW-0.1/UPW-0.1	6	22 / 100	13 / 54	4 / 10
	7	28 / 114	17 / 71	5 / 15
SD-1.0/SD-1.0	6	19 / 127	11 / 66	5 / 19
	7	31 / 235	12 / 60	5 / 21
SD-0.35/SD-1.0	6	26 / 174	18 / 133	11 / 81
	7	25 / 184	17 / 123	11 / 76
SD-0.35/SD-0.35	6	– / –	– / –	– / –
	7	– / –	13 / 109	5 / 37
C/UPW-∞	6	– / –	– / –	15 / 28
	7	66 / 143	39 / 75	15 / 27
C/UPW-1.0	6	147 / 330	59 / 130	14 / 28
	7	61 / 130	36 / 75	13 / 26
C/UPW-0.1	6	160 / 336	64 / 156	9 / 25
	7	39 / 128	21 / 109	8 / 35
C/SD-1.0	6	660 / 1128	373 / 843	73 / 300
	7	532 / 1673	306 / 968	81 / 368

Table 3.20: Nonlinear convergence results (I) for 'Lid Driven Cavity' at $Re = 10,000$: Different stopping criterions and levels; weighted, adaptive

work in the solution process of the linear subproblems, here $\epsilon_{lin} = 10^{-3}$. The results show that in most cases the use of stronger stopping criterions for the linear subproblems makes no sense: even if the number of nonlinear steps is slightly decreased, the total number of linear multigrid sweeps is massively increased.

Before we draw final conclusions from our numerical experience with non-linear iteration schemes for the incompressible Navier–Stokes equations, we additionally present some results which examine the behaviour of the proposed nonlinear techniques with respect to anisotropies in the underlying spatial mesh.

The following Tables show results for channel flow around a squared cylinder. Figure 3.27 shows the coarse mesh which can be modified by adapting the first inner line near to the square. The resulting *aspect ratios* are $AR = 10$ (this Figure, called S1), $AR = 10^3$ (called S2) and $AR = 10^5$ (called S3). The

Discret./Solver	$\epsilon_{lin}(MG)$	$Re = 10,000$ $\epsilon_{non} = 10^{-3}$ NL / MG	$Re = 10,000$ $\epsilon_{non} = 10^{-2}$ NL / MG	$Re = 10,000$ $\epsilon_{non} = 10^{-1}$ NL / MG
UPW-∞/UPW-∞	10^{-1}	20 / **53**	11 / **26**	4 / **7**
	10^{-3}	19 / **202**	11 / **125**	4 / **44**
UPW-1.0/UPW-∞	10^{-1}	21 / **54**	12 / **27**	4 / **7**
	10^{-3}	21 / **236**	12 / **137**	4 / **43**
UPW-1.0/UPW-1.0	10^{-1}	20 / **56**	12 / **29**	4 / **7**
	10^{-3}	20 / **236**	12 / **136**	4 / **47**
UPW-0.1/UPW-∞	10^{-1}	24 / **66**	14 / **36**	6 / **12**
	10^{-3}	25 / **281**	14 / **153**	6 / **54**
UPW-0.1/UPW-1.0	10^{-1}	25 / **71**	13 / **35**	5 / **10**
	10^{-3}	28 / **339**	16 / **197**	6 / **61**
UPW-0.1/UPW-0.1	10^{-1}	22 / **100**	13 / **54**	4 / **10**
	10^{-3}	20 / **385**	12 / **224**	4 / **59**

Table 3.21: Nonlinear convergence results (II) for 'Lid Driven Cavity' ($Re = 10,000$): Different stopping criterions for linear multigrid; control of weighted residuals, level 6

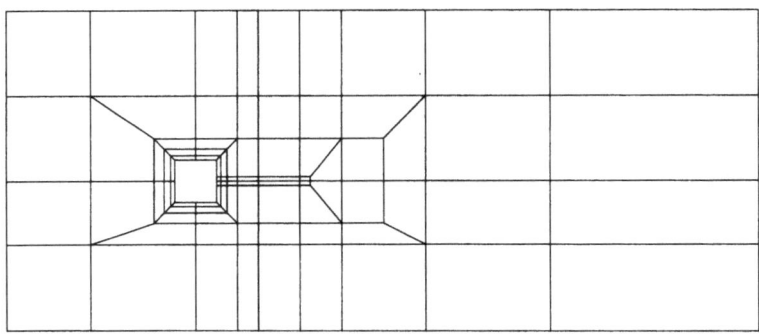

Figure 3.27: Typical (coarse) mesh S1 for nonlinear tests

following configuration is identical with the framework in Section 4.1 and also 3.1.4. We perform simulations for the viscosity parameters $1/\nu = 5, 50, 500$, hereby calculating the solution of the corresponding stationary Navier–Stokes equations. For more details about the used flow configurations, see Section 4.1.

The following Tables present the total number of nonlinear iterations (NNL), the total number of performed linear multigrid steps (NMG) for solving the auxiliary linear Oseen problems and the averaged number of multigrid steps (AMG) for one nonlinear iteration. For solving the resulting Oseen problems in each nonlinear step, we prescribe that the corresponding initial residual is damped by 0.1, that means we enforce to win one digit! In this linear multigrid, we applied 4 postsmoothing and no presmoothing steps. All cal-

LEV	S1 ($AR = 10^1$) NNL / NMG / AMG			S2 ($AR = 10^3$) NNL / NMG / AMG			S3 ($AR = 10^5$) NNL / NMG / AMG		
				$1/\nu = 5$					
3	4	8	2.0	4	8	2.0	4	16	4.0
4	4	12	3.0	4	12	3.0	4	16	4.0
5	5	24	5.0	5	25	5.0	5	22	4.5
				$1/\nu = 50$					
3	8	12	1.5	6	10	1.5	6	23	4.0
4	6	12	2.0	6	11	2.0	6	29	5.0
5	6	22	3.5	6	23	4.0	6	33	5.5
				$1/\nu = 500$					
3	10	19	2.0	10	19	2.0	10	30	3.0
4	11	21	2.0	12	21	2.0	11	35	3.0
5	11	25	2.0	12	23	2.0	11	48	4.5

Table 3.22: Nonlinear convergence results (I) for 'Flow around a square' with adaptive weighted residual control ($\epsilon = 10^{-3}$)

LEV	S1 ($AR = 10^1$) NNL / NMG / AMG			S2 ($AR = 10^3$) NNL / NMG / AMG			S3 ($AR = 10^5$) NNL / NMG / AMG		
				$1/\nu = 5$					
3	5	11	2.0	5	10	2.0	5	10	2.0
4	5	15	3.0	5	15	3.0	5	15	3.0
5	5	19	4.0	5	20	4.0	5	20	4.0
				$1/\nu = 50$					
3	9	13	1.5	8	12	1.5	7	11	1.5
4	6	11	2.0	6	10	2.0	6	11	2.0
5	7	22	3.0	6	19	3.0	6	19	3.0
				$1/\nu = 500$					
3	12	32	2.5	12	31	2.5	12	31	2.5
4	12	23	2.0	12	23	2.0	12	23	2.0
5	11	21	2.0	11	21	2.0	11	21	2.0

Table 3.23: Nonlinear convergence results (II) for 'Flow around a square' with adaptive l_2-euclidian control of the defect vector ($\epsilon = 10^{-8}$)

culations have been performed with streamline–diffusion discretization for the convective term.

These tests demonstrate that the nonlinear convergence behaviour is mainly independent of the mesh width and the shape of the mesh, here the aspect ratio. Only the Reynolds number seems to have influence on the nonlinear iteration scheme.

We finish this section about our numerical experience with nonlinear iteration techniques for solving (stationary) incompressible Navier–Stokes equations with the following conclusions and (personal) advises:

- The direct solution of the stationary Navier–Stokes equations is possible: The fixed point techniques allow the "direct" solution for a wide range of discretization approaches. The combination of outer defect correction with multigrid for the resulting linear Oseen problems as preconditioning tool leads to robust and efficient methods.

- We recommend the use of the adaptive step length control and the control of the weighted finite element residuals.

- Our results show that it is preferrable to apply the strategy "gain one digit (or similar)" for the solution of the linear subproblems. The number of necessary nonlinear steps may increase, but the total number of resulting linear sweeps seems to be better. Additionally, the use of "more stable" discretizations for the linear subproblems as preconditioners seems to be preferrable (even in the case of the central solution!).

- The nonlinear convergence rates seem to be independent of mesh size and shape of the mesh, only in the case of the central schemes a finer mesh improves the results!

3.3.3   Linearization techniques for nonstationary flows

In the nonstationary case, the typical approach is to perform first a semi-discretization in time which leads to generalized Navier–Stokes equations, with right hand side $\mathbf{f} = \mathbf{f}(t_{n+1}, t_n, \mathbf{u}^n)$ and $k = t_{n+1} - t_n$ according to the chosen time stepping scheme; additionally some parameter θ may be involved:

$$\mathbf{u} - k\nu\Delta\mathbf{u} + k\mathbf{u}\cdot\nabla\mathbf{u} + k\nabla p = \mathbf{f} \quad , \quad \nabla\cdot\mathbf{u} = g \qquad (3.184)$$

Hence, the fixed point techniques from the previous subsection can be directly applied in each time step. In contrast to the pure stationary case, now the nonlinear iteration rates can be improved if smaller time steps k are employed. However, there is another possibility which can be applied in the nonstationary context for the Navier–Stokes equations or related problems as the Burgers equations.

An easy modification of the proposed adaptive fixed point defect correction method is to choose $n = N_{max} = 1$ and $\omega^n = 1$. Then, with a given velocity field $\tilde{\mathbf{u}}$, the resulting linearized fixed point approach at time level t_{n+1} reads (with $\mathbf{u}^{n+1} = \mathbf{u}(t_{n+1})$ and $p^{n+1} = p(t_{n+1})$):

$$\begin{bmatrix} M + k\nu L + kN(\tilde{\mathbf{u}}) & kB \\ B^T & 0 \end{bmatrix} \begin{bmatrix} \mathbf{u}^{n+1} \\ p^{n+1} \end{bmatrix} = \begin{bmatrix} \mathbf{f} \\ g \end{bmatrix} \qquad (3.185)$$

This simplified approach can be interpreted as linearization of the convective term by constant extrapolation backwards in time ($\tilde{\mathbf{u}} := \mathbf{u}^n$), e.g., replacing

$$\mathbf{u}^{n+1} \cdot \nabla \mathbf{u}^{n+1} \quad \text{by} \quad \mathbf{u}^n \cdot \nabla \mathbf{u}^{n+1}. \qquad (3.186)$$

A simple improvement for obtaining fully second order accuracy with about the same numerical cost is to use a linear extrapolation in time, i.e., using

$$(2\mathbf{u}^n - \mathbf{u}^{n-1}) \cdot \nabla \mathbf{u}^{n+1} \quad \text{as approximation for} \quad \mathbf{u}^{n+1} \cdot \nabla \mathbf{u}^{n+1}, \qquad (3.187)$$

that means we set $\tilde{\mathbf{u}} := 2\mathbf{u}^n - \mathbf{u}^{n-1}$. For both schemes we have to assemble the new system matrix in each time step. The corresponding linear systems are nonsymmetric and require special solution techniques. An even simpler and very common possibility is a fully explicit treatment of the nonlinearity, i.e., replacing

$$\mathbf{u}^{n+1} \cdot \nabla \mathbf{u}^{n+1} \quad \text{by} \quad \tilde{\mathbf{u}} \cdot \nabla \tilde{\mathbf{u}}, \qquad (3.188)$$

with $\tilde{\mathbf{u}}$ one of the both previous possibilities, and considering this advection term as part of the right hand side. The resulting linear systems correspond to generalized symmetric Stokes equations, and the iteration scheme reads:

$$\begin{bmatrix} M + k\nu L & kB \\ B^T & 0 \end{bmatrix} \begin{bmatrix} \mathbf{u}^{n+1} \\ p^{n+1} \end{bmatrix} = \begin{bmatrix} \mathbf{f} - kN(\tilde{\mathbf{u}})\tilde{\mathbf{u}} \\ g \end{bmatrix} \qquad (3.189)$$

These linearization techniques look very nice from a computational point of view since only one (nonlinear) iteration in each time step has to be performed while the time stepping scheme seems to preserve its second order accuracy in time. In fact, potentially even cubic schemes (with extrapolation backwards in time involving two older time levels) can be derived and are applied by some authors.

However, our numerical experience with such high-order accurate semi-explicit schemes is somewhat disadvantageous:

- As soon as an adaptive time step control is applied, this approach results generally in varying time steps. If we perform a linear extrapolation involving $\mathbf{u}(t - \Delta t_1)$ and $\mathbf{u}(t - \Delta t_2)$ for approximating $\mathbf{u}(t)$ with second order accuracy, we derive the formula

$$\tilde{\mathbf{u}} = \frac{\Delta t_2}{\Delta t_2 - \Delta t_1}\, \mathbf{u}(t - \Delta t_1) - \frac{\Delta t_1}{\Delta t_2 - \Delta t_1}\, \mathbf{u}(t - \Delta t_2)\,. \qquad (3.190)$$

Since our performed time step control is based on 1 macro-step with $3\Delta t$ and 3 substeps with Δt, we obtain usually for the macro-prediction step with $3\Delta t$ the relation that $\Delta t_2 = \frac{4}{3}\Delta t_1$. This results in

$$\tilde{\mathbf{u}} = 4\,\mathbf{u}(t - \Delta t_1) - 3\,\mathbf{u}(t - \Delta t_2)\,.$$

If the new time step was doubled by the time step control, we even obtain the relation $\Delta t_2 = \frac{7}{6}\Delta t_1$, resp.,

$$\tilde{\mathbf{u}} = 7\,\mathbf{u}(t - \Delta t_1) - 6\,\mathbf{u}(t - \Delta t_2)\,.$$

In general, weighting factors $c(\Delta t_1, \Delta t_2)$ are the results for the linear extrapolation process such that the scheme performs the setting

$$\tilde{\mathbf{u}} = (c(\Delta t_1, \Delta t_2) + 1)\,\mathbf{u}(t - \Delta t_1) - c(\Delta t_1, \Delta t_2)\,\mathbf{u}(t - \Delta t_2)\,.$$

The constants $c(\Delta t_1, \Delta t_2)$ can become of size 2 – 20 in realistic applications. Consequently, large amplification factors for certain components of the solution vector may arise which can lead to instabilities during the evolution in time.

- The even more critical point for these linear extrapolation techniques of second order accuracy is that all our numerical calculations show a strong coupling between robustness of the resulting time stepping scheme with the underlying shape of the spatial mesh. As soon as the computational grid contains elements with largely varying sizes or with large aspect ratios AR ($AR > 10$), numerical instabilities occur! The resulting time step sizes coming from the adaptive time step control may oscillate by a factor of 1000 and more or, with equidistantly chosen time steps, the calculation breaks down if Δt is not chosen sufficiently small. We refer to the computational examples in Section 4.1 which demonstrate explicitly this numerical behaviour.

  While on isotropic meshes without complex geometrical details these linearization techniques lead to excellent results with respect to efficiency and accuracy, they fail on strongly varying meshes. This explains the fact that these techniques work for instance for "unit square" calculations (*driven cavity*) as proposed by some authors, but our experience is that one should not claim to use them for complex CFD applications.

- From a computational point of view it is not imperative that the fully nonlinear iteration is more expensive than the linearized versions with exactly one nonlinear sweep. The reason is that in the nonlinear approach the stopping criterions for the linear subproblems may be very weak ("gain one digit with respect to the initial iterate in each step"), while in the linearized versions restrictions due to accuracy requirements for momentum and continuity equations have to be satisfied in exactly one step. In that case that the solver process is dominating the CPU cost, both schemes may lead to about the same computational cost for one complete nonlinear solution step, but the resulting accuracy and particularly the robustness of the nonlinear approach is in general higher.

Our numerical tests in Chapter 4 show that the combination of the *adaptive fixed point defect correction method* with linear multigrid tools as preconditioners lead to very robust and efficient solution approaches for the stationary as well as the nonstationary incompressible Navier–Stokes or Burgers equations. Additionally, for the highly nonstationary case we can perform a second order accurate linearization by extrapolation in time which however should be only applied in the case that a priori knowledge about the topology of the grid is available.

It is obvious that in general the fully nonlinear iteration schemes cause more numerical cost in each time step than the linearization techniques (with only one iteration). Additionally, the fully explicit treatment consumes the least effort. However, the question not included in these considerations is that for the necessary size of the time step. The fully nonlinear iteration is expected to be more stable and accurate while the price to be paid for the more explicit schemes is a smaller time step size. So, asking for the total CPU time, which schemes are the more efficient ones? This question will be one of the major topics in our numerical tests in Chapter 4.

3.3.4   Other nonlinear techniques for the Navier–Stokes equations

Beside our preferred *adaptive fixed point defect correction method* which belongs to the class of quasi–Newton approaches for treating the nonlinear Navier–Stokes or Burgers equations, there are some other prominent schemes which are often proposed in the literature. One of them is the *nonlinear least square CG method* which was introduced by Glowinski and Periaux [40]. In this approach, a sequence of approximate solutions $\{\mathbf{u}^n, p^n\}$ is obtained by minimizing the least square functional

$$\|\nabla \mathbf{v}\|^2 \to \min! \tag{3.191}$$

where $\mathbf{v} = \mathbf{v}(\mathbf{u}^n, p^n)$ is determined through the (Stokes-) relation

$$\left[\begin{array}{cc} M + k\nu L & kB \\ B^T & 0 \end{array} \right] \left[\begin{array}{c} \mathbf{v} \\ q \end{array} \right] = \text{"nonlinear residual of } \{\mathbf{u}^n, p^n\} \text{ "}. \tag{3.192}$$

This minimization process is embedded into a CG ('conjugate gradient') algorithm which requires the solution of three generalized Stokes problems in each nonlinear step. Hence, efficient multigrid solvers are required which can be provided by the *MPSC* schemes. In our experience this method is very robust as it is based on the minimization of a positive functional but the convergence behaviour for higher Reynolds numbers, especially in direct stationary approaches, drastically slows down. This failure is due to the fact that only Stokes problems are used as nonlinear preconditioners which are insufficient for dominating convective terms.

In fact, if we could derive a combination of our gradient–type schemes involving Oseen problems as preconditioners with such generalized nonlinear CG algorithms for nonsymmetric problems, this solver might be a candidate for the desired final "Black Box" solver for the resulting nonlinear problems.

Another scheme which is often included in CFD tools is the *characteristics Galerkin method* which was first introduced by Pironneau [78]. It works for the nonstationary equations only and may be viewed as an *implicit upwind* in space–time. The starting point is a Lagrangian formulation of the acceleration terms in the momentum equations such that the time discretization is then performed in the resulting characteristic flow direction.

Again the incompressibilty constraint requires the solution of a generalized Stokes equation in each time step which can be performed by the proposed MPSC–type Stokes solvers. However, the characteristic velocity on the involved right hand side must be traced back to the preceeding time level through solving an ODE system for each spatial degree of freedom. Within a finite element framework this process contains the L^2-projection between different spatial meshes which requires special care for preserving the full accuracy of the method. Further, the principal unconditional stability of this scheme may be lost by using inaccurate numerical quadrature.

Since the *characteristics Galerkin method* is potentially of second order in space and time it looks like a very promising approach to CFD applications in the range of large Reynolds numbers. However, there are still several technical questions to be answered and there is no computational evidence yet that this approach is superior over simple stabilization techniques of upwind or (spatial) streamline–diffusion type. Unfortunately, we still lack of any own numerical experience and up to now no method involving this approach participated at the *1995 DFG Benchmark* such that the quality of these schemes, at least for low or medium Reynolds numbers, is still not clear to us (see at least [48]).

3.4   Linear multigrid techniques

Multigrid techniques have become one of the most important subjects in Numerical Linear Algebra. In particular for the numerical solution of discretized partial differential equations (PDE's) they play a key role due to their excellent behaviour with respect to convergence rates and computational complexity. Since the corresponding stiffness matrices, arising from PDE's, have usually a very sparse structure, iterative schemes which are essentially based on matrix-vector multiplications and simple vector modifications only, can be much superior to classical direct solvers. Since beside the smaller computational complexity (with respect to storage requirements and floating point operations) a very fast and robust convergence behaviour can be obtained in multigrid approaches, they are actually the most efficient solvers for partial differential equations in general and particularly for CFD problems.

There are many aspects of today's multigrid: robustness and efficiency on complicated domains and anisotropic meshes, adaptive multigrid, parallelism and vectorization on supercomputers, multigrid for complex systems of problems, etc. We do not plan to give an overview about all these topics: there are numerous publications and the research is still going on. For an overview concerning theoretical and practical aspects of multigrid we recommend the "classical" book of Hackbusch [49], and for recent results in multigrid/multilevel research the reader is advised to look, for instance, at [15], [18], [69] and the literature cited therein.

Chapter 2 has already introduced the basic multigrid ideas and components. However, we have exclusively concentrated on the *Multilevel Pressure Schur Complement* (MPSC) schemes for the solution of Stokes-like systems. In addition, we now provide the reader with more details for the numerical solution of accompanying subproblems which arise if MPSC schemes are applied to solve discrete incompressible flow problems. These are:

- multigrid for resulting subproblems with respect to the velocity (scalar transport–diffusion equations) and the pressure (scalar Pressure–Poisson equations).

- components as grid transfer, coarse grid operators, step-length control of correction, etc., for the resulting scalar subproblems as well as for the MPSC approach.

We start with formulating a standard multigrid algorithm for the solution of the following (discrete) linear system of equations

$$A_N u_N = f_N. \tag{3.193}$$

Further, we assume a *hierarchy of levels* $i, i = 1, \ldots, N$, which may be strongly connected to a *mesh size* parameter h_i, for instance. On each of these levels, we have to be able to define the discrete problem matrix A_i and corresponding right hand sides f_i. f_N is given on the finest level N only, while all other terms f_i are generated during the multigrid run. Then, a standard N-level multigrid algorithm $MG(N, \cdot, \cdot)$ for the solution of problem (3.193) reads:

The k-level iteration $MG(k, u_k^0, f_k)$:

The k-level iteration with initial guess u_k^0 yields an approximation to u_k, the solution of the problem

$$A_k u_k = f_k.$$

One step can be described in the following way:

For $k = 1$, $MG(1, u_1^0, f_1)$ is the exact solution: $\quad MG(1, u_1^0, f_1) = A_1^{-1} f_1$.

For $k > 1$, there are four steps:

1) m-Presmoothing steps

Apply m **smoothing steps** to u_k^0 with a *basic iteration* to obtain u_k^m.

---

2) Correction step

Calculate the restricted residual (with **restriction operator** I_k^{k-1})

$$f_{k-1} = I_k^{k-1}(f_k - A_k u_k^m),$$

and let u_{k-1}^i $(1 \le i \le p,\, p \ge 1)$ be defined recursively by

$$u_{k-1}^i = MG(k-1, u_{k-1}^{i-1}, f_{k-1}),\ 1 \le i \le p,\ u_{k-1}^0 = 0.$$

3) Step-length control of correction

Calculate u_k^{m+1} (with **prolongation operator** I_{k-1}^k) via

$$u_k^{m+1} = u_k^m + \alpha_k I_{k-1}^k u_{k-1}^p,$$

where α_k may be a fixed value or chosen adaptively so as to minimize
the error $u_k^{m+1} - u_k$ in an appropriate norm.

4) n-Postsmoothing steps

Analogously to step 1), apply n **smoothing steps** to u_k^{m+1} and obtain
u_k^{m+n+1}.

Performing one iteration step $MG(N, u_N^0, f_N)$ yields, for a given initial u_N^0,
the new approximate u_N^{m+n+1}, which may be written as

$$MG(N, u_N^0, f_N) = u_N^{m+n+1}.$$

Each run of $MG(k, \cdot, \cdot)$ is called one *cycle* of the multigrid iteration, and the
application of sufficiently many *cycles* on level N ensures the (approximate)
solution of the problem (3.193). If $p = 1$ in the correction step, we say
V–cycle, $p = 2$ denotes a *W–cycle*. In our applications we prefer something
inbetween, the so-called *F–cycle*. The value m, resp., n defining the number of
smoothing steps is small in general, between 1 (or even 0) and 4 for instance.
The components of such a typical multigrid algorithm are:

1. the **smoothing operator**, resp., the performed *basic iteration* on each level.

2. the **grid transfer operator** (*prolongation* I_{k-1}^k and *restriction* I_k^{k-1}).

3. the construction of **coarse grid operators** A_k, $k < N$.

4. the (adaptive) **step–length control** of coarse grid corrections.

5. the **matrix–vector multiplication** with matrix A_k (for defect calculations).

The topics we consider in the subsequent sections can be formulated as follows:

- Simple iterative *basic schemes* for scalar transport–diffusion and Pressure–Poisson problems and some of their properties.

- Acceleration of such *basic iterations* as preconditioners in Krylov–methods.

- Application of such *basic iterations* as smoothing operators in multigrid.

- Robust and efficient grid transfer operators.

- Construction of appropriate coarse grid matrices.

- Control of coarse grid corrections.

The typical result of optimal multigrid can be characterized by the following thumb rules:

Thumb rules for expected multigrid behaviour for (second order) PDE's:

1. The convergence rates are independent of the mesh size and additionally of the shape of the underlying mesh.

2. The convergence rates are of size $0.1 - 0.5$ for $1 - 4$ (total) smoothing steps.

3. The convergence rates are directly related to the number of smoothing steps, i.e., a doubling of the number $m + n$ of total smoothing steps leads to halved rates.

Hence, if multigrid is "optimally" performed, the typical cost for gaining one digit accuracy are usually related to a small number of matrix-vector multiplications, with a very sparse matrix only. And this relation is true independent of the complexity and size of the problem. Based on this philosophy, *multigrid* has become a "must" in current numerical schemes and is one of the "magic" key words which tends to be already sufficient to guarantee the efficiency of proposed solution schemes. However, the straightforward application of standard multigrid techniques is in general not sufficient. In fact, it is a hard job to provide optimal tools for a given class of problems, taking into account the following practical needs:

- Smoothing and/or grid transfer operators on complex meshes with large aspect ratios.

- Grid transfer routines adapted to non-standard finite elements.

- Optimal run-time behaviour on parallel or superscalar architectures.

Summing up, there are many specific aspects to take care if the typical excellent behaviour of multigrid approaches, which is known from the treatment of simple model problems on unit squares, shall be obtained in general with respect to accuracy, robustness and efficiency. The hard work is to adapt the singular components to the given problem, if possible through an adaptive procedure during the calculation. In the following sections, we demonstrate some approaches into this direction which can satisfy some of the requirements.

3.4.1   Linear basic iterations and their properties as smoothers

We concentrate on solving a (standard) linear problem, $Ax = b$, which may be thought to arise from the discretization of a partial differential equation. The underlying problem may be a generalized transport–diffusion problem for a velocity component u_i (with given $\mathbf{U}$),

$$\alpha u_i - \nu \Delta u_i + \mathbf{U} \cdot \nabla u_i = g \,, \qquad (3.194)$$

or a Pressure–Poisson problem,

$$-\Delta p = g \,. \qquad (3.195)$$

In general, the resulting matrix A is obtained by applying an appropriate discretization process to the continuous differential operator. In addition, the matrix A can also be directly constructed on discrete level via matrix–vector multiplications,

$$A := B^T M_l^{-1} B .$$
(3.196)

In that case, A denotes the resulting *reactive Schur complement* preconditioner from Section 2.3.1 which is a discrete analogue to the generalized Pressure–Poisson operator in the *global MPSC* approach. A very general way to solve iteratively such problems is to perform a *preconditioned Richardson* iteration. We call it in the following the **basic iteration**, with C^{-1} being an appropriate preconditioner for A.

Linear basic iteration:

Given: Iterate x^{l-1}

Perform: One relaxation step to obtain x^l

$$x^l = x^{l-1} - \omega^{l-1} C^{-1} (A x^{l-1} - b)$$

Since the convergence behaviour is directly related to the *iteration matrix* M^l,

$$M^l := I - \omega^l C^{-1} A ,$$
(3.197)

one is typically faced with the problem to find a compromise between:

- $\omega^l C^{-1} \approx A^{-1}$.

- C^{-1} is easily applicable.

The following typical schemes can be found in the literature (with D denoting the diagonal of A, L and U the lower, resp., the upper part from the matrix A):

1. **Classical Richardson** scheme: $\omega^l \leq 1/\lambda_{max}(A)$, $C = I$

2. **Jacobi** iteration: $\omega^l \geq 0$, $C = D$

3. **Gauß–Seidel/SOR** iteration: $\omega^l \in (0,2)$, $C = D + \omega^l L$

4. **ILU** scheme: $\omega^l \geq 0$, $C = \tilde{L}\tilde{U}$
        (with $A = \tilde{L}\tilde{U} + R$)

5. Many other modifications of these schemes:
                    SSOR, MILU, line–Gauß–Seidel, etc.

A major problem for all these basic iteration schemes is that in general they are insufficient solvers whose convergence rates are mainly depending on the problem size, resp., the underlying mesh width h. In fact, the typical result for the resulting convergence rate ρ_C is

$$\rho_C \sim 1 - O(h^\alpha) \quad , \quad \alpha \geq 0. \tag{3.198}$$

While the optimal case, $\alpha = 0$, is reached in general only by multigrid schemes, the typical range for the proposed classes of problems is $\alpha \in [0,2]$, for instance,

$\alpha = 2$ for Jacobi and Gauß–Seidel iteration,

$\alpha = 1$ for ILU and SOR (with optimal ω^l).

There is one important feature of these schemes: during the iteration process the high frequencies of A are damped very fast while the low frequencies are insufficiently reduced which results in the bad convergence behaviour. However, this fast damping of some parts of the complete spectrum of A is one essential property of these schemes for their use as smoothers in multigrid which is responsible (among others) for the resulting excellent performance. So, even being "bad solvers", they are in general "good smoothers". Therefore, we list some properties of the typical basic iterations if applied as smoothers in multigrid.

Some important properties of typical basic iterations:

Jacobi iteration:

- "worst" convergence behaviour among all described schemes, but excellent computational performance on vector- or parallel computers.

- no additional matrix for the preconditioner $C^{-1} = D^{-1}$ is necessary.
- invariant convergence rates with respect to renumbering strategies.
- working for rather general matrices if ω^l is sufficiently small.
- **Personal numerical experience:**

The Jacobi iteration should be exclusively performed on rather isotropic meshes. If elements with "large" aspect ratios (≥ 5) are contained, the rates deteriorate significantly, more smoothing steps are needed, the relaxation parameter ω^l has to be chosen smaller ($\omega^l \leq 0.7$) and the convergence behaviour deteriorates with decreasing mesh width h. In contrast, excellent computational performance can be obtained on vector computers which may come near to the Peak performance. As a conclusion, use this method only for meshes containing isotropic (macro) elements.

Gauß–Seidel/SOR iteration:

- convergence behaviour better than Jacobi, but hard to vectorize and parallelize in general.
- no additional matrix for the preconditioner $C^{-1} = (L + \omega D)^{-1}$ is needed.
- (almost) invariable convergence rates with respect to renumbering strategies in the case of pure diffusive problems (Poisson operator).
- "optimal" convergence rates can be obtained for convection-dominated problems if almost lower triangular matrices are generated by appropriate renumbering strategies (see the *downwind numbering* by Hackbusch [50]).
- working as smoother for rather general matrices, for $\omega^l = 1$ even.
- **Personal numerical experience:**

The Gauß–Seidel/SOR iteration should be performed on isotropic meshes which contain elements with moderately sized aspect ratios (≤ 10). Otherwise, the rates may deteriorate significantly, more smoothing steps are needed and the convergence behaviour deteriorates with decreasing mesh width h. This method is our favourite on most typical meshes and for workstation simulations.

ILU iteration:

- shows the most robust convergence behaviour compared with the other schemes, but hard to vectorize and parallelize.

- one additional matrix for the preconditioner C^{-1} is needed.

- renumbering strategies are absolutely necessary for convection-dominated problems as well as for pure diffusive problems to achieve "optimal" convergence rates.

- working for rather general matrices, robust even on very anisotropic meshes.

- **Personal numerical experience:**

The ILU iteration is our favourized scheme if robust multigrid tools are needed in the context of *Black Box* solvers, particularly if complicated geometries and triangulations are taken into account. However, the convergence behaviour is very sensitive with respect to applied renumbering strategies, the computational cost are significantly larger and the application to systems of equations is not straightforward.

Since these basic iterations are poor solvers they have to be accelerated. In general, there are two possibilties:

1. As **preconditioners** in Krylov-space methods, resp., conjugate gradient like schemes (*adaptive setting of optimal damping parameters ω^l*).

2. As **smoothing operators** in multigrid (*damping of high frequencies*).

The major idea with regard to their application as efficient preconditioners in conjugate gradient like schemes is described in the following (see also [3]). It is well-known that the convergence behaviour (ρ_{CG}) of such Krylov-space methods is mainly due to the condition number $cond(A)$ of the matrix A,

$$\rho_{CG} \sim 1 - O(\frac{1}{\sqrt{cond(A)}}), \qquad (3.199)$$

which typically results in estimates of type $\rho_{CG} \sim 1 - O(h)$ for the considered classes of problems. However, this (theoretically based) result is often somewhat pessimistic since additionally the distribution of the eigenvalues over the complete spectrum of A influences the convergence behaviour, too. Consequently, the essential idea of preconditioning – in fact, the algorithmic realization of preconditioning may look different – is to solve a *scaled* problem

$$C^{-1}Ax = C^{-1}b \quad \text{instead of} \quad Ax = b, \qquad (3.200)$$

with resulting matrix $\tilde{A} := C^{-1}A$. Altogether, the philosophy behind this approach is to satisfy that:

- $cond(\tilde{A}) < cond(A)$,
- the eigenvalues of $\tilde{A}$ are *clustered*.

Typical candidates for Krylov-space methods used in (CFD-) practice are the following:

Preconditioned conjugate gradient method (PCG):

- works for symmetric positive definite problems only (Pressure–Poisson!).

GMRES:

- is applicable for general matrices.
- can be performed only as truncated versions since otherwise the sequence of all iteration vectors has to be stored. This modification is necessary for CFD problems, particularly in the 3D case which leads to vectors with several millions of components.

BiCGSTAB:

- works for rather general matrices.
- involves 2 matrix-vector multiplications and 5 auxiliary vectors only.
- our favourized scheme in the context of transport–diffusion problems (as coarse grid solver).

Our numerical experience with these solution schemes, if applied to CFD problems, can be concluded in the following statements:

Numerical experience with Krylov-space methods:

1. The convergence behaviour of such PCG-like methods can be improved
   in general only up to

$$\rho_{PCG} \sim 1 - O(\frac{1}{cond(A)^{1/4}}) \sim 1 - O(h^{1/2}).\qquad (3.201)$$

   Consequently, these schemes are too slow for the solution of the positive
   definite Pressure–Poisson problems if the mesh width is fine as typically
   needed in realistic applications.

2. The same behaviour, and hence the same conclusions can be stated for
   the BiCGSTAB. However, in the context of fully nonstationary problems this scheme may be viewed as alternative to multigrid for the
   solution of resulting transport–diffusion problems if the time step k is
   very small.

3. Both schemes can be employed as coarse grid solvers in multigrid. In
   fact, their range of application is due to the following thumb rule (with
   NEQ denoting the number of equations), at least with respect to the
   computer technology in 1998:

   Use **Gaussian elimination** or **CG-like** methods for $NEQ \leq 1,000$.
   Use **CG-like** or **multigrid** methods for $1,000 \leq NEQ \leq 10,000$.
   Use **multigrid** methods for $10,000 \leq NEQ$.

4. An unsolved problem (at least for us) is the question of the play-
   together of large aspect ratios and the convergence behaviour of preconditioned Krylov-space methods. While multigrid is rather sensitive,
   eventually through the involved coarse grid problems and operators, an
   analogous behaviour for CG-like methods is not clear at all and requires
   further numerical studies.

3.4.2   Grid transfer, coarse grid operators and control of correction

The second important component in multigrid are the grid transfer operators
which *transfer* coefficient vectors (as discrete analogues to finite element functions, resp., finite element residuals) from a given mesh onto another finer or
coarser grid. Since the vector components represent the degrees of freedom
of the associated finite element function, the acting on the components of a
solution vector is equivalent with treating directly the corresponding finite

element function. Hence, we can formulate all grid transfer routines in terms of finite element notation.

Let H_{2h} and H_h be two finite element spaces associated with meshes according to a mesh width $2h$ and h, meaning that T_h is, in general, generated from T_{2h} by a refining process. Then, the *prolongation operator* $I_{2h}^h : H_{2h} \to H_h$ transfers functions from T_{2h} to T_h. In the following, we always denote by V_h the coefficient vector directly associated with the finite element function $v_h \in H_h$. Then, in matrix-vector notation, we can define a corresponding *prolongation matrix* P_{2h}^h which is associated with I_{2h}^h. The *restriction operator* $I_h^{2h} : H_h \to H_{2h}$ (to be precise: the dual spaces according to H_h, resp., H_{2h}, which however makes no difference for the discrete functions here) works analogously and is in general defined as the adjoint operator to I_{2h}^h. Translated in matrix-vector notation, the corresponding *restriction matrix* R_h^{2h} can be (up to a constant weighting factor) represented as $(P_{2h}^h)^T$. Hence, it is sufficient if we consider the *prolongation operator* I_{2h}^h only in the following. The restriction operator can be always obtained as adjoint operator to a certain prolongation operator.

In the following, let $(\cdot, \cdot)_h$ denote a discrete scalar product. Let $||| \cdot |||_n, n = 0, 1, 2$, be a discrete L^2-, H^1- or H^2- norm, and let v_{2h}, resp., v_h be arbitrary functions in H_{2h}, resp., H_h. Typically, there are two approaches for the construction of prolongation operators I_{2h}^h, via *interpolation* or *(discrete) L^2-projection*:

1. **(discrete) L^2-projection** operator I_{2h}^h:

$$(I_{2h}^h v_{2h} - v_{2h}, v_h)_h = 0 \qquad (3.202)$$

2. **interpolation** operator I_{2h}^h:

$$
\begin{aligned}
|||I_{2h}^h v_{2h} - v_{2h}|||_0 \;&\leq\; ch \; |||v_{2h}|||_1 \quad \text{(constant interpolation)} &(3.203)\\
&\leq\; ch^2 \; |||v_{2h}|||_2 \quad \text{(linear interpolation)} &(3.204)
\end{aligned}
$$

The (discrete) L^2-projection operator $I_{2h}^{h,P}$:

The prolongation operator $I_{2h}^h := I_{2h}^{h,P}$ is defined via

$$(I_{2h}^{h,P} v_{2h}, v_h)_h = (v_{2h}, v_h)_h \quad \forall v_h \in H_h \,, \ \forall v_{2h} \in H_{2h} \,.$$

If we associate with $\tilde{V}_h \in H_h$ the corresponding coefficient vector for $I_{2h}^{h,P} v_{2h}$, we derive in equivalent matrix-vector notation

$$M_h \tilde{V}_h = N_{h,2h} V_{2h} \,, \qquad\qquad (3.205)$$

and for the resulting *prolongation matrix* $P_{2h}^{h,P}$ the relation

$$\tilde{V}_h = P_{2h}^{h,P} V_{2h} := M_h^{-1} N_{h,2h} V_{2h} \,. \qquad\qquad (3.206)$$

The essential technique for the definition of the discrete scalar product $(\cdot,\cdot)_h$ is to perform a special quadrature rule which leads to a diagonal mass matrix M_h only. The matrix $N_{h,2h}$ is rectangular due to the basis function of H_{2h}, resp., H_h. The components $M_h^{(i)}$ correspond to the size of the local support of the corresponding i-th basis function. Then, the prolongation process can be performed element by element, and for standard finite elements (constant, linear, quadratic) it is equivalent to the following interpolation approach, at least on isotropic meshes. However, for divergence–free finite elements both approaches lead to different results.

The (macro-elementwise) interpolation operator $I_{2h}^{h,I}$:

The (macro-elementwise) interpolation algorithm for the nonconforming rotated bilinear finite element functions can be described as follows. We show explicitly the construction process for both edges u_5 and u_6 in Figure 3.28, then all remaining coefficients are calculated analogously. Consider Figure 3.28 which contains a (macro) element on level $2h$ and the correspondingly refined smaller elements according to level h. We assume in the following the "classical" refinement process: each coarse element is divided into 4 elements by connecting opposite midpoints. For a precise definition of the FEM spaces H_h^a and H_h^b we refer to Section 3.1.2. It should be remarked that on highly deformed meshes the "true" coefficients for the following interpolation may slightly change, but these differences are very small and all numerical tests have proven that the exclusive use of the "frozen" constants is absolutely sufficient. The corresponding weighting factors in the 3D case can be found in [93].

(Macro-elementwise) interpolation for rotated bilinear finite elements:

1. Calculate on all fine edges the corresponding interpolant values due to the given function v_{2h} which is defined via the coefficients u_1, u_2, u_3, u_4:

   - **full** interpolation for $H_h = H_h^a$ (mean-values on edges as d.o.f.'s)

   $$u_5 = u_1 - \frac{1}{4}u_2 + \frac{1}{4}u_4 \quad , \quad u_6 = \frac{5}{8}u_1 + \frac{1}{8}(u_2 + u_3 + u_4)$$

   - **full** interpolation for $H_h = H_h^b$ (midpoint-values on edges as d.o.f.'s)

   $$u_5 = \frac{15}{16}u_1 - \frac{3}{16}u_2 - \frac{1}{16}u_3 + \frac{5}{16}u_4 \,,$$
   $$u_6 = \frac{9}{16}u_1 + \frac{3}{16}u_2 + \frac{1}{16}u_3 + \frac{3}{16}u_4$$

   - **constant** interpolation for $H_h = H_h^{(a/b)}$

   $$u_5 = u_1 \quad , \quad u_6 = u_1$$

2. For the values on edges which belong to (macro) edges from T_{2h} (for instance u_5) take a mean-value by simply averaging or by taking into account the size/volume of the two adjacent macro-elements as weighting factors.

In the context of the discrete Pressure–Poisson equation with matrix

$$A := B^T M_l^{-1} B \,, \tag{3.207}$$

we can perform analogous steps, but now for piecewise constant ansatz functions. A further possiblity is to interpolate first the piecewise constant functions into the space of the nonconforming rotated bilinear finite elements. Hence, the complete grid transfer process for the piecewise constant pressure functions is formulated as follows. We assume that p_{2h} denotes a piecewise constant pressure function due to T_{2h}.

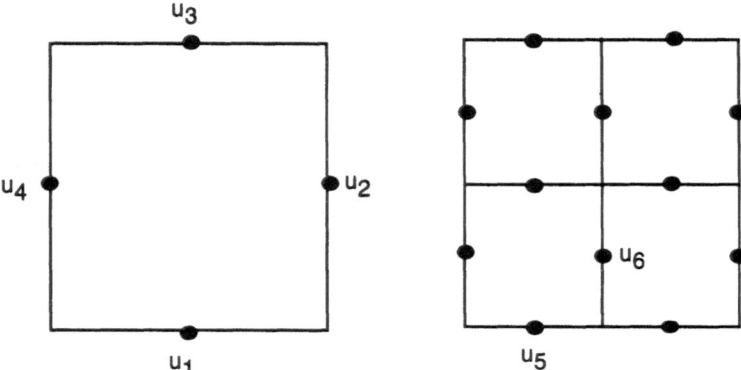

Figure 3.28: Configuration for local interpolation

**(Macro-elementwise) interpolation for piecewise constant
finite elements:**

1. Let p_{2h} be piecewise constant or calculate the nonconforming rotated bilinear finite element interpolant $\tilde{p}_{2h}$.

2. Perform the described (macro-elementwise) interpolation algorithm, with constant evaluation (for p_{2h}) or full interpolation (for $\tilde{p}_{2h}$).

3. If the (macro-elementwise) interpolation was applied to the rotated bilinear finite elements, calculate the corresponding piecewise constant interpolant p_h on T_h. The processes of interpolating between piecewise constant and rotated bilinear finite elements can be locally performed.

We allow either the full or constant interpolation on each macro-element. From theoretical considerations one prefers the full approach which corresponds to a linear/bilinear interpolation. This operator is chosen due to the well-known rule that the order of restriction and prolongation operators should be together at least two, corresponding to the order of the underlying differential operator. This requirement is not satisfied by the constant approach. Nevertheless we perform the piecewise constant interpolation, however in an adaptive setting. The criterion for deciding which local interpolation has to be taken is due to the local shape of the mesh, resp., the macro-element. In fact, if the actual macro-element or the neighboured macro-elements show large aspect ratios, or large jumps in the corresponding mesh areas/volumes, we apply the locally constant interpolation. This selection is directed by a user-defined threshold parameter.

Adaptive prolongation for rotated bilinear finite elements:

1. Calculate on all fine edges the corresponding interpolant values due to the given function v_{2h} which is defined via the coefficients u_1, u_2, u_3, u_4. There are two possibilities which may depend on the aspect ratios (AR) or the relation between the sizes (SR) of the actual macro-element and its neighboured cells.

2. If these quantities are below a certain threshold parameter, then perform the **full** (rotated bilinear) interpolation which reads:

   - for $H_h = H_h^a$ (mean-values on edges as d.o.f.'s)

   $$u_5 = u_1 - \frac{1}{4}u_2 + \frac{1}{4}u_4 \quad , \quad u_6 = \frac{5}{8}u_1 + \frac{1}{8}(u_2 + u_3 + u_4)$$

   - for $H_h = H_h^b$ (midpoints on edges as d.o.f.'s)

   $$u_5 = \frac{15}{16}u_1 - \frac{3}{16}u_2 - \frac{1}{16}u_3 + \frac{5}{16}u_4 \, ,$$
   $$u_6 = \frac{9}{16}u_1 + \frac{3}{16}u_2 + \frac{1}{16}u_3 + \frac{3}{16}u_4$$

3. If these quantities AR and SR are larger than a prescribed threshold parameter, then perform the **locally constant** interpolation which reads:

   $$u_5 = u_1 \quad , \quad u_6 = u_1$$

4. For the values on edges which belong to (macro) edges from T_{2h} (for instance u_5) take a mean-value by simply averaging or by taking into account the size of the two adjacent macro-elements as weighting factors.

Our numerical tests show that up to aspect ratios of size $10 - 100$ the full interpolation works fine, however for even higher mesh anisotropies the convergence fails while the **elementwise** application of the constant interpolation performs well. This knowledge is new (at least for us) and will be illustrated in the following numerical examples.

Construction of coarse grid operators:

In finite element approaches, the typical technique to derive the stiffness matrices corresponding to a given mesh is to apply the discrete variational formulation separately on each level. Let T_{h_i} be a given triangulation which is associated with the finite element space H_{h_i}. Further, denote by $\varphi_{h_i}^j$, $j = 1, \ldots, N(i)$, the corresponding basis functions. Then, in the **standard finite element approach** the matrix entries of the related stiffnes matrix A_{h_i} are determined via

$$A_{h_i}^{(k,l)} := a_{h_i}(\varphi_{h_i}^k, \varphi_{h_i}^l), \tag{3.208}$$

in which case $a_{h_i}(\cdot, \cdot)$ denotes the discrete bilinear form corresponding to the continuous differential operator. The same approach is performed in the context of the discrete Pressure–Poisson problems, in which case A_{h_i} is defined as

$$A_{h_i} := B_{h_i}^T M_{h_i,l}^{-1} B_{h_i}. \tag{3.209}$$

Another possibility which is often used in the multigrid context is the so-called **Galerkin approach** which is in fact equivalent to the described finite element technique for many finite element spaces (for example, for the conforming linear or bilinear spaces). Using the notation P_{2h}^h for the *prolongation* and R_h^{2h} for the *restriction matrix*, we can construct the coarse grid matrix A_{2h} depending on A_h via

$$A_{2h} := R_h^{2h} A_h P_{2h}^h. \tag{3.210}$$

These matrix operations can be performed locally (element by element) such that only singular rows/columns of A_h, resp., A_{2h} have to be treated, one after each other.

While both construction processes are identical for simple conforming elements and for problems with constant coefficients, there may arise differences if transport–diffusion operators with solution-depending coefficients are treated. Anyway, both techniques lead to completely different results if applied to the nonconforming finite elements. If the second approach (*Galerkin approach*) is performed with grid transfer matrices due to the full local interpolation, the construction process results in an expanding stencil for A_{2h}. This loss of sparsity is further continued if this approach is recursively applied

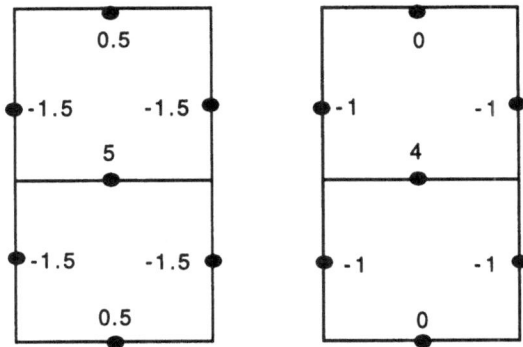

Figure 3.29: Canonical and modified matrix stencil

up to the lowest grid level. Hence, if the second order interpolation operators shall be utilized to obtain the full multigrid performance, one is restricted to apply the typical finite element approach.

However, analogously to the described situation of the adaptive prolongation which enforces us to perform locally the constant interpolation for large grid anisotropies, one should analogously modify the entries of the coarse grid matrix due to elements with large aspect ratios. As a result, if we apply the corresponding grid transfer operators based on this constant approach, the resulting matrix A_{2h} can be calculated via the *Galerkin approach*.

The typical effect is the following. While the stencil obtained by the usual finite element approach looks like in Figure 3.29 (left), the modified stencil is (at least on cartesian meshes) the usual 5-point stencil.

In fact, we perform these modifications of A_{2h} in the same way as described in the *adaptive prolongation algorithm*, by replacing only those rows of A_{2h} which are associated with edges which fail the threshold parameters in the adaptive selection process. Both approaches together, modifying the construction of the grid transfer and the coarse grid matrices via an adaptive switching between constant and full local interpolation, seem to guarantee the typical multigrid efficiency on very anisotropic meshes even. These statements contradict to some traditional experience, but they can be confirmed by the following numerical results.

Adaptive step-lenght control of the coarse grid correction:

There is another component in multigrid algorithms which is often needed to ensure a robust numerical behaviour, namely the step-lenght control of the correction.

$$u_k^{m+1} = u_k^m + \alpha_k I_{k-1}^k u_{k-1}^p . \tag{3.211}$$

The parameter α_k may be arbitrarily fixed or chosen adaptively so as to minimize the error $u_k^{m+1} - u_k$ in an appropriate norm. The original idea for this supplementary step comes from the treatment of nonconforming finite elements. In contrast to conforming finite elements which guarantee a best approximation result of the coarse grid solution u_{k-1}^p with respect to the energy norm (and hence allow $\alpha_k = 1$), this orthogonality relation is lost if nonconforming spaces are employed. However, the same failure may happen with conforming finite elements if full multigrid with approximative solution of the coarser problems only is applied such that a perturbed u_{k-1}^p is the result.

In an anlogous manner as for the nonlinear solvers, the optimal value α_k can be determined via energy minimization,

$$\alpha_k = \frac{(f_k - A_k u_k^m, I_{k-1}^k u_{k-1}^p)_k}{(A_k I_{k-1}^k u_{k-1}^p, I_{k-1}^k u_{k-1}^p)_k} , \tag{3.212}$$

or via minimization in the defect norm,

$$\alpha_k = \frac{(f_k - A_k u_k^m, A_k I_{k-1}^k u_{k-1}^p)_k}{(A_k I_{k-1}^k u_{k-1}^p, A_k I_{k-1}^k u_{k-1}^p)_k} , \tag{3.213}$$

In both cases, $(\cdot, \cdot)_k$ represents the euclidian scalar product on level k. While from a theoretical point of view the energy minimization is restricted to symmetric positive definite matrices A_k only, the defect minimization is applicable for (almost) all types of matrices. However, in contrast to the nonlinear iteration schemes in which case the numerical results show that the minimization in the defect norm is preferrable, linear multigrid seems to require the energy minimization. Even in the case of transport–diffusion problems with highly nonsymmetric matrices A_k, the strict application of the proposed energy minimization formulation often leads to improved results although the theoretical derivation of this formula is not correct.

We will demonstrate the importance of these three non-standard aspects,

- **adaptively** chosen grid transfer routines,

- **adaptive** construction of coarse grid operators,

- **adaptive** step-lenght control of the coarse grid correction,

by the following numerical examples. Together with appropriate smoothing operators, we can show that "Black Box" multigrid solvers for the auxiliary "inner" subproblems for the pressure and velocity components can be obtained such that together with the "outer" MPSC approach very efficient and robust solvers for discretized incompressible Navier–Stokes equations are possible.

3.4.3   Numerical examples for multigrid performance

We concentrate on the following aspects to confirm the theoretical considerations from the previous section:

1. **Smoothing:** How does the resulting multigrid convergence behaviour depend on the specified smoothers of Jacobi (JAC), Gauß–Seidel (GS), SOR and ILU type, with regard to the complexity of the computational mesh? What is the influence of certain renumbering strategies on the global convergence behaviour in Pressure–Poisson or transport–diffusion problems? How is the sensitivity of the multigrid rates with respect to large aspect ratios and complicated small-scale geometrical structures?

2. **Intergrid transfer:** What is the resulting difference if we strictly apply locally the full or constant interpolation in comparison with the adaptive grid transfer? Are the results different if we apply these techniques to pressure subproblems (piecewise constant spaces) or to velocity subproblems (rotated bilinear finite elements) on the other hand?

3. **Coarse grid matrix:** Does it pay to apply the adaptive construction of coarse grid matrices which uses the locally constant interpolation operator for the construction of certain rows in the matrix? Or is it absolutely necessary?

4. **Coarse grid correction:** What is the influence of the adaptive step-length procedure for controlling the calculated coarse grid correction? Does it work for transport–diffusion problems?

We perform test calculations for the same configurations as in Chapter 4 which contains examples for the computational solution of incompressible flow problems. Hence, this section can be seen as supplementary part which

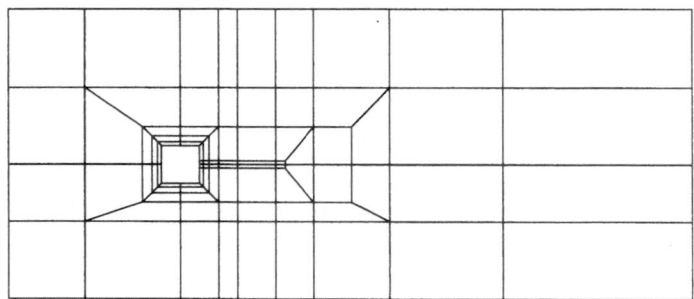

Figure 3.30: Typical (coarse) mesh S1 for linear multigrid tests

level	elements(for pressure)	edges (for velocity)
1	86	190
3	1,448	2,824
4	5,504	11,152
5	22,016	44,320

Table 3.24: Geometrical information and degrees of freedom for various levels

describes the efficient and robust multigrid solution of the resulting auxiliary subproblems in the context of the complete Navier–Stokes equations.

We start with the 'channel flow around a squared cylinder'. Figure 3.30 shows again the coarse mesh which can be modified by adapting the first inner line near to the square. The resulting *aspect ratios* are $AR = 10$ (this mesh, called S1), $AR = 10^3$ (called S2) and $AR = 10^5$ (called S3). The following configuration is identical with the framework in Section 4.1 and also 3.1.4 in which more details about the used flow simulation can be found.

We employ different viscosity parameters, such that in combination with the described nonlinear iterative techniques (see Section 3.2) our task here consists of the solution of linear convection–diffusion problems, resp., Poisson problems. In the following tables and diagrams, we use the following notation for classifying the performed solution methods; the precise explanation of the specific tools is described in the beginning of this Section. We do not only explain the applied methods, additionally we also provide the reader with some conclusions with respect to their properties: we define those multigrid components which have proven in all our test calculations to be candidates for the desired "Black Box" multigrid: for the auxiliary scalar subproblems in the velocity as well as the pressure component.

Notation and rating of the performed multigrid components:

1. **Applied smoothing operators: JACn/GSn/ILUn(s1,s2)**

   - **JACn(s1,s2)**: Jacobi iteration with damping (in most cases: $\omega^l \in [0.6, 0.8]$)

   - **GSn(s1,s2)**: SOR scheme, resp., Gauß–Seidel iteration (if $\omega^l = 1$)

   - **ILUn(s1,s2)**: Incomplete LU decomposition, with $\omega^l \in [0.8, 1.0]$ in most cases

     – 'n' denotes the renumbering strategy: '0' means **no** renumbering ("two level ordering", as usual in finite elements), while '1', resp.,'2' mean sorting the x, resp., the y–coordinate. Version '3' denotes a Cuthill-McKee-like method.

     – 's1' denotes the number of presmoothing steps, 's2' the post-smoothing steps.

     – in all cases we perform the F–cycle.

   - Candidates for the **"Black Box"** solver: **ILUn(0,s2)**

We recommend to use the Gauß–Seidel iteration as smoother (typically 2 – 8 smoothing steps!) for the pressure as well as the velocity subproblems as long as the mesh is more or less isotropic (aspect ratios smaller than 10 – 50!), with any numbering of the unknowns. Only in the case of convection–dominated problems, streamwise–renumbering techniques should be applied to improve essentially the convergence behaviour.

For meshes containing cells with even larger aspect ratios, the ILU scheme should be applied as smoother (mostly with 1 – 4 smoothing steps, slightly damped; in most cases we prefer postsmoothing only). However, for the use of ILU it is absolutely necessary to renumber appropriately the unknowns! In the case of convection–dominated problems, again the streamwise numbering might be optimal. However, for diffusive problems, resp., for the pressure subproblems, certain renumbering strategies (see [106]) have to be applied, too. By computational reasons, the Cuthill-McKee-like methods may be the favourized schemes!

2. **Applied grid transfer operators: INT(type)(eps)**

- 'type' denotes the kind of interpolation, for velocity, resp., pressure functions:

  - <u>velocity:</u> 'type=c' means locally constant, 'type=l' means rotated bilinear interpolation, but with simply averaging the coarse edge values. In contrast, 'type=f' denotes the bilinear interpolation with weighting factors according to the element sizes (corresponds to the idea of L^2-projection). We note that **INT(c) = INT(l)(0) = INT(f)(0)**

  - <u>pressure:</u> 'type=c' means locally constant, 'type=f' denotes the rotated bilinear interpolation with weighting factors according to the element sizes. This procedure requires first the interpolation from the constant into the rotated bilinear space, and vice versa. Both constant–linear interpolation processes can be easily applied on local element level.

  - in the case 'type=c', the value for 'eps' is obsolete.

- 'eps' denotes the threshold parameter for adaptively switching between elementwisely constant and rotated bilinear grid transfer; that means: for elements, resp., edges belonging to elements with aspect ratios larger than the parameter 'eps', we use the locally constant interpolation, otherwise the rotated bilinear version.

- Candidates for the **"Black Box"** solver: **INT(f)(50)**, resp., **INT(f)(0)**

We recommend for the velocity subproblems to use the rotated bilinear interpolation ('**type=f**') with weighting factors according to the element sizes. This interpolation can be interpreted as a special variant of a discrete L^2-projection between coarse and fine mesh. While the purely constant interpolation ('c') leads to much worse results in almost all cases, the special weighting factors ('f') guarantee the typical convergence rates even on very anisotropic meshes, in contrast to the simple averaging process ('l') !

For the pressure subproblems, the decision between constant or rotated bilinear grid transfer is not so clear. The local interpolation to the rotated bilinear ansatz spaces is clearly preferrable, but then the simple constant interpolation for the bilinear functions leads already to excellent results (version **INT(f)(0)**).

Working on highly anisotropic meshes, the adaptive switching between locally constant and fully bilinear interpolation on element level seems to be a must for the nonconforming rotated bilinear finite elements! Numerical tests show that the threshold parameter 'eps' should be chosen in the range of 'eps' $\in [50, 100]$!

3. **Construction of coarse grid matrices: MAT(eps)**

   - 'eps' denotes the threshold parameter for adaptively switching between the typical finite element approach and the locally rowwise Galerkin-type construction of the coarse grid matrix for the velocity subproblems; that means: for elements, resp., edges belonging to elements with aspect ratios larger than the parameter 'eps', we use the locally constant interpolation procedure for the calculation of specific rows in the coarse grid matrix. Otherwise the standard finite element approach as described above is employed (that means: no modification!).

   - For the pressure matrices $(P := B^T M_l^{-1} B)$, due to the piecewise constant ansatz functions, we can always apply the standard finite element approach.

   - Candidates for the **"Black Box"** solver: **MAT(50)**

The numerical tests show the "surprising(?)" result that the adaptive switching between both described construction processes is absolutely necessary, if the rotated bilinear nonconforming finite elements are utilized. All calculations indicate that analogously to the grid transfer process a threshold parameter of size 'eps' $\in [50, 100]$ should be used!

4. **Coarse grid correction: CORR(type)**

- 'type' denotes the kind of coarse grid correction:
  - 'type=adap1' means the adaptive calculation of the damping factor for the coarse grid correction via the (discrete) energy functional.
  - 'type=adap2' denotes the adaptive calculation of the damping factor for the coarse grid correction via the (discrete) defect norm.
  - 'type=value' denotes the precribed setting (via 'value') of the corresponding damping factor for the coarse grid correction.

- Candidates for the **"Black Box"** solver: CORR(adap1)

We recommend the use of the adaptive control for the damping parameter via optimization by the discrete energy functional. Even for nonsymmetric velocity matrices, but generated with streamline–diffusion or upwinding for stabilization, this adaptive process may help to stabilize the multigrid run. The result is **not** that the convergence rates dramatically improve, but the typical quality is preserved for almost arbitrary meshes and parameter configurations, without the need for determining manually the "right" damping factors!

If we summarize these recommendations, our final "Black Box" multigrid solver for the auxiliary velocity problems (nonconforming rotated bilinear finite element) has the form

$$\text{ILU}_{ren}(0,\text{s2})\text{-INT(f)(50)-MAT(50)-CORR(adap1)}$$

while the preferred combination for the Pressure–Poisson problems (piecewise constant ansatz function) is:

$$\text{ILU}_{ren}(0,\text{s2})\text{-INT(f)(0)-CORR(adap1)}$$

In both combinations, the value for 's2' is typically 2 or 4. The aim of the following calculations is to "prove" these recommendations. Hereby, the reader should always keep in mind the meaning of being a "Black Box multigrid solver" in our context:

1. In many cases, the emphasized combinations lead to the best convergence results and optimal computational cost, or the results are at least not far away from the best possible.

2. Our preferred combinations lead in (almost) all cases to a robust and efficient convergence behaviour, **without** the need for calibrating manually the multigrid parameters for each configuration!

We start with the hardest case – calculations on mesh S3 with maximum aspect ratios $AR = 10^5$ – and demonstrate first that **only** the described combinations lead to satisfying results. In addition, we show "negative" results for other combinations. The results for this extreme mesh are followed by calculations on the more isotropic meshes S2 and particularly S1 which allow the efficient multigrid treatment with almost all defined combinations of multigrid components. However, the reader should keep in mind that even mesh S1 is still quite anisotropic ($AR = 10$), compared to other typical meshes which are usually employed in numerical simulations. This explains the somewhat "bad" convergence behaviour for the Jacobi and Gauß–Seidel scheme which may further improve on more regular meshes.

The following Tables present the observed multigrid convergence rates for a sequence of test calculations. To be precise, we perform the following settings:

- We use the control via the weighted finite element residuals (see Section 3.3) for $H_h^{(a)}$.

- The stopping criterion is $\epsilon_{abs} = 10^{-5}$. Additionally, we prescribe to "gain three digits", that means $\epsilon_{rel} = 10^{-3}$.

- We apply a total number of 50 linear multigrid sweeps in the maximum.

- We apply the F–cycle in all computations.

The following multigrid rates are averaged rates, for a specific test configuration, to examine the linear convergence behaviour. Therefore, the resulting rates are often somewhat pessimistic, particularly compared with the "real" convergence behaviour in the complete Navier–Stokes context. This may explain the fact that we seem to be satisfied by rates of order $\rho_{mg} \sim 0.3 - 0.4$,

ILU1(0,s2)			
MG-LEVEL	s2=2	s2=4	s2=8
3	0.36	0.23	0.17
4	0.39	0.23	0.17
5	0.43	0.24	0.18
GS0(0,s2)			
MG-LEVEL	s2=4	s2=16	s2=64
3	0.74	0.33	0.19
4	0.94	0.75	0.33
5	0.97	0.95	0.76
JAC0(0,s2)			
MG-LEVEL	s2=16	s2=64	s2=256
3	0.61	0.25	0.17
4	0.93	0.63	0.24
5	0.96	0.93	0.64

Table 3.25: MG results for Poisson problem (rotated bilinear velocity) on mesh S3 with adaptive grid transfer components **INT(f)(50)-MAT(50)-CORR(adap1)**

or that a doubling of smoothing steps does not lead to halved rates: The reason is simply that often the initial values were chosen in such a way that (unfortunately) the first iteration increased the residual, before the typical multigrid damping was started. As pointed out, in the complete Navier-Stokes context with applications in the MPSC schemes, the analogous linear multigrid rates are often much better.

First we start with examining the multigrid solution of Poisson problems, discretized with the nonconforming rotated bilinear finite elements. We show the results for our "Black Box" multigrid candidate **ILU1(0,s2)-INT(f)(50)-MAT(50)-CORR(adap1)** - that means we apply our locally adaptive grid transfer and coarse grid techniques - and we concentrate first on the choice of different smoothing operators and different numbers of smoothing steps.

As can be seen, only ILU as smoother leads to "good" convergence rates while in contrast both Jacobi and Gauß–Seidel schemes show two typical deficiencies:

1. the multigrid rates are "bad".

2. the convergence rates deteriorate with increasing the number of unknowns, and require a coupling between the number of smoothing steps and the mesh level.

For both schemes (GS, JAC), multigrid is far away from being optimal in such a sense that "good" convergence rates are obtained independent of the

	ILU1(0,s2)					
MG-LEVEL	s2=2	s2=4	s2=8	s2=2	s2=4	s2=8
3	–	0.21	–	–	0.23	–
4	–	0.23	–	–	0.25	–
5	–	0.25	–	–	0.31	–
	GS0(0,s2)					
MG-LEVEL	s2=4	s2=16	s2=64	s2=4	s2=8	s2=16
3	0.71	0.29	0.18	0.57	0.34	0.24
4	0.94	0.71	0.29	0.67	0.43	0.26
5	0.99	0.94	0.71	0.74	0.55	0.31
	JAC0(0,s2)					
MG-LEVEL	s2=16	s2=64	s2=256	s2=16	s2=32	s2=64
3	0.57	0.23	0.14	0.36	0.25	0.21
4	0.88	0.58	0.20	0.41	0.27	0.23
5	0.98	0.88	0.59	0.56	0.28	0.24

Table 3.26: MG results for Poisson problem (velocity) on meshes S2 (left) and S1 (right)

mesh level. These results are also confirmed by the following Table 3.26 which shows analogous results on the meshes S1 and S2. As explained above, the reader should keep in mind that even mesh S1 is rather anisotropic (aspect ratio $AR = 10$). In contrast, the observed bad convergence behaviour for the Jacobi and Gauß–Seidel schemes further improves on more regular meshes.

Additionally, we can demonstrate the influence of renumbering strategies. While the Jacobi iteration is by construction independent of the kind of numbering, this is not clear for the Gauß–Seidel iteration. However, as already discussed in [106], the convergence rates for Poisson–like problems are (more or less) independent of the numbering (compare the results for **GS1(0,4)** in Table 3.27 with **GS0(0,4)** in Table 3.25). For Poisson–like problems, too, the strategy of renumbering the unknowns in combination with ILU has no great influence, as long as any "clever" renumbering is performed at all (it is evident that for special cases as tridiagonal matrices certain strategies are optimal, but here we discuss the more general framework!). However, as Table 3.27 shows, **no renumbering**, i.e., *two level ordering* as in **ILU0(0,4)**, deteriorates all nice properties of ILU smoothing. Hence, for the subsequent calculations we restrict us to the smoother **ILU1(0,4)**.

Further, Tables 3.27 and 3.28 show the results if we modify the grid transfer (**INT**), the coarse grid operator (**MAT**) and the coarse grid control (**CORR**). As can be seen – we work with fully weighted interpolation operators **INT(f)**, but vary the threshold parameters to change between locally constant and rotated bilinear interpolation – the choice of "too small" parameters (0 or 5), and particularly "too large" parameters ('∞' means that we always apply the fully bilinear interpolation) leads to much worse results, if we compare with **INT(f)(50)-MAT(50)**. Additionally, the following results

INT(f)()-CORR(adap1)

	ILU0(0,4) INT(50) MAT(50)	GS1(0,4) INT(50) MAT(50)	ILU1(0,4) INT(0) MAT(0)	ILU1(0,4) INT(5) MAT(5)	ILU1(0,4) INT(∞) MAT(∞)	ILU1(0,4) INT(50) MAT(∞)	ILU1(0,4) INT(∞) MAT(50)
3	0.48	0.72	0.67	0.49	0.92	0.92	0.92
4	0.81	0.94	0.80	0.68	0.98	0.98	0.97
5	0.94	0.97	0.83	0.77	0.99	0.99	0.91

Table 3.27: MG results for Poisson problem (velocity) on mesh S3 for different parameters

ILU1(0,4)-INT(f)()

	INT(50) MAT(50) CORR(0.8)	INT(50) MAT(50) CORR(1.0)	INT(0) MAT(0) CORR(1.0)	INT(5) MAT(5) CORR(1.0)	INT(∞) MAT(50) CORR(0.8)
3	0.21	0.09	0.68	0.40	0.57
4	0.26	0.12	0.81	0.62	0.42
5	0.31	0.18	0.87	0.79	0.48

ILU1(0,4)-INT(l)(50)-MAT(50)

	CORR(0.6)	CORR(0.8)	CORR(1.0)	CORR(adap1)
3	0.51	0.57	0.69	0.92
4	0.60	0.41	0.50	0.97
5	0.68	0.48	0.85	0.91

Table 3.28: MG results for Poisson problem (velocity) on mesh S3 for different configurations

in Table 3.28 concern the case of adaptive step length control of the correction (**CORR(adap1)**) or not (**CORR(0.8)**, resp., **CORR(1.0)**). In combination with the fully adaptive intergrid operator **INT(f)(50)-MAT(50)**, the adaptive step–length control is only slightly worse, but promises more robustness at the same time. The prescribed setting of the damping parameter (0.8 or 1.0) seems to improve the results for other intergrid operators **INT(f)()-MAT()**, but this improvement is not significant. Finally, we also show the corresponding results for the applied grid transfer operator **INT(l)()** which uses simply averaging on macro–edges only.

As a first conclusion, we can state the following results for the nonconforming rotated bilinear FEM spaces, at least for Poisson–like problems:

- **ILU** with renumbering is a good candidate for smoothing. In most cases, we prefer the case of postsmoothing only (**ILU1(0,s2)**).

- As grid transfer operator the **adaptive fully weighted** interpolation **INT(f)(eps1)** should be used (with 'eps1' denoting the threshold parameter for local interpolation).

- As coarse grid operator the **locally adaptive** matrix **MAT(eps2)** should be taken.

Mesh S2:	ILU1(0,4)-INT()()-CORR(adap1)					
	INT(f)(0) MAT(0)	INT(f)(50) MAT(50)	INT(f)(∞) MAT(∞)	INT(l)(0) MAT(0)	INT(l)(50) MAT(50)	INT(l)(∞) MAT(∞)
3	0.69	0.21	0.90	0.69	0.33	0.91
4	0.81	0.23	0.91	0.81	0.29	0.91
5	0.86	0.25	0.94	0.86	0.29	0.93
Mesh S1:	ILU1(0,4)-INT()()-CORR(adap1)					
	INT(f)(0) MAT(0)	INT(f)(50) MAT(50)	INT(f)(∞) MAT(∞)	INT(l)(0) MAT(0)	INT(l)(50) MAT(50)	INT(l)(∞) MAT(∞)
3	0.69	0.23	0.23	0.69	0.23	0.23
4	0.81	0.25	0.25	0.81	0.25	0.25
5	0.86	0.31	0.31	0.86	0.31	0.31

Table 3.29: Some results for Poisson problem (velocity) on meshes S1 and S2

- In both cases, the parameters 'eps1' and 'eps2' should be of size 50 – 100. Smaller threshold values lead to essentially worse results, while larger values, resp., the consequent application of the bilinear interpolation only in the limit case, may even lead to divergence.

- The adaptive step–length control **CORR(adap1)** works only in combination with **INT(f)(50)-MAT(50)**. The results have not to be better than compared with fixed parameter settings (**CORR(1.0)** or **CORR(0.8)**), but they are determined in an adaptive framework without any a priori knowledge which is essential if problems with solving on coarser levels should occur!

We finish this first part - numerical results for Poisson–like problems - with calculations on the more regular meshes S1 and S2 (see Table 3.29). On mesh S2 (aspect ratios $AR \leq 1,000$) most of the conclusions remain valid; it is "new" that combination **INT(l)(50)-MAT(50)** (simply averaging instead of weighted averaging on macro edges!) works fine in combination with the adaptive step–length control process **CORR(adap1)**.

Looking at the analogous calculations on mesh S1 (aspect ratios $AR = 10$), we see that almost all combinations lead to satisfying results. Only the choice of constant interpolation operators (**INT(c)=INT(l)(0)=INT(f)(0)**, **MAT(0)**) leads to worse results: their application seems to lead to mesh independent, but unfortunately to "bad" convergence results in multigrid.

Next, we perform similar tests for convection–diffusion problems, with varying viscosity parameters ν,

$$-\nu\Delta u + \mathbf{U} \cdot \nabla u = \text{ right hand side},$$

with the given vector $\mathbf{U}$ which is taken from the solution of a corresponding Stokes problem. Both, $\mathbf{U}$ and the diameter of the domain Ω, are of order 1, such that the viscosity parameter ν can be identified with the Reynolds number in the complete Navier–Stokes model. We neglect the case of an additional reactive term αu which typically arises from a time discretization process. Since the resulting complete stiffnes matrix is modified by a positive mass matrix which may be assumed to be even diagonal (*lumping*), the corresponding convergence results essentially improve in general.

For discretizing the convective term, we either apply the upwinding variants UPW-∞ (simple first order upwind) and UPW-0.1 (adaptively damped upwinding, see Section 3.1.4), or stabilization via streamline–diffusion, SD-0.5 (again see Section 3.1.4). As a result from the previous tests, we use the locally adapted intergrid operators **INT(f)(50)-MAT(50)** in the following tests. Then, the remaining main topics which we examine are:

1. How do certain numbering strategies for the GS and the ILU smoothing influence the multigrid rates? Are there differences between upwinding and streamline–diffusion?

2. How does the quality of the stabilization technique (the more sophisticated techniques UPW-0.1 and SD-0.5 in contrast to the simpler, but more stable UPW-∞) influence the algebraic properties, that means the multigrid convergence behaviour?

3. Does the adaptive step–length control **CORR(adap1)** works as minimization process, even for nonsymmetric matrices which allow no rigorous justification?

4. And finally: Are "Black Box" multigrid components possible for discrete convection–dominated problems, independent of the viscosity parameter and the shape of the mesh?

We start in Table 3.30 with results on mesh S1, hereby examining carefully the difference between both upwind variants UPW-∞ and UPW-0.1, with respect to the viscosity parameter ν, the kind of numbering in Gauß–Seidel smoothing and the influence of the step–length control **CORR()**. We always apply the adaptive grid transfer techniques **INT(f)(50)-MAT(50)**. **GS1** corresponds to "renumbering from left to right" in contrast to the other described possibilities ("two level" (**0**), "up to down" (**2**) and Cuthill-McKee (**3**)).

The results from Table 3.30 can be summarized as follows:

1. In combination with the first order upwinding UPW-∞ which potentially may lead to M–matrices, resp., to lower triangular matrices in the

UPW-∞

	GS1(0,4) $1/\nu = 10^2$	GS1(0,4) $1/\nu = 10^4$	GS1(0,4) $1/\nu = 10^6$	ILU1(0,2) $1/\nu = 10^2$	ILU1(0,2) $1/\nu = 10^4$	ILU1(0,2) $1/\nu = 10^6$
3	0.32	0.12	0.10	0.21	0.07	0.17
4	0.43	0.21	0.14	0.32	0.08	0.15
5	0.51	0.29	0.22	0.48	0.18	0.20

UPW-0.1

	GS1(0,4) $1/\nu = 10^2$	GS1(0,4) $1/\nu = 10^4$	GS1(0,4) $1/\nu = 10^6$	ILU1(0,2) $1/\nu = 10^2$	ILU1(0,2) $1/\nu = 10^4$	ILU1(0,2) $1/\nu = 10^6$
3	0.36	0.09	0.09	0.24	0.23	0.13
4	0.46	0.61	0.62	0.35	0.12	0.11
5	0.53	div	div	0.49	0.23	0.16

UPW-∞   + $1/\nu = 10^6$

	GS0(0,4) CORR(adap1)	GS1(0,4) CORR(adap1)	GS2(0,4) CORR(adap1)	GS3(0,4) CORR(adap1)	GS1(0,4) CORR(1.0)
3	0.35	0.10	0.58	0.40	div
4	0.28	0.14	0.68	0.46	0.36
5	0.36	0.22	0.77	0.48	0.24

UPW-0.1   + $1/\nu = 10^6$

	GS0(0,4) CORR(adap1)	GS1(0,4) CORR(adap1)	GS2(0,4) CORR(adap1)	GS3(0,4) CORR(adap1)	GS1(0,4) CORR(1.0)
3	0.43	0.09	0.59	0.49	0.22
4	0.52	0.61	0.98	0.64	0.42
5	0.62	div	div	0.90	div

Table 3.30: MG results for convection–diffusion problem (velocity) on mesh S1

limit case of vanishing diffusion, the Gauß–Seidel scheme with renumbering "from left to right" (according to the Stokes flow **U**) leads to excellent multigrid rates which even further improve for decreasing viscosity! However, in the case of the more accurate upwind discretization UPW-0.1 the excellent algebraic properties are weakened, and due to the loss of lower triangularity, the ILU smoother is a must!

2. As expected leads the GS1 smoother (that means with renumbering "from left to right") in the case of dominating convection to the best results, if the upwind scheme UPW-∞ is applied. However, due to the more diffusive character of the "better" scheme UPW-0.1, this property is lost for the more accurate discretization UPW-0.1.

3. The adaptive step–length control seems to work well, even for nonsymmetric matrices!

Next, we show the results for the Gauß–Seidel scheme if convection-dominated problems come together with anisotropic meshes (large aspect ratios, for instance for the resolution of boundary layers!). While for moderate meshes the stiffness matrices tend to lower triangular matrices, at least in combination with special upwinding and renumbering, the limit case of

Mesh S1

	UPW-∞ $1/\nu = 10^2$	UPW-∞ $1/\nu = 10^4$	UPW-∞ $1/\nu = 10^6$	UPW-0.1 $1/\nu = 10^2$	UPW-0.1 $1/\nu = 10^4$	UPW-0.1 $1/\nu = 10^6$
3	0.32	0.12	0.10	0.35	0.69	0.08
4	0.43	0.21	0.14	0.46	div	0.61
5	0.51	0.29	0.22	0.53	div	0.90

Mesh S2

	UPW-∞ $1/\nu = 10^2$	UPW-∞ $1/\nu = 10^4$	UPW-∞ $1/\nu = 10^6$	UPW-0.1 $1/\nu = 10^2$	UPW-0.1 $1/\nu = 10^4$	UPW-0.1 $1/\nu = 10^6$
3	0.60	0.56	0.41	0.69	0.75	0.42
4	0.90	0.86	0.77	0.92	div	0.86
5	0.98	0.96	0.92	0.98	div	div

Mesh S3

	UPW-∞ $1/\nu = 10^2$	UPW-∞ $1/\nu = 10^4$	UPW-∞ $1/\nu = 10^6$	UPW-0.1 $1/\nu = 10^2$	UPW-0.1 $1/\nu = 10^4$	UPW-0.1 $1/\nu = 10^6$
3	0.63	0.62	0.60	0.75	0.81	0.61
4	0.92	0.92	0.89	0.95	div	0.90
5	0.98	0.97	0.97	div	div	div

Table 3.31: MG results for convection–diffusion problem on all meshes for the GS–smoother (**GS1(0,4)-INT(f)(50)-MAT(50)-CORR(adap1)**)

UPW-0.1   +   CORR(adap1)

	ILU1(0,2) $1/\nu = 10^2$	ILU1(0,2) $1/\nu = 10^4$	ILU1(0,2) $1/\nu = 10^6$	ILU1(0,4) $1/\nu = 10^2$	ILU1(0,4) $1/\nu = 10^4$	ILU1(0,4) $1/\nu = 10^6$
3	0.25	0.22	0.05	0.15	0.04	0.01
4	0.35	0.27	0.09	0.19	0.10	0.01
5	0.42	0.22	0.19	0.20	0.10	0.04

UPW-0.1   +   ILU1(0,4)

	CORR(0.8) $1/\nu = 10^2$	CORR(0.8) $1/\nu = 10^4$	CORR(0.8) $1/\nu = 10^6$	CORR(1.0) $1/\nu = 10^2$	CORR(1.0) $1/\nu = 10^4$	CORR(1.0) $1/\nu = 10^6$
3	0.15	0.02	0.01	0.05	0.01	0.01
4	0.20	0.09	0.01	0.11	0.03	0.01
5	0.23	0.16	0.04	0.18	0.08	0.02

Table 3.32: MG results for convection–diffusion problem on mesh S3 with ILU–smoothing

highly anisotropic meshes, as for instance S3 ($AR = 10^5$), includes tridiagonal matrices! However, as can be seen in Table 3.31, then the standard Gauß–Seidel schemes may fail as smoother!

In contrast to Gauß–Seidel smoothing, the ILU scheme solves exactly in the limit case both lower triangular as well as tridiagonal matrices (see Table 3.32). While the case of balanced diffusive and convective parts ($1/\nu = 10^2$) seems to be the hardest (4 smoothing steps), the convergence rates improve for convective dominance (with the right numbering!) such that 2 smoothing steps are sufficient. It is remarkable that even in this highly nonsymmetric case the adaptive step–length control of the coarse grid correction **CORR(adap1)** works fine.

		ILU1(0,4) $1/\nu = 10^2$	ILU1(0,4) $1/\nu = 10^4$	ILU2(0,4) $1/\nu = 10^2$	ILU2(0,4) $1/\nu = 10^4$
	3	0.13	div	0.10	0.33
Mesh S1:	4	0.15	div	0.15	0.31
	5	0.18	div	0.18	0.39
	3	0.13	div	0.07	0.23
Mesh S2:	4	0.15	0.45	0.12	0.28
	5	0.20	0.38	0.19	0.36
	3	0.14	div	0.10	0.24
Mesh S3:	4	0.16	0.45	0.16	0.29
	5	0.18	0.64	0.21	0.36

Table 3.33: MG results for convection–diffusion problem (velocity) for SD-0.5

Next, we show analogous results if streamline–diffusion techniques are applied to discretize the convective term (here: SD-0.5, see also Section 3.1.4). While the upwinding variants UPW lead in the limit case of "pure" convection to lower triangular matrices if we renumber from "left to right" (**ILU1**), the streamline–diffusion approaches lead to tridiagonal matrices, but with numbering from "up to down" (**ILU2**). The corresponding results, for all meshes S1 – S3 and different viscosity parameters, can be found in Table 3.33. They clearly show the need for appropriate renumbering strategies, well adapted to the performed discretization scheme.

The analogous tests as for the velocity subproblems (neglecting the convective terms) can be performed for the Pressure–Poisson equation. And in fact, the results are:

1. Jacobi and Gauß–Seidel smoothing for Pressure–Poisson problems lead to satisfying results on regular meshes only (moderate aspect ratios!). While their convergence rates are independent of the kind of numbering, ILU smoothing requires again any kind of "clever" numbering of the unknowns. We will show in the following Tables, that then the ILU smoothing (in combination with appropriate grid transfer) leads again to excellent multigrid rates, on quite arbitrary meshes.

2. If we apply rotated bilinear interpolation (after interpolating the pressure variable from the constant into the bilinear space, and vice versa), we should use again the fully weighted version **INT(f)(eps)** with a certain threshold parameter 'eps' to switch locally between constant and bilinear interpolation. However, we have to test again for several values for 'eps' and to compare with the "natural" constant grid transfer **INT(c)** for the pressure.

3. We always apply the explicit product formula $P := B^T M_l^{-1} B$ for calculating the Pressure–Poisson matrix, for each grid level separately.

	ILU3(0,2) INT(c) adap1	ILU3(0,2) INT(c) 1.0	ILU3(0,2) INT(f)(0) adap1	ILU3(0,2) INT(f)(0) 1.0	ILU3(0,2) INT(f)(50) adap1	ILU3(0,2) INT(f)(50) 1.0
3	0.67	0.56	0.39	0.33	0.38	0.39
4	0.69	0.57	0.30	0.31	0.41	0.41
5	0.70	0.59	0.28	0.29	0.42	0.42
	ILU3(0,4) INT(c) adap1	ILU3(0,4) INT(c) 1.0	ILU3(0,4) INT(f)(0) adap1	ILU3(0,4) INT(f)(0) 1.0	ILU3(0,4) INT(f)(50) adap1	ILU3(0,4) INT(f)(50) 1.0
3	0.59	0.30	0.17	0.12	0.20	0.16
4	0.61	0.32	0.11	0.10	0.17	0.17
5	0.63	0.33	0.10	0.10	0.18	0.18

INT(f)(50)-CORR(adap1)			
	ILU1(0,4)	ILU2(0,4)	ILU3(0,4)
3	0.17	0.18	0.20
4	0.16	0.17	0.17
5	0.18	0.18	0.18

Table 3.34: MG results for Pressure–Poisson problem on mesh S3

4. The minimization property of the adaptive step–length control **CORR(adap1)** is justified since the stiffness matrix is positive (semi-) definite.

Table 3.34 shows the results on the hardest mesh S3 for different smoothing steps, for three grid transfer operators **INT(c)**, **INT(f)(0)** and **INT(f)(50)** and for adaptive step-length control (**CORR(adap1)**) versus fixed setting (**CORR(1.0)**). Additionally, we provide the reader with a comparison of different renumbering strategies (version **0 - 3** as before).

While the purely constant interpolation operator **INT(c)** leads obviously to the worst results, the use of components from the bilinear interpolation process indicates clear improvements. However, it is surprising that the simpler variant **INT(f)(0)**, which first interpolates the piecewise constant pressure functions into the space of rotated bilinear functions and then uses constant interpolation for the bilinear finite elements, gives the best convergence results, and hence the best computational efficiency rates, too.

The kind of numbering is not so decisive; it is more important to apply any clever renumbering in contrast to the usual "two level ordering" at all. In the same way, adaptive step–length control via **CORR(adap1)** leads to no improvement, but also no deterioration in comparison to a manually fixed value (**CORR(1.0)**). Due to the small additional numerical cost and our positive experience with this adaptive tool, we suggest to apply this additional control to ensure a wide range of robustness. But this recommendation is not mandatory!

		ILU3(0,2) INT(c) adap1	ILU3(0,2) INT(c) 1.0	ILU3(0,2) INT(f)(0) adap1	ILU3(0,2) INT(f)(0) 1.0	ILU3(0,2) INT(f)(50) adap1	ILU3(0,2) INT(f)(50) 1.0
	3	0.61	0.49	0.18	0.18	0.29	0.28
Mesh S1:	4	0.66	0.57	0.21	0.21	0.36	0.36
	5	0.69	0.62	0.25	0.25	0.44	0.44
	3	0.66	0.55	0.28	0.28	0.38	0.37
Mesh S2:	4	0.67	0.57	0.25	0.26	0.39	0.39
	5	0.69	0.59	0.26	0.25	0.44	0.44
	3	0.67	0.56	0.39	0.33	0.38	0.39
Mesh S3:	4	0.69	0.57	0.30	0.31	0.41	0.41
	5	0.70	0.59	0.28	0.29	0.42	0.42

Table 3.35: MG results for Pressure–Poisson problem on all meshes for different grid transfers

To come to an end, we show the corresponding results for all meshes S1 – S3 which demonstrate the robust behaviour of our preferred combinations as Pressure-Poisson solver

$$\boxed{\text{ILU3(0,s2)-INT(f)(0)-CORR(adap1)}}$$

respectively,

$$\boxed{\text{ILU3(0,s2)-INT(f)(0)-CORR(1.0)}}$$

with respect to local mesh anisotropies.

We finish this section, concerning our numerical experience with (linear) multigrid for velocity and pressure subproblems in the framework of the incompressible Navier–Stokes equations, with some test calculations on a unit square, but with anisotropic mesh refinement near the boundary. These simulations are thought to be representative for *driven cavity*-like configurations, resp., for situations where strongly nonuniform refinement in normal direction to the walls is required. In contrast to the previous 'flow in channel around obstacle' configuration, here we miss a characteristic flow direction ("from left to right") which causes additional difficulties for the applied smoothers.

Figure 3.31 shows the coarse meshes for configuration DC0, and also for DC1 which is generated from DC0 by moving all interior and boundary points into

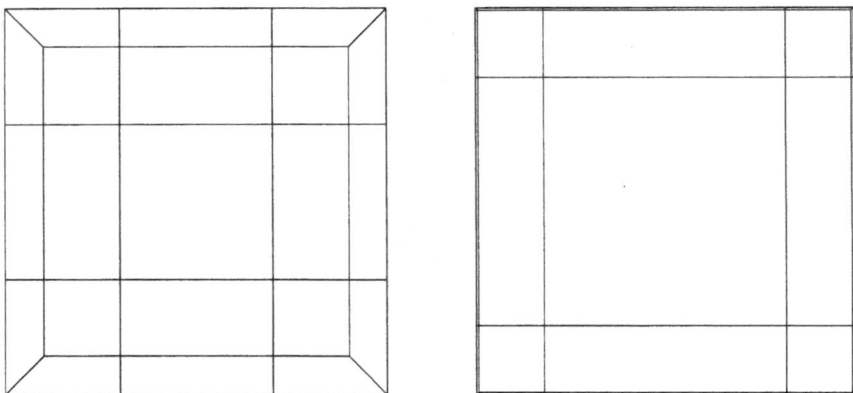

Figure 3.31: Coarse meshes DC0 and DC1 for linear multigrid tests

level	elements(for pressure)	edges (for velocity)
1	21	48
4	1,344	2,736
5	5,376	10,848
6	21,504	43,200

Table 3.36: Geometrical information and degrees of freedom for various levels

direction of the corners. While mesh DC0 can be called to be "regular", with very weak alignement of the grid points towards the boundary lines, mesh DC1 has to handle grid cells with a maximum aspect ratio $AR = 100$. We omit to show graphically the grids DC2 ($AR = 10^4$) and DC3 ($AR = 10^6$) which are thought to simulate the case of extreme mesh adaptation in order to approximate boundary layers for high Reynolds numbers.

Again, we treat a sequence of Poisson, resp., Pressure–Poisson problems, followed by linear convection–diffusion equations with different viscosity parameters ν, discretized via our preferred upwinding UPW-0.1, resp., via the streamline–diffusion approach SD-0.5.

The aim of the following calculations is to demonstrate the ability of our introduced parameter combinations for the "Black Box" multigrid solver, namely for the velocity problems:

ILU3(0,s2)-INT(f)(50)-MAT(50)-CORR(adap1)

		ILU3(0,2) INT(c)	ILU3(0,2) INT(f)(0)	ILU3(0,2) INT(f)(50)	ILU3(0,4) INT(f)(0)
	4	0.42	0.13	0.22	0.04
Mesh DC0:	5	0.47	0.16	0.27	0.04
	6	0.50	0.17	0.29	0.04
	4	0.48	0.26	0.30	0.08
Mesh DC1:	5	0.56	0.35	0.41	0.12
	6	0.68	0.49	0.58	0.24
	4	0.53	0.28	0.37	0.09
Mesh DC2:	5	0.56	0.31	0.39	0.09
	6	0.56	0.31	0.41	0.10
	4	0.62	0.37	0.45	0.14
Mesh DC3:	5	0.64	0.38	0.48	0.15
	6	0.66	0.40	0.51	0.18

Table 3.37: MG results for Pressure–Poisson problem on meshes DC10 – DC13

respectively for the pressure subproblems:

> **ILU3(0,s2)-INT(f)(0)-CORR(1.0)**

In contrast to the previous tests, we now perform the **ILU3** scheme (with a Cuthill-McKee-like technique for sorting the unknowns; however there is no significant difference to the other described numbering strategies) for treating the auxiliary velocity problems. Further, we use in the following tests for the Pressure–Poisson problem the non–adaptive version of the step–length control **CORR(1.0)** since most tests have demonstrated that the differences compared to **CORR(adap1)** are neglegible for this kind of problem.

In the following tables, we utilize the same notation as before, and we start with considering the results for the Pressure–Poisson problem on all meshes DC0 – DC3, hereby varying the grid transfer routines **INT** and the number of smoothing steps.

Again, the combination **ILU** smoothing (2 or 4 smoothing steps) with the mixed grid transfer operator **INT(f)(0)** leads to excellent convergence rates which even seem to be independent from the local anisotropies in the employed meshes DC0 – DC3 (keep in mind that the aspect ratio changes from $AR = 4$ to $AR = 10^6$!).

In our final test sequence, we perform analogous calculations for the rotated bilinear velocity ansatz functions, for Poisson as well as for convection–

Poisson problem

	DC0	DC1	DC2	DC3
4	0.05	0.11	0.12	0.16
5	0.05	0.16	0.16	0.22
6	0.05	0.33	0.21	0.27

Convection–diffusion problem

		UPW-0.1 $1/\nu = 500$	UPW-0.1 $1/\nu = 5,000$	SD-0.5 $1/\nu = 500$	SD-0.5 $1/\nu = 5,000$
	4	0.06	0.20	0.10	0.65
Mesh DC0:	5	0.05	0.19	0.06	0.44
	6	0.06	0.21	0.06	0.21
	4	0.13	0.19	0.23	0.93
Mesh DC1:	5	0.28	0.30	0.30	0.73
	6	0.54	0.46	0.54	0.58
	4	0.10	0.11	0.11	div
Mesh DC2:	5	0.13	0.13	0.14	div
	6	0.16	0.17	0.16	div

Table 3.38: MG results for Poisson, resp., convection–diffusion problems

diffusion problems. The used parameter combination is **ILU3(0,4)-INT(f)(50)-MAT(50)-CORR(adap1)**.

Most of the previous conclusions remain valid, however some new observations have to be stated which are typical for many other calculations:

- The case of moderately anisotropic meshes (DC1) in combination with a moderately small viscosity parameter ($1/\nu = 500-5,000$) leads often to worse multigrid behaviour than for the extreme cases. The explanation may be that most components are designed to be more or less exact solution tools in such limit cases which consequently have problems in the "intermediate range" when all difficulties have about same weight.

- We cannot neglect that there are still numerical problems with streamline–diffusion approaches. At the moment, we cannot state to have found the "final Black Box" multigrid configuration for these discretization techniques. That is why we still prefer the upwinding schemes.

- The shown results are – as explained above – rather pessimistic. In "real" Navier–Stokes simulations the observed multigrid rates are often better!

We finish this section about linear multigrid techniques with some concluding remarks. Even though the auxiliary subproblems for pressure and velocity components – and also the complete Navier–Stokes equations as shown in Chapter 4 – can be numerically solved with special multigrid components which have a certain "Black Box" character, all proposed schemes have some disadvantages. While Jacobi and Gauß–Seidel iteration as smoothers are restricted to regular meshes only, the ILU scheme runs into numerical and computational problems if applied to systems of equations or if performed on vector or parallel computers.

An alternative approach which may overcome these numerical and computational problems seems to be the technique which has been applied in the context of the *local MPSC* schemes. Here, we apply some (adaptively chosen) blocking techniques which aim to hide all numerical anisotropies in *patches*. The corresponding local matrices may be very ill-conditioned but they can be nevertheless "inverted" in a very efficient and robust manner since direct solution techniques and the high performance of processor caches can be exploited. Then, the outer iteration is based on a Jacobi/Gauß–Seidel-like approach which performs well in the case of isotropic macro-patches. Hence, the aim is to switch between such Jacobi-like techniques which allow pipelining and vectorization facilities, and direct subproblem solvers which exploit the fast floating point and cache units. Beside the gain in computational efficiency, further robustness is obtained via constructing subproblems which contain most of the local anisotropies. Since this approach is quite easily applicable to complex systems, too, this method is our favourized technique for the future (see the description of the FEAST project [1]).

3.5 Boundary conditions

Most flow problems of scientific or engineering interest, such as flows past obstacles, around corners, or through pipes or apertures, are first conceptualized in unbounded domains. This is an idealization intended to focus on a phenomenon of interest, free of the effects of distant boundaries. We follow the ideas from the paper [56] which is a common work together with John Heywood and Rolf Rannacher and start with reviewing the mathematical formulations for unbounded domains of a class of problems that involve the prescription of pressure drops and/or net flux conditions. These formulations are suggestive of analogous formulations for bounded domains, which are appropriate when a bounded domain is obtained as the truncation of an unbounded domain for the purpose of making a numerical computation.

We focus particularly on variational formulations, rather than on their classical counterparts. The principal issue concerning variational formulations is not a choice of boundary conditions, but a choice of function spaces. This choice of function spaces, however, seems relatively straightforward in comparison to choosing boundary conditions. We accept what seems to be the simplest and most natural choice for these function spaces, namely that which leaves functions as free as possible, and investigate the consequences through numerical experiments and by drawing out the relationship between the resulting variational problems and the "artificial" boundary conditions that are implicit in them.

We often refer to the boundary conditions arrived at this way as *do nothing* boundary conditions, since we do not try to achieve any special effect or boundary condition through restrictions in the function spaces. As it turns out, these boundary conditions are the same as those that have already been recommended by Gresho, see, e.g., [43], for use along "open" boundaries (for more recent references see [44]). In our variational formulations, these boundary conditions are implicitly combined with *net flux* and/or *pressure drop* conditions, and apply equally along both inflow and outflow boundaries. This allows us to consider two types of problems that we show to be dual to each other:

- Find certain net fluxes (say through individual pipes in a network of pipes) from prescribed pressure drops.

- Find the pressure drops that produce these net fluxes.

There are currently many possible choices of outflow boundary conditions under consideration by the CFD community, without any completely clear criteria for preferring one over another. Gresho's contention that these "do nothing" boundary conditions are probably the best possible general purpose boundary conditions for use along outflow boundaries seems to be supported from the mathematical point of view by their simplicity and elegance within the variational framework. Interestingly, they have not received much attention from the mathematical community. However, as it now seems clear that they have a great practical importance, it is apparent that they deserve serious mathematical investigation as part of the general Navier-Stokes theory.

To that end, the article [56] offers what we can provide in the way of Lemmas of existence, uniqueness, continuous dependence, and stability, and draws attention to several points of difficulty that limit our Lemmas in comparison to what is known in the case of Dirichlet boundary conditions.

Our interest in these matters was stimulated by our experience in testing an older 2D finite element code which is based on the use of discretely divergence–free finite elements. Because it uses divergence–free elements, it is natural to formulate problems for this code in the same way as they are usually formulated by mathematicians, namely as pressure–free variational

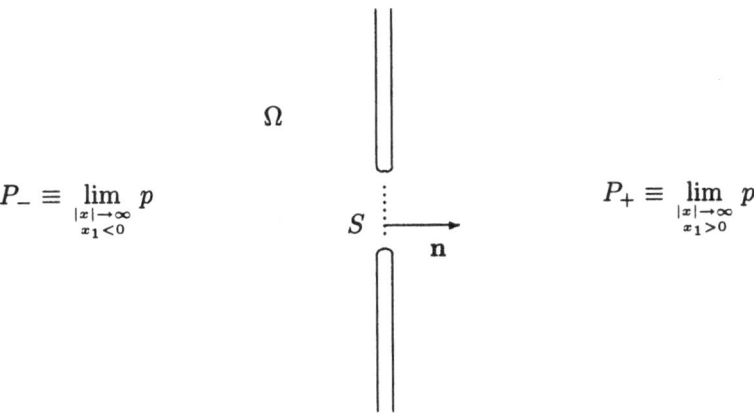

Figure 3.32: Notation for flow through an aperture in an infinite wall

problems for the velocity, using a test space of divergence–free functions. Our analysis begins with formulations of this type. However, we also derive equivalent formulations in terms of both the velocity and pressure as primary variables, for two- as well as threedimensional situations. The "do nothing" approach leads to the same result in each case. Of course, the mathematical questions studied are relevant to all methods of simulating viscous incompressible flow subject to these boundary conditions, whatever the means of enforcing them.

3.5.1 Variational formulations in unbounded domains

As mentioned above, many problems in fluid dynamics are conceptualized and studied mathematically in unbounded domains. For example, in studying flow past an obstacle, one would usually like to determine the asymptotic structure of the wake and the force on the obstacle, free of the influence of distant boundaries. Another example, the one that we are particularly interested in here, concerns fluid jets. To fix ideas, consider a plane wall which has a hole in it, and let the flow region be its complement. We call this an aperture domain (Figure 3.32).

It is a natural problem in an aperture domain to study a jet of fluid that is driven through the aperture by a drop in the pressure, from one side of the wall to the other. To make the drop in pressure quantitatively precise, one may prove first that the pressure must tend to a limit at infinity, in each half space, and then consider the difference in these limits.

Thus, the problem of finding a jet through an aperture may be formulated in classical terms by supplementing the usual initial boundary value problem for the Navier-Stokes equations (with Dirichlet boundary conditions),

$$\mathbf{u}_t + \mathbf{u} \cdot \nabla \mathbf{u} - \nu \Delta \mathbf{u} + \nabla p = 0 \quad , \quad \nabla \cdot \mathbf{u} = 0 , \tag{3.214}$$

$$\mathbf{u}|_{t=0} = \mathbf{u}_0 \quad , \quad \mathbf{u}_{|\partial\Omega} = 0 \quad , \quad \mathbf{u}(x,t) \to 0 \quad \text{as} \quad |x| \to \infty , \tag{3.215}$$

with an auxiliary condition on the pressure for prescribed $P(t)$,

$$\lim_{\substack{|x| \to \infty \\ x_1 < 0}} p(x,t) - \lim_{\substack{|x| \to \infty \\ x_1 > 0}} p(x,t) = P(t) . \tag{3.216}$$

As it happens, there is another equally good way of determining such jets. Instead of prescribing the drop in pressure, one may prescribe the net flux through the aperture. That is, one can replace the auxiliary pressure condition (3.216) by the auxiliary flux condition

$$\int_S \mathbf{u} \cdot \mathbf{n} \, ds = F(t) , \tag{3.217}$$

where $F(t)$ is prescribed.

Let us turn now to the variational formulations of these problems. Consider first the initial boundary value problem (3.214), (3.215) without the auxiliary conditions (3.216) or (3.217). In the older mathematical literature it is often posed for arbitrary domains Ω, bounded or unbounded, as follows:

Find $\mathbf{u}(t)$ *satisfying the initial condition* $\mathbf{u}_{|t=0} = \mathbf{u}_0$, *such that, for all* $t > 0$,

$$\mathbf{u}(t) \in \mathbf{J}_1^*(\Omega) \equiv \left\{ \varphi \in \mathbf{W}_2^1(\Omega) : \varphi_{|\partial\Omega} = 0, \nabla \cdot \varphi = 0 \right\} , \tag{3.218}$$

$$\nu(\nabla \mathbf{u}, \nabla \varphi) + (\mathbf{u}_t + \mathbf{u} \cdot \nabla \mathbf{u}, \varphi) = 0 \quad , \quad \forall \varphi \in \mathbf{J}_1^*(\Omega) . \tag{3.219}$$

Here, $(\cdot, \cdot)$ denotes the inner-product in $\mathbf{L}^2(\Omega)$, and $\mathbf{W}_2^1(\Omega)$ denotes the Sobolev space consisting of functions that belong to $\mathbf{L}^2(\Omega)$ and have first order spatial derivatives in $\mathbf{L}^2(\Omega)$. We are using bold face to indicate $\mathbf{R}^n$-valued functions and function spaces. Elsewhere in the older literature, the same problem (3.214), (3.215) is formulated slightly differently, as follows:

Find $\mathbf{u}(t)$ *satisfying the initial condition* $\mathbf{u}_{|t=0} = \mathbf{u}_0$, *such that, for all* $t > 0$,

$$\mathbf{u}(t) \in \mathbf{J}_1(\Omega) \equiv \text{Completion of } \mathbf{D}(\Omega) \text{ in } \mathbf{W}_1^2(\Omega), \qquad (3.220)$$

$$\nu(\nabla \mathbf{u}, \nabla \varphi) + (\mathbf{u}_t + \mathbf{u} \cdot \nabla \mathbf{u}, \varphi) = 0 \quad, \quad \forall \varphi \in \mathbf{J}_1(\Omega). \qquad (3.221)$$

Here, $\mathbf{D}(\Omega) \equiv \{\varphi \in \mathbf{C}_0^\infty(\Omega) : \nabla \cdot \varphi = 0\}$, where $\mathbf{C}_0^\infty(\Omega)$ is the set of all smooth functions with compact supports in Ω (i.e., smooth functions vanishing near the boundary and near infinity). The completion of $\mathbf{D}(\Omega)$ in $\mathbf{W}_1^2(\Omega)$ consists of those elements of $\mathbf{W}_1^2(\Omega)$ which can be approximated arbitrarily closely by elements of $\mathbf{D}(\Omega)$ in the norm

$$\|\mathbf{u}\|_{\mathbf{W}_1^2(\Omega)} = \left(\int_\Omega (|\mathbf{u}|^2 + |\nabla \mathbf{u}|^2) \, dx \right)^{1/2}.$$

It was originally thought that the spaces $\mathbf{J}_1^*(\Omega)$ and $\mathbf{J}_1(\Omega)$ are the same and that consequently these two formulations of problem (3.214) are the same. In fact, for certain classes of domains, the three-dimensional aperture domain being a prototypical example, these function spaces are different, and neither of the formulations (3.218) or (3.220) correctly represents problem (3.214). Instead, hidden within the formulation (3.218) is an auxiliary condition on the pressure, namely that the pressure drop must be zero in the sense of condition (3.216). And hidden within the formulation (3.220) is another different auxiliary condition, namely that the net flux through the aperture must be zero in the sense of condition (3.217).

Indeed, if we take an element φ of $\mathbf{D}(\Omega)$ and apply the divergence Lemma to it in the left half-space, we see that

$$\int_S \varphi \cdot \mathbf{n} \, dS = \int_{x_1 < 0} \nabla \cdot \varphi \, dx = 0.$$

It follows that elements of $\mathbf{J}_1(\Omega)$, being limits of functions in $\mathbf{D}(\Omega)$, must also have zero net flux through the aperture, and thus this condition is also contained in the formulation (3.220).

The analysis of the formulation (3.218) is more involved. Let us consider only the case of a three-dimensional aperture domain. Then, one may construct an explicit function $\mathbf{b}$ in $\mathbf{J}_1^*(\Omega)$ that carries a nontrivial net flux through the aperture, and normalize it by requiring that

$$\int_S \mathbf{b} \cdot \mathbf{n} \, dS = 1.$$

This, of course, establishes that the two spaces $\mathbf{J}_1(\Omega)$ and $\mathbf{J}_1^*(\Omega)$ are different. Further, it can be proven that the only real difference between these two function spaces is the single flux carrier $\mathbf{b}$. More precisely, $\mathbf{J}_1(\Omega)$ is contained in $\mathbf{J}_1^*(\Omega)$ while, on the other hand, every element φ of $\mathbf{J}_1^*(\Omega)$ can be written as $\varphi = F\mathbf{b} + \psi$, where ψ is some element of $\mathbf{J}_1(\Omega)$ and $F = \int_S \varphi \cdot \mathbf{n}\, dS$.

The original intention for setting the condition (3.219) in posing the Dirichlet problem was to insure that there is a scalar function p such that $-\nabla p = \mathbf{u}_t + \mathbf{u} \cdot \nabla \mathbf{u} - \nu \Delta \mathbf{u}$. However, for that, it is enough to test with test functions φ belonging to $\mathbf{J}_1(\Omega)$, or even to its dense subset $\mathbf{D}(\Omega)$. When we test with all φ in $\mathbf{J}_1^*(\Omega)$, that includes a test with the flux carrier $\mathbf{b}$. This extra test is in fact a test of the pressure drop. It can be shown that (3.219) holds with $\varphi = \mathbf{b}$ if and only if the pressure drop is zero. Thus, the variational formulation (3.219) of problem (3.215) actually contains the "hidden" condition that the pressure drop, from one side of the wall to the other, must be zero.

It remains now to generalize the variational formulations (3.219) and (3.221) so as to intentionally incorporate prescribed values of the pressure drop $P(t)$ in (3.216) or of the net flux $F(t)$ in (3.215). We will refer to the literature for the rigorous analysis and simply state here the final results. The correct variational formulation of the prescribed pressure drop problem (3.214) – (3.216) is:

Find $\mathbf{u}(t)$ *satisfying the initial condition* $\mathbf{u}_{|t=0} = \mathbf{u}_0$, *such that, for all* $t > 0$,

$$\mathbf{u}(t) \in \mathbf{J}_1^*(\Omega), \tag{3.222}$$

$$\nu(\nabla \mathbf{u}, \nabla \varphi) + (\mathbf{u}_t + \mathbf{u} \cdot \nabla \mathbf{u}, \varphi) = -P(t) \int_S \varphi \cdot \mathbf{n}\, ds, \quad \forall \varphi \in \mathbf{J}_1^*(\Omega). \tag{3.223}$$

The correct variational formulation of the prescribed net flux problem (3.214), (3.215), (3.217) is:

Find $\mathbf{u}(t) = F(t)\mathbf{b} + \mathbf{v}(t)$ *satisfying the initial condition* $\mathbf{u}_{|t=0} = \mathbf{u}_0$, *such that, for all* $t > 0$,

$$\mathbf{v}(t) \in \mathbf{J}_1(\Omega), \tag{3.224}$$

$$\nu(\nabla \mathbf{u}, \nabla \varphi) + (\mathbf{u}_t + \mathbf{u} \cdot \nabla \mathbf{u}, \varphi) = 0, \quad \forall \varphi \in \mathbf{J}_1(\Omega). \tag{3.225}$$

Perhaps it should be pointed out that, having constructed $\mathbf{b}$, the real unknown in (3.223) is $\mathbf{v}$, and that equation (3.225) can be equivalently written as

$$\nu(\nabla \mathbf{v}, \nabla \varphi) + (\mathbf{v}_t + \mathbf{v} \cdot \nabla \mathbf{v} + \mathbf{b} \cdot \nabla \mathbf{v} + \mathbf{v} \cdot \nabla \mathbf{b}, \varphi) = -\nu(\nabla \mathbf{b}, \nabla \varphi) - (\mathbf{b}_t + \mathbf{b} \cdot \nabla \mathbf{b}, \varphi).$$

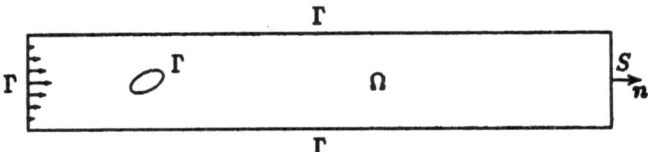

Figure 3.33: Notation for a flow region having an artificial boundary at the outlet S

The results that we have described in this section are from the papers [52], [54] of Heywood, which initiated a general study of the relationship between the geometry of unbounded domains and of the auxiliary conditions that are needed to formulate well posed problems for the Navier-Stokes equations. These investigations all depend in an essential way on an analysis of the function spaces that enter into the variational formulations of these problems. One notable result, already given in [52], is that $\mathbf{J}_1^*(\Omega) = \mathbf{J}_1(\Omega)$ in the case of an exterior domain. Consequently, pressure drops cannot be prescribed in an exterior domain, and solutions of the initial value (Dirichlet) problem (3.214), (3.215) are uniquely determined without them. In particular, flow past an obstacle in an exterior domain must be driven by the prescription of a nonzero limit for the velocity at infinity. Thus, there are fundamentally different mechanisms that drive nontrivial flows in different types of unbounded domains.

The next section concerns the truncation to bounded domains of flows which, in the idealization of an unbounded domain, are driven by pressure drops. What we are going to do now is change the point of view to that of the computational practitioner, and use the theory from this section as guidance in formulating problems.

3.5.2 Variational formulations in bounded domains

To fix ideas in a familiar setting with which we can make later comparisons, let us begin by considering a common test problem, that of calculating nonsteady flow past an obstacle (here taken as an inclined ellipse) situated in a rectangle (see Figure 3.33).

The velocity $\mathbf{u}(t)$ is required to be zero on the upper and lower boundaries and on the surface of the ellipse, while a parabolic "Poiseuille" inflow profile is prescribed on the upstream boundary. We denote by Γ the union of those portions of the boundary on which Dirichlet conditions are imposed. Rather than giving serious thought to the downstream boundary condition on S, in seeking a variational formulation one can simply decide to "do nothing", in other words, leave the solution and the test space free on that portion of the boundary.

To give a variational formulation of this problem using *solenoidal spaces* (meaning spaces consisting of solenoidal functions), the first step is to construct a solenoidal extension $\mathbf{b}$ of the prescribed Dirichlet boundary values into the whole of the domain Ω. Note, that since $\mathbf{b}$ is to be solenoidal, it must carry the incoming flux at the left boundary through the domain and out across the downstream boundary. The construction of such a function $\mathbf{b}$ might appear to be a difficult task. For a construction in the case of continuous functions see Galdi [35].

Fortunately, in computations involving the discretely divergence free finite elements, the construction of $\mathbf{b}$ can be achieved by simply prescribing the appropriate nodal values for the streamfunction related basis functions along the boundary Γ. This procedure automatically generates a discretely divergence–free extension $\mathbf{b}$ of the boundary values having support in a one element wide strip along the boundary. While in practice one need not be concious of this, we need to realize that it is being done in order to analyze the method and the variational formulations behind it.

Having constructed a solenoidal extension $\mathbf{b}$ of the boundary values, a variational formulation of the test problem, using solenoidal vector fields, is obtained by requiring that $\mathbf{u}(t) = \mathbf{b} + \mathbf{v}(t)$, where, for all t,

$$\mathbf{v}(t) \in \mathbf{J}_1^*(\Omega) \equiv \left\{ \boldsymbol{\varphi} \in \mathbf{W}_2^1(\Omega) : \boldsymbol{\varphi}|_\Gamma = 0, \nabla \cdot \boldsymbol{\varphi} = 0 \right\}, \qquad (3.226)$$

$$\nu(\nabla \mathbf{u}, \nabla \boldsymbol{\varphi}) + (\mathbf{u}_t + \mathbf{u} \cdot \nabla \mathbf{u}, \boldsymbol{\varphi}) = 0, \qquad \forall \boldsymbol{\varphi} \in \mathbf{J}_1^*(\Omega). \qquad (3.227)$$

In order to discuss both steady and nonsteady problems simultaneously, we have omitted the initial condition in writing (3.226). Thus, (3.226), (3.227) represent the Navier-Stokes equations along with boundary conditions. The initial boundary value problem is formulated by adding the initial condition $\mathbf{u}|_{t=0} = \mathbf{u}_0$. The stationary problem is formulated by adding the condition that $\mathbf{u}_t = 0$. What we are mainly interested in is how the equations are combined with boundary conditions and other "hidden" auxiliary conditions in variational formulations.

Corresponding to (3.226), (3.227), which is a pressure free formulation using solenoidal spaces, there is also an equivalent *standard* formulation which is expressed without reference to solenoidal spaces. For this formulation the extension $\mathbf{b}$ needs not be solenoidal (again, it can be constructed by simply assigning nodal values along the boundary). The requirement is then that $\mathbf{u}(t) = \mathbf{b} + \mathbf{v}(t)$, where, for all t,

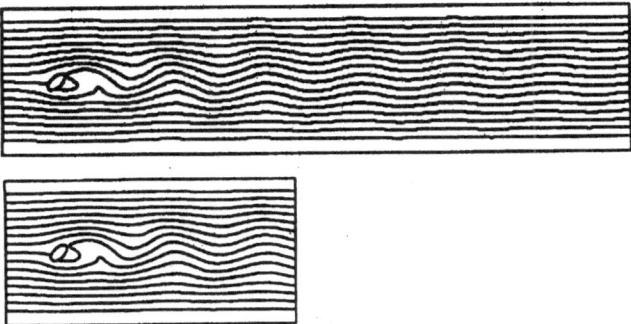

Figure 3.34: Streamlines, at $Re = 500$, after 100 time steps, starting from Stokes flow with constant Poiseuille inflow, computed in domains of different lengths on the basis of the variational formulation (3.227). The flow is nearly identical in the shorter common region, indicating a satisfactory treatment of the downstream artificial boundary.

$$\mathbf{v}(t) \in \mathbf{V}_1^*(\Omega) \equiv \left\{ \varphi \in \mathbf{W}_2^1(\Omega) : \varphi|_\Gamma = 0 \right\}, \quad p(t) \in L^2(\Omega), \qquad (3.228)$$

$$\nu(\nabla\mathbf{u}, \nabla\varphi) + (\mathbf{u}_t + \mathbf{u} \cdot \nabla\mathbf{u}, \varphi) - (p, \nabla \cdot \varphi) = 0, \quad \forall \varphi \in \mathbf{V}_1^*(\Omega), \quad (3.229)$$
$$(\chi, \nabla \cdot \mathbf{u}) = 0, \quad \forall \chi \in L^2(\Omega). \quad (3.230)$$

The results of our computations based on (3.227), like those reported by others on the basis of (3.229), (3.230), show a truely remarkable "transparency" of the downstream boundary when it is handled in this way (Figures 3.34 – 3.36). Testing, by doubling the length of the computational domain, is seen to make almost no discernable difference in the flow in the shorter common region. Figures 3.34 – 3.36 are different representations of exactly the same computations.

As these results appear highly satisfactory, there seems little reason to ask about the boundary conditions that must be implicit in these variational formulations. But now, to motivate such questions, let us consider low Reynolds number flow through a junction in a system of pipes, again prescribing a Poiseuille inflow upstream. Figure 3.37 shows steady streamlines for computations based on the same variational formulations as above, each with the same inflow, but with varying lengths of pipe beyond the junction.

There seems to be something of a puzzle here, in that the flow through the junction is seen to be highly dependent on the positions of the artificial boundaries, even if they are far from the junction. One might wonder whether the variational formulation (3.226), (3.227) or (3.228), (3.229) has

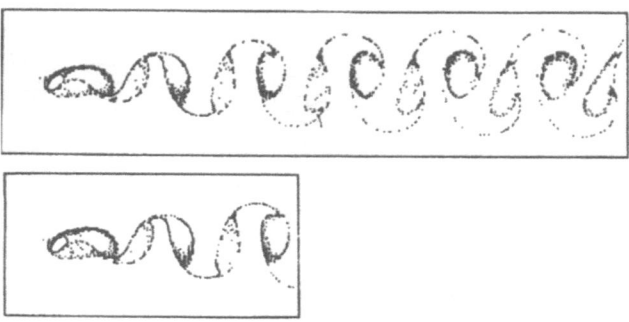

Figure 3.35: The same computations represented by particle tracing, showing von Kármán vortex streets as usually visualized in physical experiments by smoke or similar.

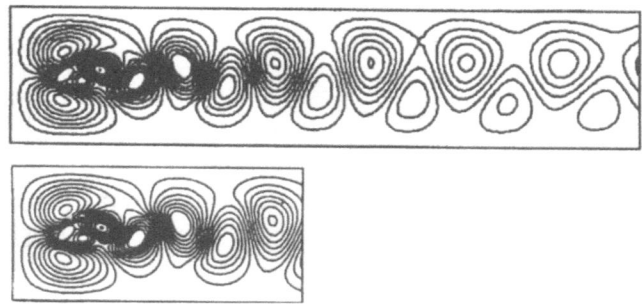

Figure 3.36: Relative streamlines for the same computations, showing the difference $\mathbf{u} - \bar{\mathbf{u}}$ between the nonlinear solution $\mathbf{u}$ and the solution $\bar{\mathbf{u}}$ of the *stationary* Stokes problem on the same domain Ω.

some "hidden" condition within it, analogous to the "hidden" pressure condition (3.216) in the variational formulation (3.219) of problem (3.215). That is the point of this discussion.

It does have a precisely analogous "hidden" pressure condition. This can be seen by examining the natural boundary conditions that are associated with the variational formulations (3.227) and (3.229), as we shall show later. In particular, it will be seen that they imply that the mean pressure on each free section S_i (cf. Figure 3.38) is zero:

$$\frac{1}{|S_i|} \int_{S_i} p \, ds = 0 \quad , \quad \text{for} \quad i = 2, 3 .$$

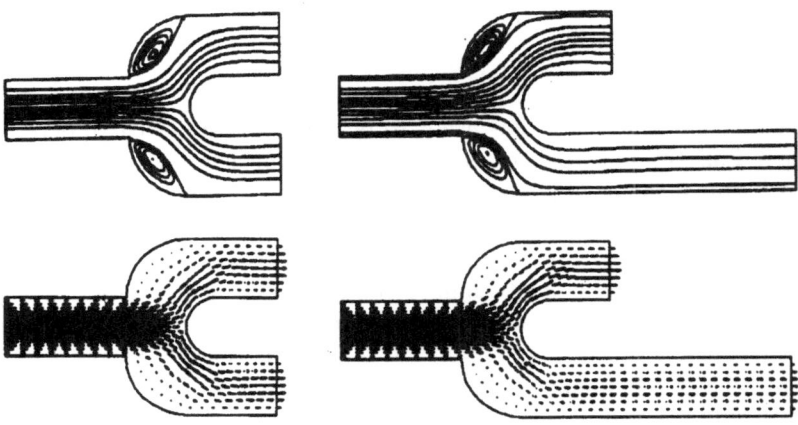

Figure 3.37: Streamlines and velocity plot of flow at $Re = 50$ with constant Poiseuille inflow, computed using the same variational formulation (3.227) as in Figure 3.34. The net flux through each outlet is highly dependent upon the relative lengths of the downstream sections.

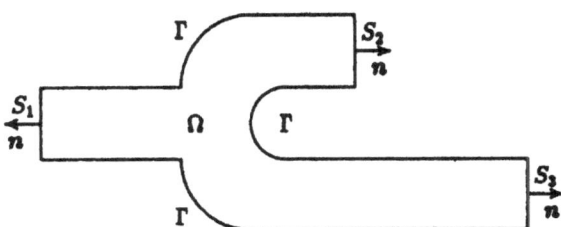

Figure 3.38: Notation for regions having artificial boundaries at multiple inlet/outlets S_i.

Thus, in Figure 3.37, the pressure gradient is greater in the shorter of the two outflow sections, which explains why there is a greater flow through that section. This example suggests that we might consider formulating problems more generally, in terms of *prescribed pressure drops*, and that we need not distinguish between sections of inflow and outflow, or even know which are which. For a flow region with multiple inlet/outlets, as indicated in Figure 3.38, it seems natural to seek solutions for which the mean pressure over each outlet section is prescribed. Therefore, let us consider the following

Prescribed Pressure Drop Problem:

For any prescribed $P_i(t)$, find $\mathbf{u}(t)$ and $p(t)$ such that

$$\mathbf{u}_t + \mathbf{u} \cdot \nabla \mathbf{u} - \nu \Delta \mathbf{u} + \nabla p = 0 \quad , \quad \nabla \cdot \mathbf{u} = 0, \qquad (3.231)$$

$$\mathbf{u}_{|\Gamma} = 0 \quad , \quad \frac{1}{|S_i|} \int_{S_i} p \, ds = P_i(t) \, . \qquad (3.232)$$

It is this type of problem that needs to be considered in order to determine the net flux through each of various inlets or outlets, given the pressure drops between them. One can even ask whether there will be a positive net inflow or outflow through some particular duct, for given prescribed values of the mean pressures. Notice, that at this point, we do not want to committ ourselves to any particular boundary conditions along the outlets S_i. Our only stated objective is to achieve prescribed differences between the mean pressures across the various outlets. It is implicit that we want to achieve this by whatever boundary conditions work best, in some vague sense. We hope to find these by posing the problem variationally in the most natural possible way. What would that formulation be? For guidance we look to the analogous problem (3.214) – (3.216) for unbounded domains. Its variational formulation (3.222), (3.223) can be copied word for word.

Variational Pressure Drop Problem (with solenoidal spaces):

Find $\mathbf{u}(t)$ such that, for all t,

$$\mathbf{u}(t) \in \mathbf{J}_1^*(\Omega) \equiv \left\{ \boldsymbol{\varphi} \in \mathbf{W}_2^1(\Omega) : \boldsymbol{\varphi}|_{\Gamma} = 0, \nabla \cdot \boldsymbol{\varphi} = 0 \right\} \, , \qquad (3.233)$$

$$\nu(\nabla \mathbf{u}, \nabla \boldsymbol{\varphi}) + (\mathbf{u}_t + \mathbf{u} \cdot \nabla \mathbf{u}, \boldsymbol{\varphi}) = - \sum_i P_i(t) \int_{S_i} \boldsymbol{\varphi} \cdot \mathbf{n} \, ds \quad , \quad \forall \boldsymbol{\varphi} \in \mathbf{J}_1^*(\Omega) \, .$$
$$(3.234)$$

It is easy to see that such a variational formulation is mathematically well posed. It is somewhat more difficult to translate it, precisely, in terms of boundary conditions and the like. When one does, as in the next section, it will be seen that the conditions (3.233), (3.234) imply something more along the free boundary S than was asked for in (3.232). Therefore problem (3.231), (3.232), by itself, does not quite form a well posed problem. We note too that a more general class of functionals can be introduced to the right side of (3.234). However, the simple case considered here seems to have a very wide range of useful applications. It is interesting that this problem, in which conditions for the pressure are prescribed, is so easily set in a pressure free variational formulation. The analogue of problem (3.234) in terms of both primary variables is posed as follows:

Variational Pressure Drop Problem (without solenoidal spaces):

Find $\mathbf{u}(t)$ such that, for all t,

$$\mathbf{u}(t) \in \mathbf{V}_1^*(\Omega) \equiv \left\{ \boldsymbol{\varphi} \in \mathbf{W}_2^1(\Omega) : \boldsymbol{\varphi}|_{\Gamma} = 0 \right\}, \qquad p(t) \in L^2(\Omega) \, , \qquad (3.235)$$

$$\nu(\nabla \mathbf{u}, \nabla \boldsymbol{\varphi}) + (\mathbf{u}_t + \mathbf{u} \cdot \nabla \mathbf{u}, \boldsymbol{\varphi}) - (p, \nabla \cdot \boldsymbol{\varphi}) = -\sum_j P_j(t) \int_{S_j} \boldsymbol{\varphi} \cdot \mathbf{n} \, ds, \ \forall \boldsymbol{\varphi} \in \mathbf{V}_1^*(\Omega),$$

$$(3.236)$$

$$(\chi, \nabla \cdot \mathbf{u}) = 0 \quad , \quad \forall \chi \in L^2(\Omega). \qquad (3.237)$$

The prescription of pressure drops is not the only natural way of posing problems for flow through a system of ducts, like that of Figure 3.38. Indeed, one may wish to *find* the pressure drops that are required to achieve a desired *net flux* through each of various ducts. Thus, we also consider the

Prescribed Net Flux Problem:

For any prescribed $F_i(t)$ satisfying $\sum_i F_i(t) = 0$, find $\mathbf{u}(t)$ and $p(t)$ such that

$$\mathbf{u}_t + \mathbf{u} \cdot \nabla \mathbf{u} - \nu \Delta \mathbf{u} + \nabla p = 0 \quad , \quad \nabla \cdot \mathbf{u} = 0, \qquad (3.238)$$

$$\mathbf{u}_{|\Gamma} = 0 \quad , \quad \int_{S_i} \mathbf{u} \cdot \mathbf{n} \, ds = F_i(t). \qquad (3.239)$$

To incorporate these flux conditions into a variational formulation of the Navier-Stokes equations, we look again to the analogous problem (3.214), (3.215), (3.217) for unbounded domains. Its variational formulation (3.225) can be copied exactly. One first constructs solenoidal flux carriers $\mathbf{b}_i, i \geq 2$, carrying a unit net flux from an arbitrarily chosen reference inlet/outlet S_1 to each of the others. If there are three inlet/outlets (cf. Figure 3.38), let $\mathbf{b}_i, i = 2, 3$, satisfy

$$\mathbf{b}_i \in \mathbf{J}_1^*(\Omega) \quad , \quad \int_{S_1} \mathbf{b}_i \cdot \mathbf{n} \, ds = -1 \quad , \quad \int_{S_j} \mathbf{b}_i \cdot \mathbf{n} \, ds = \delta_{ij} \quad , \quad j = 2, 3.$$

$$(3.240)$$

Then, an appropriate variational formulation is the following:

Variational Net Flux Problem (with solenoidal spaces):

Find $\mathbf{u}(t) = F_2(t)\mathbf{b}_2 + F_3(t)\mathbf{b}_3 + \mathbf{v}(t)$ such that, for all t,

$$\mathbf{v}(t) \in \mathbf{J}_1(\Omega) \equiv \left\{ \boldsymbol{\varphi} \in \mathbf{W}_2^1(\Omega) : \boldsymbol{\varphi}_{|\Gamma} = 0, \nabla \cdot \boldsymbol{\varphi} = 0, \int_{S_i} \boldsymbol{\varphi} \cdot \mathbf{n} \, ds = 0, \forall i \right\},$$

$$(3.241)$$

$$\nu(\nabla \mathbf{u}, \nabla \boldsymbol{\varphi}) + (\mathbf{u}_t + \mathbf{u} \cdot \nabla \mathbf{u}, \boldsymbol{\varphi}) = 0 \quad , \quad \forall \boldsymbol{\varphi} \in \mathbf{J}_1(\Omega). \qquad (3.242)$$

If one is not using solenoidal spaces, the functions $\mathbf{b}_i$ are not required to be solenoidal, and the appropriate formulation is:

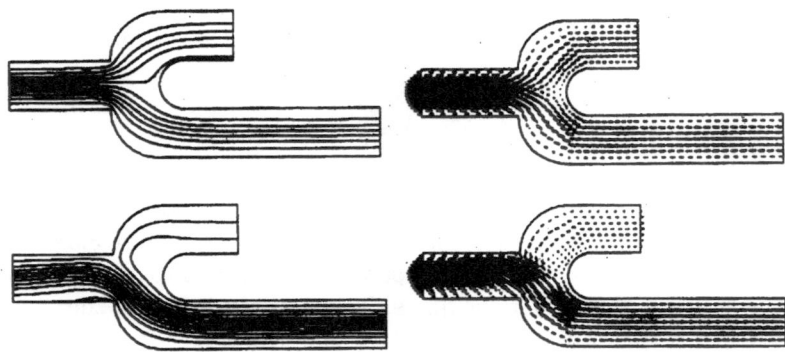

Figure 3.39: Typical steady computations based on the pressure drop formulation (3.234). Let P_i denote the prescribed mean pressure over the inlet/outlet S_i, numbered as in Figure 3.38. For the computation on the top, $P_1 = 0$, $P_2 = 1.5$, $P_3 = 2$. This produces inflow across S_2 and S_3. For the other computation on the bottom, $P_1 = 0$, $P_2 = 0.5$, $P_3 = 2$, which produces outflow through both S_1 and S_2. The Reynolds number is approximately $Re = 50$.

Variational Net Flux Problem (without solenoidal spaces):

Find $\mathbf{u}(t) = F_2(t)\mathbf{b}_2 + F_3(t)\mathbf{b}_3 + \mathbf{v}(t)$ *and* $p(t)$ *such that, for all* t,

$$\mathbf{v}(t) \in \mathbf{V}_1(\Omega) \equiv \left\{ \boldsymbol{\varphi} \in \mathbf{W}_2^1(\Omega) : \boldsymbol{\varphi}_{|\Gamma} = 0, \int_{S_i} \boldsymbol{\varphi} \cdot \mathbf{n}\, ds = 0\,, \forall i \right\},\ p(t) \in L^2(\Omega)\,,$$

$$(3.243)$$

$$\nu(\nabla \mathbf{u}, \nabla \boldsymbol{\varphi}) + (\mathbf{u}_t + \mathbf{u} \cdot \nabla \mathbf{u}, \boldsymbol{\varphi}) - (p, \nabla \cdot \boldsymbol{\varphi}) = 0\,, \quad \forall \boldsymbol{\varphi} \in \mathbf{V}_1(\Omega)\,, \quad (3.244)$$

$$(\chi, \nabla \cdot \mathbf{u}) = 0\,, \quad \forall \chi \in L^2(\Omega)\,. \quad (3.245)$$

These formulations, too, are examined later. Again, it will be seen that these well posed variational problems contain further boundary conditions along the free boundary S than asked for in (3.239). So, by itself, problem (3.238), (3.239) is not quite well posed. Figures 3.39 – 3.42 present the results of several typical computations for problems with prescribed net fluxes or pressure drops, based on the formulations (3.234) and (3.242).

The problem of a jet through an aperture in a wall can be regarded as a prototypical problem for computational procedures based on the variational formulations (3.234) and (3.242). The computational results shown in Figures 3.43 – 3.45 have been the first of this type that we know of.

Before we began our numerical experiments with free inflow boundary conditions, we were concerned that such problems might be quite unstable, already

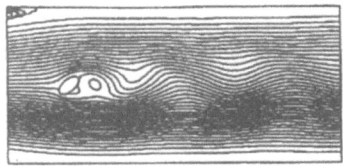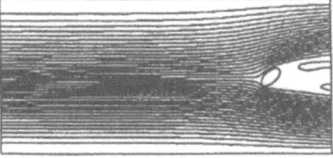

Figure 3.40: Streamlines of flow past an inclined ellipse, at $Re = 500$ based on the variational flux formulation (3.242) with both upstream and downstream boundaries free. Except for the free upstream boundary, all parameters are the same as in Figure 3.34. In the left figure, approximately 60% of the flux passes under the ellipse, compared to approximately 50% in the right figure.

on the continuous, theoretical level, and that this instability might limit their computational usefulness to very low Reynolds numbers. For instance, it seemed possible that the upstream Dirichlet condition in Figure 3.34 might be an important stabilizing factor, for lack of which the computations shown in Figures 3.39 and 3.40 would somehow collapse.

This concern was heightened by a look at the existence theory for such problems. In [56], we present the basic estimates that we know of, upon which an existence theory can be given for steady and nonsteady solutions of prescribed flux and pressure problems. It can be seen that some of these estimates require assumptions about the smallness of the data that one does not encounter in dealing with Dirichlet boundary conditions, giving the impression (we have not explicitly evaluated the constants) that these Lemmas may be valid only for very small data. Thus, anticipating difficulties that have not actually arisen in our computations, we looked at alternative variational formulations of flux and pressure problems using symmetrised "conservative" forms of the nonlinear term. Using these forms, the nonlinear term vanishes identically in the energy estimates, facilitating existence Lemmas for less restrictive data. Of course, changing the variational form also changes the problem that is being solved, and may render it unsatisfactory in other respects. That seems to be the case.

One is led to the first of the conservative forms we are referring to by using the identity $\nabla(\frac{1}{2}|\mathbf{u}|^2) = \mathbf{u} \cdot (\nabla\mathbf{u})^T$, to write the Navier-Stokes equations as

$$\mathbf{u}_t + \mathbf{u} \cdot \nabla\mathbf{u} - \mathbf{u} \cdot (\nabla\mathbf{u})^T - \nu\Delta\mathbf{u} = -\nabla(p + \frac{1}{2}|\mathbf{u}|^2) = -\nabla\bar{p}. \qquad (3.246)$$

This leads to a variational formulation in which the term $(\mathbf{u} \cdot \nabla\mathbf{u}, \varphi)$ is replaced by

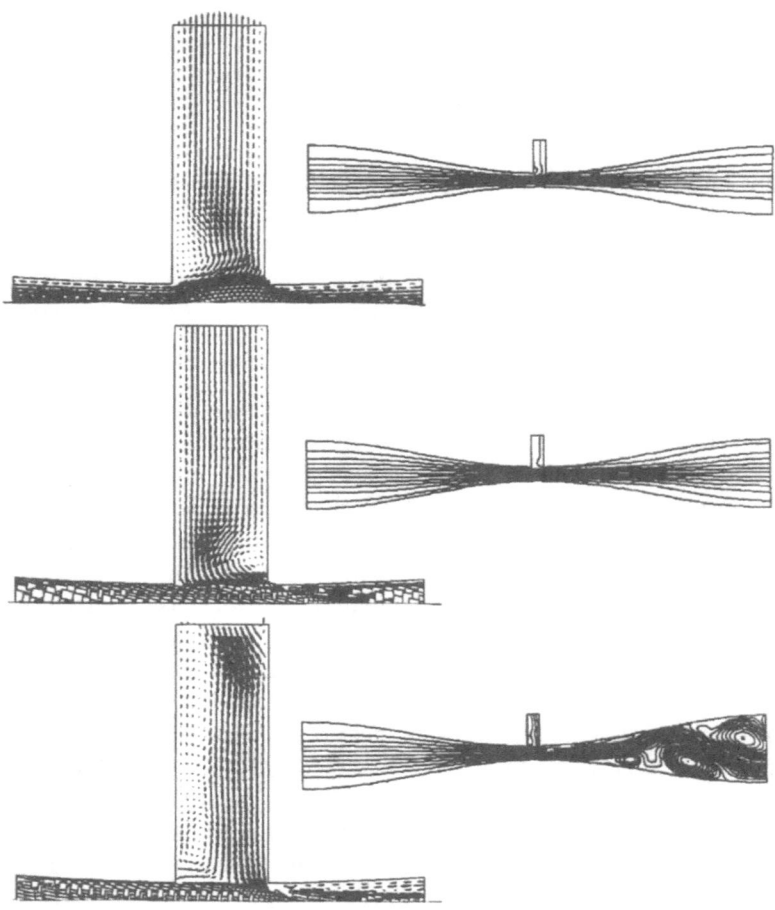

Figure 3.41: Formulations (3.234) and (3.242) are combined here (introducing a flux carrier from S_1 exiting either S_2 or S_3) in this test of the Bernoulli principle. The results of three computations are shown, with enlargements of the upper duct. In each case, an incoming net flux F_1 is prescribed across a free boundary S_1 on the left, while the mean pressure P_2 on a free boundary S_2, at the top of the small upper inlet/outlet, is prescibed to be equal to the mean pressure P_3 on a free boundary S_3 at the right. In the first case, at very low Reynolds number, $Re = 10$, there is outflow at S_2. In second case, at $Re = 50$, there is inflow at S_2, as predicted by Bernoulli's principle. In the third case, at $Re = 1,000$, there is inflow at S_2 and complex vortex structure in both the upper and downstream ducts. See Figure 3.42 for an enlargement.

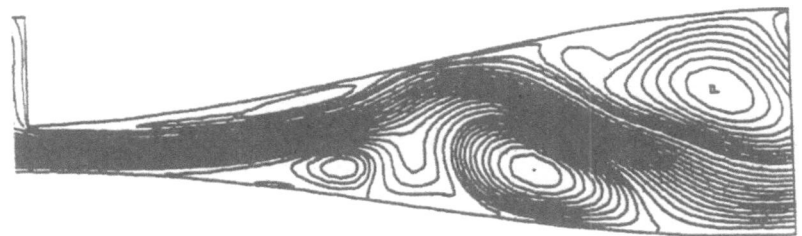

Figure 3.42: Enlargement of part of Figure 3.41.

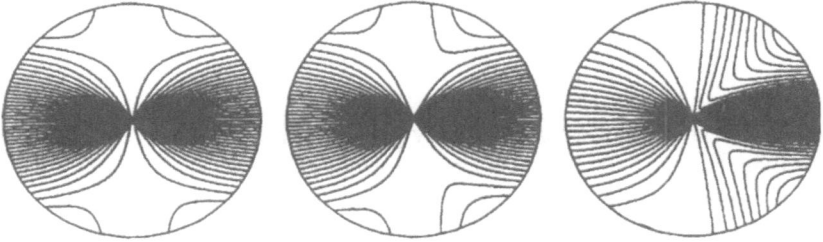

Figure 3.43: Streamlines of a steady jet through an aperture in a wall (a line segment, not visible here; see the following pictures, particularly the particle tracing), for Reynolds numbers $Re = 1, 10, 100$, based on the variational formulation (3.242) for flow with a prescribed net flux. The fluid adheres to the linear wall, while the left and right semicircles are free artificial boundaries.

$$(\mathbf{u} \cdot \nabla \mathbf{u}, \boldsymbol{\varphi}) - (\boldsymbol{\varphi} \cdot \nabla \mathbf{u}, \mathbf{u}).$$

On the right side, the additional term is absorbed into the pressure, giving what is referred to as "total pressure" or "Bernoulli pressure". The total pressure is constant along streamlines in Euler flow, and therefore it is an important quantity in some high Reynolds number situations. For example, it is the "Bernoulli principle" that explains the inflow through the central duct of Figure 3.41. The reason that the additional term on the left side of (3.246) facilitates the existence theory is that when (3.246) is multiplied through by $\mathbf{u}$, to obtain an energy estimate, the nonlinear term disappears, as if one were considering homogeneous Dirichlet data. Thus motivated, we consider the following problems:

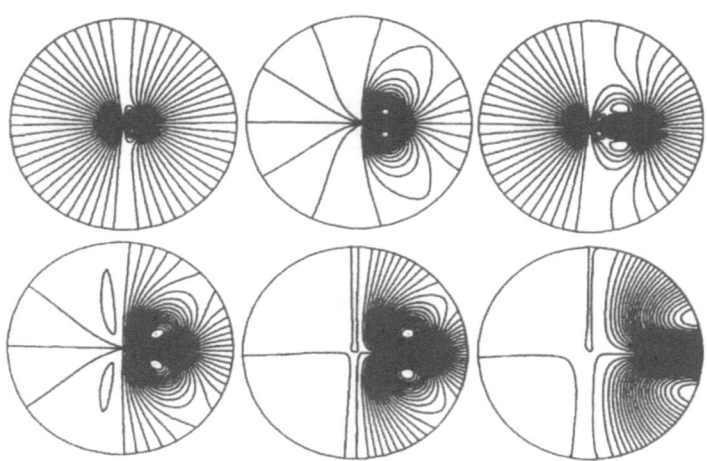

Figure 3.44: Streamlines of a nonsteady jet through an aperture in a wall (a line segment), based on the variational formulation (3.234) for flow with a prescribed time dependent pressure drop $P(t)$. The initial velocity is $\mathbf{u}_0 = 0$, and $P(t)$ is the step function $P = 1$ for $0 \leq t \leq 40$, $P(t) = 0$ for $40 < t \leq 80$, $P(t) = 1$ for $80 < t \leq 120$, and $P(t) = 0$ for $t > 121$. The figures are for $t = 20, 60, 100, 140, 200, 500$. This produces two short bursts ("puff-puff") through the hole, which are visualized by particle tracing in the next Figure.

Variational Total Pressure Drop Problem
(with solenoidal spaces):

Find $\mathbf{u}(t)$ *such that, for all* t, $\mathbf{u}(t) \in \mathbf{J}_1^*(\Omega)$ *and*

$$\nu(\nabla\mathbf{u}, \nabla\boldsymbol{\varphi}) + (\mathbf{u}_t + \mathbf{u}\cdot\nabla\mathbf{u} - \mathbf{u}\cdot(\nabla\mathbf{u})^T, \boldsymbol{\varphi}) = -\sum_j P_j(t)\int_{S_j} \boldsymbol{\varphi}\cdot\mathbf{n}\,ds\,, \ \forall\boldsymbol{\varphi} \in \mathbf{J}_1^*(\Omega)\,.$$

$$(3.247)$$

Variational Net Flux Problem Involving Total Pressure
(with solenoidal spaces):

Find $\mathbf{u}(t) = F_2(t)\mathbf{b}_2 + F_3(t)\mathbf{b}_3 + \mathbf{v}(t)$, *such that, for all* t, $\mathbf{v}(t) \in \mathbf{J}_1(\Omega)$ *and*

$$\nu(\nabla\mathbf{u}, \nabla\boldsymbol{\varphi}) + (\mathbf{u}_t + \mathbf{u}\cdot\nabla\mathbf{u} - \mathbf{u}\cdot(\nabla\mathbf{u})^T, \boldsymbol{\varphi}) = 0, \qquad \forall\boldsymbol{\varphi} \in \mathbf{J}_1(\Omega)\,. \quad (3.248)$$

It will be seen that the pressure condition corresponding to the problem (3.247) is no longer (3.232), but rather

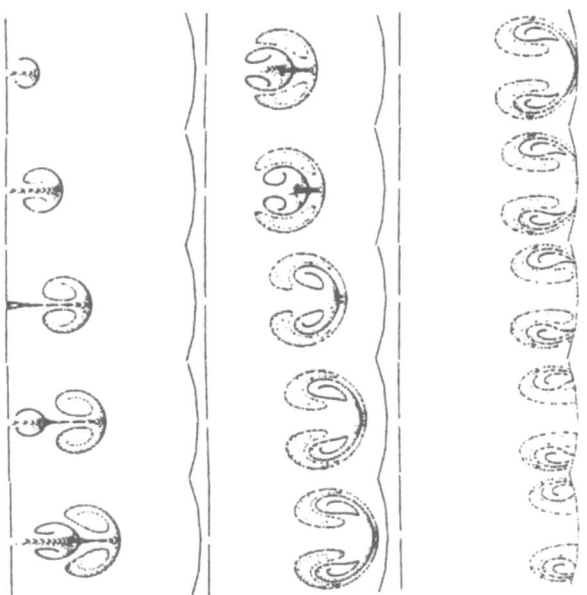

Figure 3.45: The same computations as in Figure 3.44, visualized by tracing particles that are introduced at the aperture during the "puffs". The result is two "smoke rings" that leapfrog each other and eventually exit the free artificial boundary on the right.

$$\frac{1}{|S_i|} \int_{S_i} (p + \frac{1}{2}|\mathbf{u}|^2)\, ds = P_i(t) \, . \tag{3.249}$$

Another conservative form which is often taken for convenience in analysing numerical methods is obtained by replacing $(\mathbf{u} \cdot \nabla \mathbf{u}, \varphi)$ by $\frac{1}{2}(\mathbf{u} \cdot \nabla \mathbf{u}, \varphi) - \frac{1}{2}(\mathbf{u} \cdot \nabla \varphi, \mathbf{u})$ in (3.234) and (3.242). This gives a legitimate weak form of the Navier-Stokes equations because $(\mathbf{u} \cdot \nabla \mathbf{u}, \varphi) = \frac{1}{2}(\mathbf{u} \cdot \nabla \mathbf{u}, \varphi) - \frac{1}{2}(\mathbf{u} \cdot \nabla \varphi, \mathbf{u})$ if φ is a solenoidal test function which vanishes on the boundary. When the variational equation (3.234) is changed in this way, the new pressure condition corresponding to (3.232) is

$$\frac{1}{|S_i|} \int_{S_i} (p + \frac{1}{2}|\mathbf{u} \cdot \mathbf{n}|^2)\, ds = P_i(t) \, . \tag{3.250}$$

Figure 3.46 presents the result of a typical computation based on the formulation (3.248).

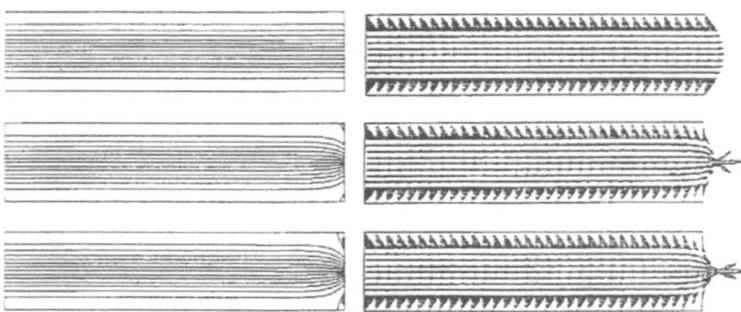

Figure 3.46: Streamlines and vector plots of pipe flow at $Re = 50$, with an artificial outflow boundary. The upper figures are based on the standard formulation (3.227), and the lower on the total pressure formulation (3.249), resp., (3.250).

Clearly, the boundary conditions that are implicit in the total pressure variational formulation (3.248) are not very satisfactory for the problems that we have been considering, although they might, perhaps, be satisfactory for some other types of problems. To reason further about this, it is neccesary to identify the boundary conditions which are implicit in the various formulations that we have been considering.

3.5.3   Associated boundary conditions of flux and pressure drop type

We present that for smooth solutions the variational formulations given above, with and without use of solenoidal spaces, are equivalent and that the prescribed pressure drop problem also admits a formulation in terms of classically prescribed boundary conditions. Solutions of the prescribed flux problem satisfy the same boundary conditions, but with an unkown pressure drop. Hence the prescribed flux problem does not have a fully equivalent formulation in terms of classical boundary conditions.

Let us consider, first, the variational pressure drop problems (3.234) and (3.236), and show that they are both equivalent to the classical problem (3.251), (3.252) below. The corresponding proofs can be found in [56].

Classical Pressure Drop Problem:

For any prescribed constants P_i, find $\mathbf{u}(t)$ and $p(t)$ such that, for all t,

$$\mathbf{u}_t + \mathbf{u} \cdot \nabla \mathbf{u} - \nu \Delta \mathbf{u} + \nabla p = 0 \quad , \quad \nabla \cdot \mathbf{u} = 0, \qquad (3.251)$$

$$\mathbf{u}_{|\Gamma} = 0 \quad , \quad (p - \nu \partial_n \mathbf{u}_n)_{|S_i} = P_i \quad , \quad \partial_n \mathbf{u}_{\tau |S_i} = 0 \,. \qquad (3.252)$$

Lemma 8 *For smooth solutions, the three formulations (3.234) and (3.236), (3.237) and (3.251), (3.252) of the prescribed pressure drop problem are all equivalent to each other.*

We make several final remarks concerning the striking success of the boundary conditions (3.252) on the artificial boundaries S_i. First, if a straight section of pipe is bounded at its ends by perpendicular sections S_i, then the unique steady solution of (3.251) is Poiseuille flow. We imagine and intend (perhaps the reader has questioned this) that the domains in Figures 3.33 – 3.39 are truncations of large domains that continue as straight sections of pipe for some distance beyond each of the S_i. Having this intention, any boundary condition which is not satisfied by Poiseuille flow would probably be found unsatifactory. Second, realizing that no artificial boundary condition can do a perfect job in nontrivial situations, we find it very satisfying that (3.252) appears to work so well in calculating flows like those of Figures 3.39 and 3.43.

Next, we consider the prescribed net flux problem and show that its variational formulations (3.242) and (3.244), (3.245) are equivalent. To identify the boundary conditions that are implicit in these problems, it is easiest to consider the $\mathbf{V}_1$-formulation (3.243), (3.244). Integrating (3.244) by parts, we obtain

$$(\mathbf{u}_t + \mathbf{u} \cdot \nabla \mathbf{u} - \nu \Delta \mathbf{u} + \nabla p, \boldsymbol{\varphi}) + \int_{\partial\Omega} (\nu \partial_n \mathbf{u} - p\mathbf{n}) \cdot \boldsymbol{\varphi}\, ds = 0, \quad \forall \boldsymbol{\varphi} \in \mathbf{V}_1(\Omega) \,.$$
$$(3.253)$$

Testing with functions $\boldsymbol{\varphi}$ vanishing on the boundary, one sees that $\mathbf{u}_t + \mathbf{u} \cdot \nabla \mathbf{u} - \nu \Delta \mathbf{u} + \nabla p = 0$, and hence that the first integral in (3.253) vanishes for all $\boldsymbol{\varphi} \in \mathbf{V}_1(\Omega)$. Then, testing with $\boldsymbol{\varphi}$ that are nonzero on $\partial\Omega$, one can conclude that

$$\mathbf{u}_{|\Gamma} = 0, \quad (p - \nu \partial_n \mathbf{u}_n)_{|S_i} = c_i \quad , \quad \partial_n \mathbf{u}_{\tau |S_i} = 0, \qquad (3.254)$$

for some constants c_i. The argument for this can be made more simply than that for (3.252) because the test functions in $\mathbf{V}_1(\Omega)$ need not be solenoidal. Notice, however, that since they are constrained by the condition $\int_{S_i} \boldsymbol{\varphi} \cdot \mathbf{n}\, ds = 0$, one can only show that $p - \nu \partial_n \mathbf{u}_n$ is constant on each S_i, and not necessarily zero. Integrating the second of conditions (3.254), we get

$$\frac{1}{|S_i|} \int_{S_i} p \, ds = c_i + \frac{\nu}{|S_i|} \int_{S_i} \partial_n u_n \, ds \,. \qquad (3.255)$$

It is evident that the boundary conditions are the same for the prescribed flux problem as for the prescribed pressure drop problem, except that the mean pressures c_i, which appear in them, are unknowns. One obtains ([56]):

Lemma 9 *For smooth solutions, the two formulations (3.242) and (3.244), (3.245) of the prescribed net flux problem are equivalent to each other. Their solutions satisfy the same boundary conditions (3.254) as solutions of the prescribed pressure drop problem, but with mean pressures $c_i(t)$ that are not known in advance of solving the problem.*

Lemma 10 *Smooth solutions of the variational total pressure drop problem (3.247) and of the variational flux problem involving total pressure (3.248) satisfy the boundary conditions*

$$\mathbf{u}_{|\Gamma} = 0 \,, \quad (p + \frac{1}{2}|\mathbf{u}|^2 - \nu \partial_n u_n)_{|S_i} = P_i(t) \,, \quad \partial_n u_\tau = 0 \quad on \ S_i \,. \quad (3.256)$$

However, for the flux problem, the pressures $P_i(t)$ are not known in advance of solving the problem.

Similarily, if one replaces the nonlinear term $(\mathbf{u} \cdot \nabla \mathbf{u}, \varphi)$ in the variational pressure and flux problems (3.234) and (3.242) by the symmetrized form $\frac{1}{2}(\mathbf{u} \cdot \nabla \mathbf{u}, \varphi) - \frac{1}{2}(\mathbf{u} \cdot \nabla \varphi, \mathbf{u})$, then the associated boundary conditions are

$$\mathbf{u}_{|\Gamma} = 0 \,, \quad (p + \frac{1}{2}|u_n|^2 - \nu \partial_n u_n)_{|S_i} = P_i(t) \,, \quad \nu \partial_n u_\tau = \frac{1}{2} u_n u_\tau \quad on \ S_i \,.$$
$$(3.257)$$

It is evident, examining the boundary conditions (3.256) and (3.257), that they are not satisfied by Poiseuille flow. Thus their poor performance in the computation shown in Figure 3.46 was to be expected. Analogous effects could be expected if the deformation tensor was used.

The previous considerations show that the right choice of variational formulations and corresponding natural boundary conditions may be in general a

Figure 3.47: Vector plots of pipe flow at $Re = 50$, with an artificial outflow boundary. The computation on the left is based on the standard variational formulation (3.227). The computation on the right is made similarly, but with the term $\nu(\nabla \mathbf{u}, \nabla \varphi)$ in (3.227) replaced by $\nu(D\mathbf{u}, D\varphi)$, where D is the deformation tensor.

very delicate task. However, our numerical experience is excellent and the given computational experiments which are mainly based on the FEATFLOW package demonstrate the wide range of possible applications. Beside the aspect of imposing natural boundary conditions, the next section concerns the implementation of Dirichlet boundary conditions due to complicated geometrical details or time dependent domains.

3.5.4   Implementation of boundary conditions

If multigrid solvers are applied, one central problem is the sufficiently accurate approximation of complex geometries and their associated boundary curves or surfaces. While there is "no" problem to design fine meshes which provide all necessary information, the construction of sequences of "coarse" meshes is the crucial point; particularly if one keeps in mind that such sequences of more or less nested meshes are necessary for multigrid.

One possible approach is to start with a coarse mesh which contains already most of geometrical fine-scale details. This technique may work efficiently in 2D cases, but for analogous 3D applications the resulting "coarsest" mesh is in general very large (much more than 10,000 mesh cells) such that the typical multigrid efficiency is lost due to a dominating coarse grid solver. In contrast, completely different – since non–nested – grids in comparison to the finest mesh may be used, but the corresponding intergrid transfer routines, which interpolate from one mesh to another, are difficult to handle. Consequently, the resulting multigrid solver spends most of its time with grid transfer routines on lower levels. Further, the convergence rates may deteriorate since they massively depend on the choice of subgrids and corresponding transfer operators.

We prefer a third approach which is embedded into a general framework for implementing boundary conditions in iterative solution techniques: the *iterative filtering technique* in combination with *fictitious boundary conditions*.

> 'Employ a (rough) boundary parametrization which sufficiently de-
> scribes all large-scale structures with regard to the boundary condi-
> tions. Treat all fine-scale features as interior objects such that the
> corresponding components in all matrices and vectors are unknown
> degrees of freedom which are implicitely incorporated into all iter-
> ative solution steps. Hence, standard tools for grid refinement in
> interior regions are easily applicable and highly accurate approx-
> imations can be obtained. Further, utilize filtering techniques to
> project the corresponding vector components onto the subspace of
> "correct" boundary conditions, before and directly after each iter-
> ative substep. This additionally ensures the typical performance
> of standard multigrid solvers without requiring additional modifi-
> cations in the software components.'

We assume that we solve the following abstract continuous problem given
in operator notation which may be linked to a typical second-order partial
differential equation (Poisson problem),

$$Au = f. \qquad (3.258)$$

For a general variational approach, let V be a Hilbert space with inner prod-
uct $(\cdot,\cdot)$ and corresponding norm $\|\cdot\|$, and let $a(\cdot,\cdot)$ be a bilinear form. We
seek a solution to the following variational problem:

$$\text{Find } u \in V, \text{ such that: } \quad a(u,\varphi) = (f,\varphi) \quad \forall \varphi \in V. \qquad (3.259)$$

This problem is approximated by a finite element method using a sequence
of finite dimensional subspaces "$V_h \subset V$" parameterized by a discretization
parameter h, such that the discrete problems read as usually:

$$\text{Find } u_h \in V_h, \text{ such that: } \quad a_h(u_h,\varphi) = (f_h,\varphi) \quad \forall \varphi \in V_h. \qquad (3.260)$$

In matrix-vector notation, we derive the discrete linear system

$$A_h U_h = F_h. \qquad (3.261)$$

Further, we assume that the continuous problem is defined with corresponding boundary conditions $B(u) = g$ on a boundary part Γ. Let be B a boundary operator which involves a combination of function values and partial

derivatives of u on Γ. In the following we imagine for simplicity that we perform Dirichlet boundary values $u = g$ on Γ. Then, there are three possiblities after the discretization process to involve these boundary conditions into the solution process of the matrix-vector problem (3.261). In all cases, we assume that the matrix A_h (and also U_h and F_h) have not yet incorporated any boundary condition which in fact corresponds to a natural boundary condition due to the partial integration involved,

$$\partial_n u = 0 \quad \text{on } \Gamma = \partial\Omega. \tag{3.262}$$

Further, let $S_h(\Gamma)$ denote all degrees of freedom, resp., all components of the solution vector U_h, which are related to the boundary Γ. For example, in the case of bilinear finite elements these are the nodes on Γ (or better: on Γ_h as an approximation to Γ), while for the nonconforming rotated bilinear elements the edges on Γ, resp., on Γ_h are associated.

Then, we can proceed as follows to apply Dirichlet boundary conditions:

1. **Fully explicit treatment:**

   We eliminate in A_h all rows and columns belonging to $S_h(\Gamma)$. Additionally, we modify all components of the right hand side vector F_h not belonging to $S_h(\Gamma)$ according to this elimination process. Further, we may prescribe the correspondingly prescribed Dirichlet values for all these components of the vectors U_h and F_h which belong to $S_h(\Gamma)$. But this step is not necessary since all components of $S_h(\Gamma)$ are treated as being well-known and hence do not longer belong to the set of unknowns.

   This process is often performed if direct solvers (Gaussian elimination) are applied. However, problems arise if the right hand sides change since corresponding modifications due to (then already) eliminated matrix elements of A_h have to be repeated. As a further consequence, two different matrices may be needed if the same operator A is required with two different boundary conditions, for instance Dirichlet values for the momentum equations and natural settings for the associated Pressure–Poisson problems.

2. **Semi-implicit treatment:**

   We replace only those rows of A_h by rows of the identity matrix which correspond to $S_h(\Gamma)$. The other rows and columns remain unchanged. Further, during an initialisation process only, we have to prescribe the given Dirichlet values for all these components of the vectors U_h and F_h which belong to $S_h(\Gamma)$. This setting guarantees in combination with iterative solvers that the resulting solution vectors always satisfy the prescribed boundary conditions.

This approach seems to be the most common in the framework of iterative solution tools. Although the components of $S_h(\Gamma)$ cannot change their value, they belong explicitly to the set of unknowns and are treated by each component of the iterative solver (e.g., smoothing, prolongation, restriction, defect calculation, preconditioning). Again, two different matrices may be needed when two different boundary conditions, for instance Dirichlet and natural conditions, are prescribed for the same operator A.

3. **Fully implicit treatment:**

   We do not modify A_h which consequently is the Neumann matrix due to natural boundary conditions. Again, we have to prescribe the given Dirichlet values for all these components of the vectors U_h and F_h which belong to $S_h(\Gamma)$, but <u>before</u> and <u>after</u> each iteration step (*filtering*). All components of the solution vectors have to be treated as unknowns by all components of the iterative solvers.

   This approach is the most general for iterative solution techniques. Even if different boundary conditions are involved with the same operator A, we can always work with exactly the same matrix A_h. Only the performed *filtering operator* has to be changed.

While the use of direct solution tools often requires the application of the *fully explicit treatment,* all three approaches are equivalent if iterative solvers are applied, and lead to exactly the same results. We prefer the *fully implicit treatment* since this is the most general approach for our employed multigrid solvers and finite element approaches. One important component inside is the *projection filter* which administrates the boundary setting for all vector components of $S_h(\Gamma)$ during the iterative procedure. This tool is the essential trick for easy implementations of complicated boundary value settings with respect to:

- several boundary components for discretely divergence-free finite elements
- small-scale details due to complex geometrical boundaries
- moving boundaries in time

We will discuss the last two items and demonstrate their efficient behaviour for some 2D and also 3D calculations. Most of these tests have been performed by P. Schreiber and the author, and they can be found, with more practical details, in the Thesis of Schreiber [93]. See also the examples at:

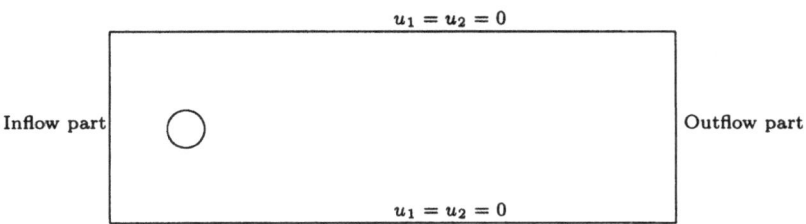

Figure 3.48: 2D flow around a circle

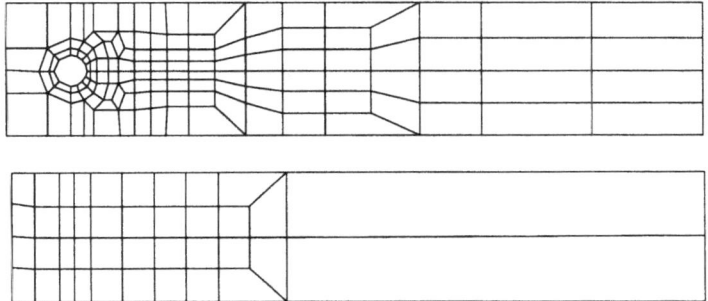

Figure 3.49: Coarse meshes for circle calculations

http://www.iwr.uni-heidelberg.de/~featflow

2D flow around a circle:

We consider the standard benchmark problem of '2D flow around a circle in a channel' (see Figure 3.48 for the geometrical description and Section 1.1 for more details).

In all previous calculations, the meshes have been adapted to the underlying geometry by prescribing two boundary components, one for the outer channel and the other for the inner circle. Now, we make additional simulations on a "pure" channel geometry, too, which does not capture the inner circle by grid points (see both coarse meshes in Figure 3.49).

For the second mesh we apply a special *filtering operator* which sets all vector components of $S_h(\Gamma)$ to zero, and $S_h(\Gamma)$ is related to all coordinates with are inside or directly on the circle. So, the resulting filtering technique is a very easy one, and the corresponding index pointers containing all vector components which have to be treated during the iterative process, can be generated

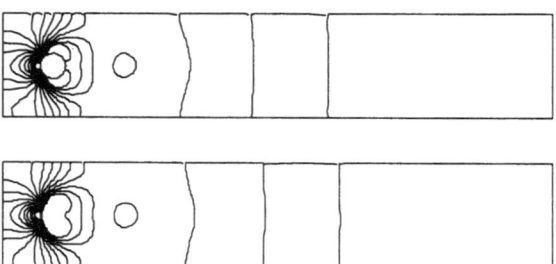

Figure 3.50: Isolines for pressure approximations

	adapted mesh			channel mesh		
NEL	8,320	33,280	133,120	10,240	40,960	163,840
Pres 1	0.1343	0.1327	0.1322	0.1299	0.1304	0.1312
Pres 2	0.1916-1	0.1661-1	0.1538-1	0.1513-1	0.1485-1	0.1468-1
Pres 3	0.2847-1	0.2812-1	0.2787-1	0.2665-1	0.2739-1	0.2761-1
Pres 4	0.2271-1	0.2272-1	0.2267-1	0.2112-1	0.2207-1	0.2237-1
ΔP	**0.1151**	**0.1161**	**0.1168**	**0.1148**	**0.1155**	**0.1165**
Velo 1	0.2156	0.2062	0.2029	0.2113	0.2051	0.2027
Velo 2	0.1372-3	0.5247-3	0.7310-3	0.4633-3	0.7053-3	0.7982-3
Velo 3	0.2569	0.2516	0.2497	0.2535	0.2508	0.2496
Velo 4	-0.3247-3	-0.2294-3	-0.1584-3	-0.3807-3	-0.2067-3	-0.1370-3

Table 3.39: Quantitative results for flow around circle

once in an initialization phase. A qualitative comparison for pressure plots
with regard to a stationary Navier–Stokes calculation is shown in Figure 3.50.

Table 3.39 provides a quantitative comparison of velocity and pressure values
at special points and of the pressure difference Δp defined on the circle. The
results for both approaches agree very well.

The other interesting aspect is the resulting multigrid convergence, in particular if we apply the *filtering process* depending on the mesh level. It is
not surprising at all that the multigrid rates are excellent if we perform the
special filter process equally on all levels. In our configuration – CC2D (=
local MPSC) with two smoothing steps – the typical result, averaged over all
nonlinear steps, is

$$\rho_{mg}^{all} \approx 0.15 \,,$$

while in that case that we allow the circle geometry on the finest level 1 only
the multigrid rates deteriorate,

$$\rho_{mg}^{1} \approx 0.9 \,.$$

However, it is sufficient to perform such filter techniques on the two or three finest levels only while on the coarser meshes a simple channel configuration (without any interior object!) is applied. In this case, the resulting multigrid rates do not differ from the case of applying this filter on all levels, $\rho_{mg}^{all} \sim \rho_{mg}^{2/3}$. This technique of modifying the geometrical details on coarser meshes has shown to be a powerful tool in the context of multilevel approaches, see also the 3D results in [93].

We do not claim that these *fictitious boundary* techniques are superior to the typical approach of resolving accurately small-scale structures by local grid adaption. But there are situations which require the combination of both techniques. Our numerical experience can be concluded in the following "thumb rules":

- If the geometry is not too complex, approximate the boundary parts by corresponding boundary parametrizations (as piecewise parametrized functions) and via adapted meshes (for instance, objects which are not to numerous and which have a piecewise smooth surface as circles, squares, etc.).

- Try to approximate the boundary parts by a "rough" boundary parametrization which contains already most of the important structure. Then, apply the *iterative filtering* techniques to resolve the fine scale–structure depending on the granularity of the mesh. This approach may be useful if large-scale objects with complicated surfaces are approximated, for instance mountains or buildings, or car shapes with special features.

- If the geometrical details are too small, perform the *fictitious boundary* techniques only. This approach is necessary if many (> 100) boundary parts are required (see [93] for the example of a 3D heating device with about 1000 small holes at the outflow and 250 apertures in the interior) or if small-scale objects are involved (for instance, cars with antennae).

Summing up, the combination of **mesh adaption** and **iterative filtering techniques** with or without **fictitious boundaries** leads to the following advantageous numerical behaviour:

1. If these techniques are applied with respect to different levels of refinement, the coarse mesh is allowed to consist of few elements only. If the accuracy of approximating the complex boundary parts is simultanously increased with refining the mesh, the full multigrid efficiency can be obtained.

2. The combination of semi–adapted meshes (near boundary parts) with these *fictitious boundary* techniques leads to accurate results. The number of unknowns is not optimal since components corresponding to some parts of the boundaries with prescribed values are explicitly involved before they are set again to the prescribed boundary values. However, the employed data structures are locally regular such that high performance rates on fast computer platforms can be the result.

3. Especially for complex 3D geometries and for moving boundaries, these techniques provide an elegant way to obtain qualitatively accurate results without paying to much for additional implementation or extended run-time behaviour.

We demonstrate some further features of these iterative techniques by the following examples which again are taken from the Thesis of P. Schreiber [93].

Flow in a moving water pump:

The geometry of the pump is a cylinder in which a piston moves up and down. The cylinder ends in a channel perpendicular to it. This channel has a valve on each side in order to allow the water to come in and to leave the pump. When the piston moves upwards the left valve is closed and the right one is open, resp., vice versa when the piston moves downwards. We control this physical situation by prescribing Dirichlet or natural boundary conditions at the channel ends, with changing position in time. The speed of the piston motion is prescribed by oscillating Dirichlet values for $S_h(\Gamma)$ which is some part of the cylinder, depending on time.

Figure 3.51 shows the coarse mesh for a 2D calculation which is <u>not</u> changed in time. The same calculation has been performed for the 3D case which shows a qualitatively similar behaviour. Figure 3.53 shows isolines for the pressure and (normalized) velocity plots for two different states. Analogous velocity plots in 3D are shown in Figure 3.54. The reader should keep in mind that all following calculations are performed on the **same** (!) mesh as shown in the Figures 3.51 and 3.52; only the components of the actual 'boundary' change in time!

Flow through moving propeller blades:

We simulate the flow in a box containing a rotating propeller. The inflow part is on the right side while on the other sides some apertures in the walls are incorporated which are controlled by natural boundary conditions. On the propeller surface (and in fact for all components belonging to $S_h(\Gamma)$)

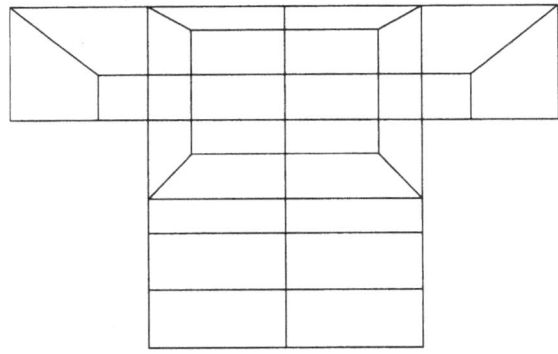

Figure 3.51: Coarse mesh for water pump in 2D

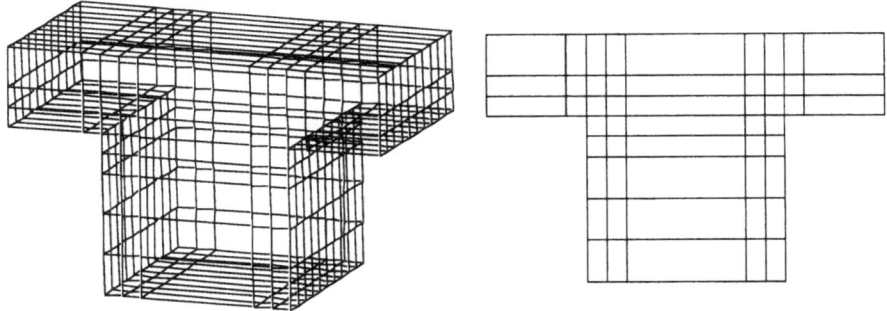

Figure 3.52: Coarse mesh for water pump in 3D

we prescribe the derivative of the propeller motion as Dirichlet values. The computational mesh is an equidistantly discretized unit square!

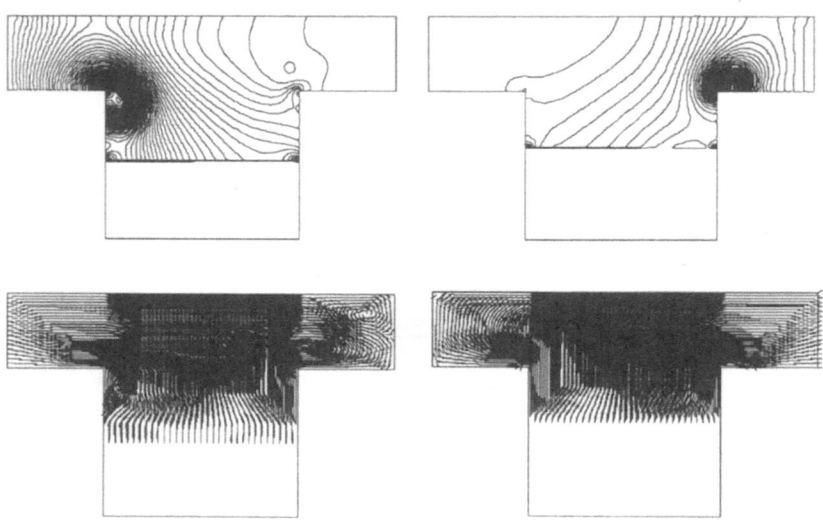

Figure 3.53: Pressure and velocity snapshots for 2D water pump

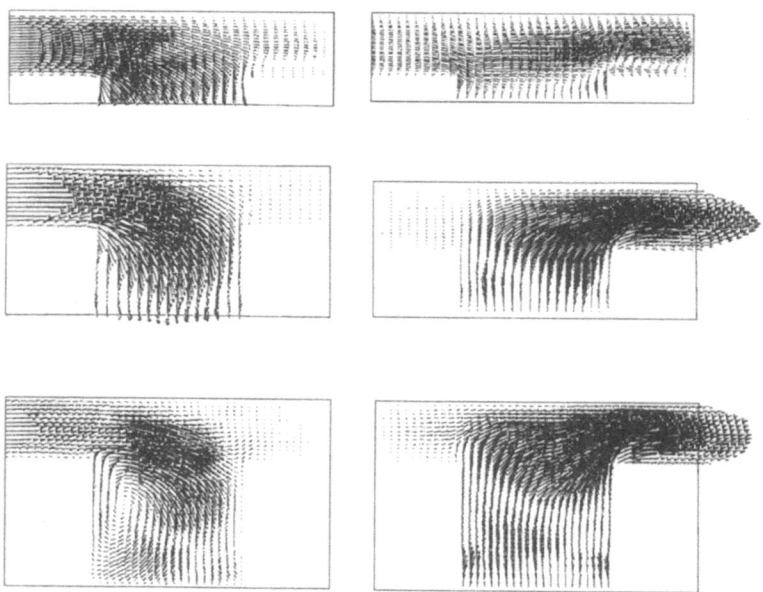

Figure 3.54: Some velocity snapshots for 3D water pump

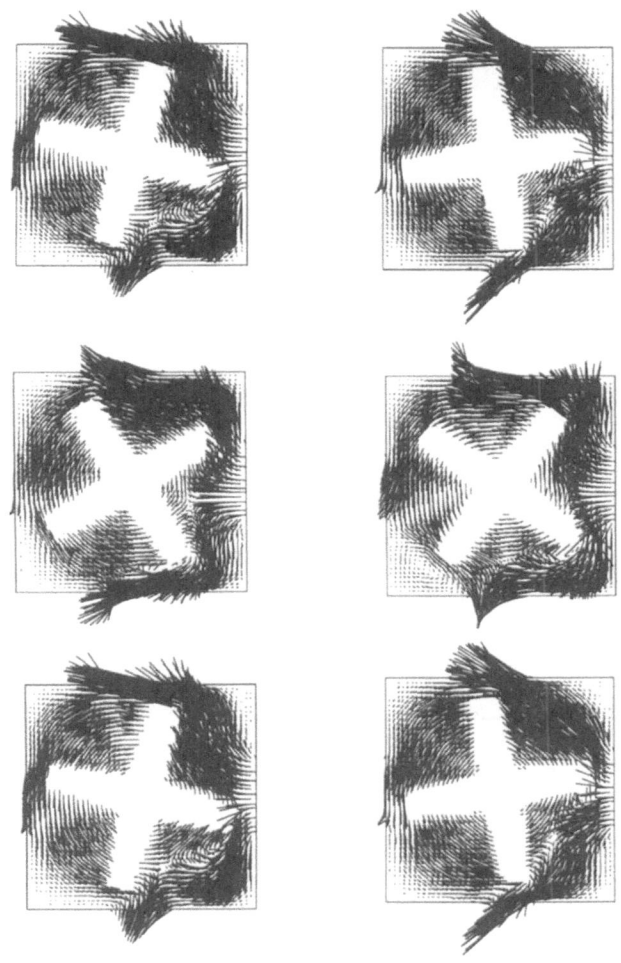

Figure 3.55: Some velocity snapshots for rotating propeller

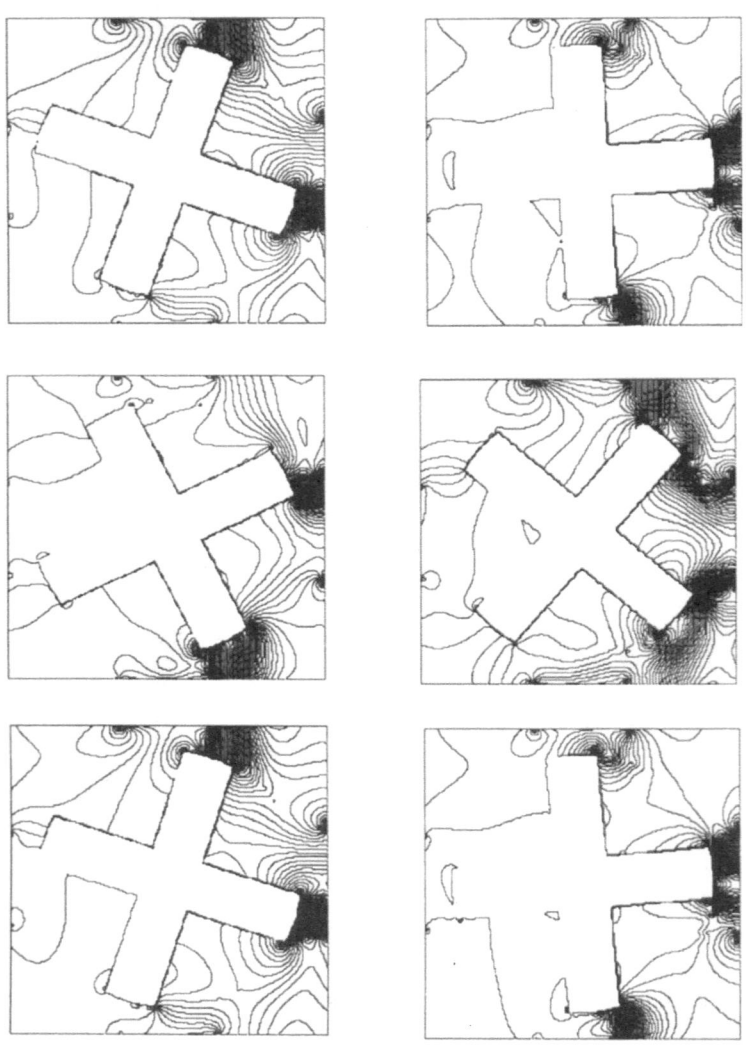

Figure 3.56: Some pressure isolines for rotating propeller

Chapter 4

Numerical comparisons of Navier-Stokes solvers

While the previous Chapter 3 has been mainly concerned with the aspect of appropriate discretization techniques, we have derived in Chapter 2 the *global* and *local MPSC* schemes for the "pure solution process" of the resulting discrete saddle point problems

$$Su + kBp = \mathbf{f}\,, \quad B^T \mathbf{u} = g \quad \text{with} \quad S := \alpha M + \nu \theta_1 kL + k\theta_2 K\,. \quad (4.1)$$

$\alpha, k, \theta_1, \theta_2$ are parameters and $\mathbf{f}$ and g are right hand sides depending on the performed discretizations in space and time. M represents the (diagonal) mass matrix and L is the discretized Laplacian. K corresponds to the convective term and may depend on a given vector $\tilde{\mathbf{U}}$ or on the solution $\mathbf{u}$ itself in the fully nonlinear case.

We have already introduced the **Galerkin** schemes which include both *local* and *global PSC* approaches. The *global SPSC* variants ('S' for 'single grid') are close to the *pressure correction* techniques and are based on modifications of classical **projection** schemes. In contrast, the *global* (multilevel) *MPSC* schemes may differ in the choice of the *pressure Schur complement* preconditioners and their embedding into the multigrid context. Consequently, the resulting methods are modified versions of SIMPLE, Uzawa and projection-like approaches.

Alternatively, the *local MPSC* approaches are based on the concept of local preconditioning and can be interpreted as certain block-Jacobi/Gauß–Seidel–like iterations applied to linearized Navier–Stokes systems. The different choices for the involved preconditioners include schemes as the **Vanka** smoother. While this approach corresponds to elementwise block-Gauß–Seidel schemes, it can be generalized to the (adaptive) *patchwise approach* which defines the corresponding size of the "subdomains" in an adaptive setting during the calculation.

Both techniques are in general applied to linearized problems (Oseen equations) while the nonlinearity is treated in an outer iteration by using the adaptive fixed point defect correction method from Section 3.1.3. The same techniques can be also applied in the case of the nonstationary Navier–Stokes equations by performing first the time and space discretization as an outer process. The resulting discretized equations correspond to (nonlinear) generalized Navier–Stokes–like problems as introduced in (4.1). Applying the **Galerkin** schemes in each time step leads to the so-called *coupled* or *Galerkin approaches* as time stepping schemes. However, in contrast to the treatment of purely steady equations, one is now allowed to weaken the solution process of such coupled nonlinear equations.

For example, we might apply linearization techniques or we solve only approximately the momentum and/or the continuity equation. This failure can be compensated by smaller time steps to guarantee a comparable accuracy in time (and space) as the fully coupled approaches may provide. Even explicit advancing of the momentum equation with a distinguished emphasis on the incompressibility constraint is applicable. One essential assumption for a fair comparison of these schemes is the utilization of the described time step control from Section 3.1.2 for selecting the appropriate time steps.

Further, we have introduced in Section 2.5 some projection-type approaches which were called **1-step projection**, with fully nonlinear, semi–implicit or semi–explicit treatment of the nonlinearity. These can be applied as classical *single grid* SPSC methods or as MPSC schemes, accelerated via multigrid. Though these projection-type schemes seem to be approaches for nonstationary problems only, they can be interpreted as **incomplete solvers** for the resulting generalized Navier–Stokes–like equations in (4.1).

Finally, we can play around with different techniques for treating the nonlinearity in the momentum equations: **fully nonlinear, semi–implicit** or **semi–explicit**. However, due to the assumed splitting between discretization and solution part there is one common issue for all resulting schemes: If the discretization is fixed (to be precise: the Crank–Nicolson- or Fractional-step-θ–scheme together with the nonconforming $\tilde{Q}1/Q0$ Stokes element) they all result in exactly the same solution on a fixed spatial mesh. This can be enforced by requiring that the time step k is sufficiently small, or some iteration counters N (for the nonlinear solver) or L (for the linear coupled solver) are sufficiently large. Since all of them can be interpreted as **special solution methods** for the same problem (4.1), they are now comparable with respect to robustness and efficiency in solving steady problems and hence with respect to the accuracy of the resulting time stepping scheme, too.

In the following, we attempt to figure out the characteristic features for the different schemes. One essential point is to perform numerical tests for problems which are in some sense representative for "real life" applications. This requirement leads to test cases which involve complex physical behaviour in space and time as well as complex geometries, both for stationary and nonstationary flows.

We examine the following major aspects:

- How important is the correct treatment of the nonlinearity? Do linearization strategies lead to satisfying results?

- How decisive is the correct treatment of the incompressibility constraint which is much more accentuated in projection-like schemes? Is it really so important to stress explicitely the "exact" mass conservation for laminar incompressible flows?

- Are explicit techniques possible for the nonlinearity, particularly on complex geometries with locally small spatial mesh widths?

Moreover, we are interested in the subsequent aspects concerning the efficiency of the "solver engine" for the resulting discrete saddle point problems:

- Which preconditioners lead to (linear) solvers working independently of the viscosity parameter, the size and shape of the triangulation and the complexity of the geometry?

- Which solution schemes are suited for direct stationary approaches, and which improve their efficiency with respect to decreasing time steps k?

- What is the influence of the solver performance on accuracy, robustness and efficiency of the overall solution process?

- How do discretization and solution techniques influence each others?

We perform tests for the following characteristic flow situations. They involve steady and nonsteady configurations in which case the temporal behaviour is self-induced (large Reynolds numbers) or initiated by nonsteady inflow boundary conditions. We examine:

1. steady Stokes flow, resp., for low Reynolds numbers,

2. steady flow for Reynolds numbers of medium size,

3. nonsteady Stokes flow, resp., nonstationary low Reynolds number problems,

4. nonsteady flow for midranged Reynolds numbers,

5. fully nonstationary high Reynolds number flow,

6. nonstationary configurations for varying Reynolds numbers with/without steady limit.

The first five flow types are thought to be "extreme cases" which represent different states of typical flow behaviour. First, our aim is to find the optimal schemes among all proposed approaches with respect to each of these flow configurations. Beside the selection of the "good schemes", we provide typical deficiencies of certain variants which may occur in practise for the given configurations. These computations will provide a characterization of the introduced schemes. Based on this knowledge we can finally show that there are indeed solution approaches which work likewise robust and efficient for all examined flow situations, even for topic 6 with varying Reynolds numbers. These special schemes will be candidates for the aimed "discrete Black Box" solver.

The quality of the direct steady solvers is analysed with respect to:

- the complexity of the domain, resp., the shape of the mesh (large aspect ratios!).

- the size of the viscosity parameter ν (assuming a characteristic velocity $\mathbf{u}_{mean} = 1$).

- the size of the performed time step k (which is simulated by $\alpha = 1/k$ in front of the zero-order reaction term – the mass matrix M – in the generalized steady equations).

The same criterions are decisive in the context of the nonstationary schemes. However, the numerical behaviour of the complete time stepping schemes is not only determined by the quality of solving the Navier–Stokes subproblems, but more generally by the complete treatment of the three topics **nonlinearity, incompressibility** and **complete outer control**.

The specific flow problems we consider in the following are:

1. **Flow around a cylinder**

   This problem corresponds exactly to the *1995 DFG Benchmark* which is explained in Section 1.1 and [87]. We concentrate on the nonstationary case and show the typical behaviour for self-induced nonsteady flow with periodic oscillations. The results demonstrate differences with respect to (temporal) **accuracy** and **efficiency**, but neglect the case of complex geometries and meshes.

2. **Channel flow around a square**

   This flow problem is similar to the *1995 DFG Benchmark* configuration, but we strongly vary the aspect ratios of the spatial mesh. These variations are thought to represent the typical numerical behaviour for flows with boundary layers which have to be sufficiently approximated by the mesh. The corresponding aspect ratios range from $AR \approx 1$ up to $AR \approx 10^5$.

   We perform intensive studies for steady flows (Stokes and Navier–Stokes) and compare the different components of the MPSC approaches and the nonlinear iterations with respect to the underlying mesh and the viscosity parameter ν. Further, we prescribe time dependent inflow conditions such that the resulting nonsteady flow behaviour can be studied for the full range of Reynolds numbers representing highly nonstationary flows ($\nu \approx 10^{-3}$) up to very low Reynolds number flow ($\nu \approx 10^3$), but all time dependent. These computations are mainly thought to figure out the **robustness** behaviour of the various components and Navier–Stokes solvers.

3. **Flow through a Venturi pipe**

   This configuration represents the case of (self-induced) highly oscillating flows without "simple" periodical structures. The corresponding flow pattern is the most complicated of all our tests and addresses the question of how to measure the solution quality for such problems. Main topic of our tests is the analysis of the resulting **total efficiency** for different time stepping schemes with respect to such complex flow behaviour.

4.1 Some exemplary numerical examples

The following numerical examples are all from [105] which contains the complete definitions for all performed test configurations and Navier-Stokes solvers. Here, we restrict exemplarily to a few characteristic flow configurations to demonstrate the most important conclusions.

1.) Nonstationary flow around a cylinder:

The following configurations are related to the *1995 DFG Benchmark: Flow around a cylinder* which is carefully explained in Section 1.1, see also [87]. Figure 4.1 shows the coarse mesh (level 1); the refined levels have been created via recursively connecting opposite midpoints. All calculations have been done on the levels 4–6.

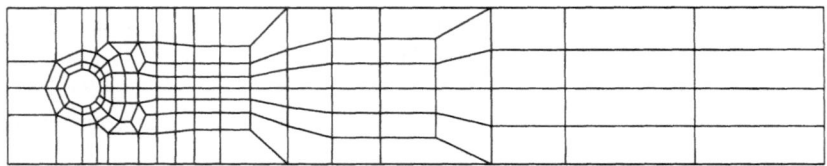

level	vertices	elements	midpoints	total unknowns
1	130	156	286	**702**
4	8,320	8,528	16,848	**42,016**
5	33,280	33,696	66,976	**167,232**
6	133,120	133,952	267,072	**667,264**

Table 4.1: Geometrical information for 'Flow around a cylinder'

We perform nonstationary tests which mainly aim to examine the corresponding (temporal) **accuracy** of the different Navier-Stokes solvers, on a fixed spatial mesh. We start here with the "simplest" (nonstationary) case of periodically oscillating flow for a medium Reynolds number, and neglect the influence of locally - in space and time - complex changes onto the numerical behaviour of the solvers.

Test configuration:

- no slip condition at the upper and lower walls and at the cylinder, parabolic inflow (left side) with a maximum velocity of $u_{max} = 1.5$, natural boundary conditions at the "outflow" (right side)

- viscosity parameter $\nu = 10^{-3}$, resulting Reynolds number $Re = 100$

- upwind discretization of the convective terms (*adaptive upwinding*), Crank–Nicolson as time discretization

- performed for $T = [0, 1]$, started with a fully developed solution

We apply the adaptive time step control from Section 3.3.1 with corresponding functionals for simultanously controlling velocity, pressure, drag and lift coefficients. For the error evaluation, we first perform reference calculations on the same spatial mesh, for each level, and store the resulting drag and lift values as function in time. The following abbreviations are employed:

- NT ~ number of performed macro time steps (1 step with $3k$ and 3 steps with k)

- l_∞ ~ maximum absolute value, resp., l_∞-error over the calculated time interval, l_2 ~ l_2-value, resp., l_2-error over the calculated time interval

- *mean* ~ meanvalue, resp., error for the meanvalue over the calculated time interval, *peak* ~ maximum value, resp., error for the maximum value over the time interval

We start with showing the corresponding drag (c_w) and lift (c_a) plots for the reference solutions, with respect to each grid level 4 – 6.

We provide our subjective ratings of the schemes which are described more carefully in [105]. In the following, "OK" stands for an "acceptable error", while "PERF" means that the approximate and the reference solution are almost identical. Additionally, "CPU" means the total CPU time on an IBM RS6000/590, followed by the percentage of the "solver" part (first item) and the "matrix generation" part (second). The idea is the following: the "solver" part, that means mainly the multigrid routines, can be expected to exploit all features of new processor technologies into direction of Peak performance, at least if an appropriate implementation is carefully done. In contrast, the "matrix generation" denotes the memory access intensive part which is in general only hard to accelerate. Therefore, an optimal scheme for the future should spend most of the elapsed time with the solution process, that means the "solver" part. And the hope is that these schemes will be essentially improved by our next generation of PDE software (see [1]).

We use the following abbreviations for the employed schemes:

- **n-Gal** nonlinear *full Galerkin* approach with the global MPSC approach for Oseen subproblems and "exact" solution of the nonlinear steady equations in each time step

- **l-Gal** similar to the **n-Gal** scheme but with 2nd order linearization of the convective direction via linear extrapolation in time

- **n-Pro** *nonlinear 1–step projection* with diffusive preconditioner; corresponds to classical 2nd order projection-schemes ('Van Kan', 'Gresho–2')

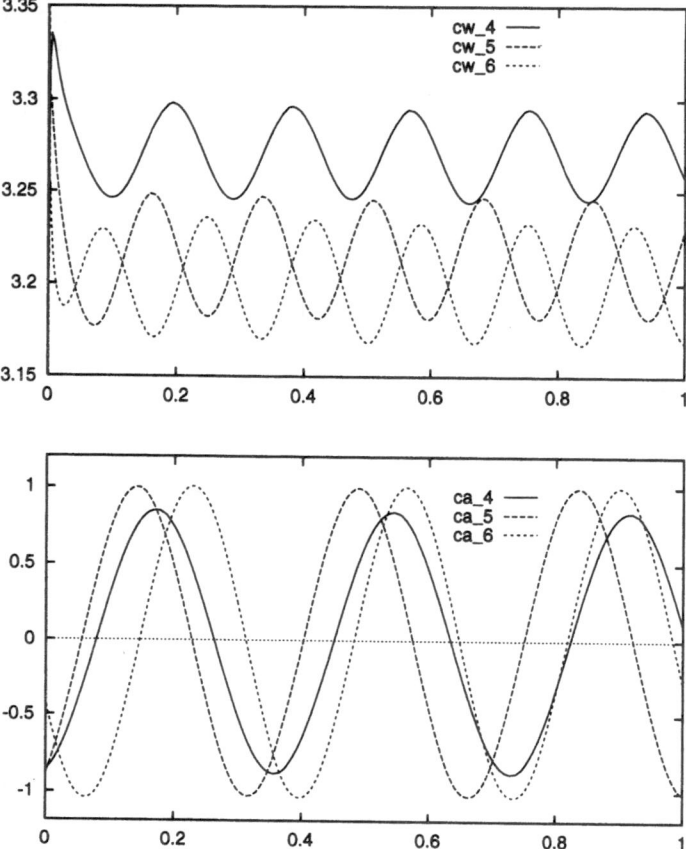

Figure 4.1: Reference results for drag (cw) and lift (ca)

- **l-Pro**   *semi-implicit 1–step projection* with linear extrapolation (2nd order) of the convective direction

- **c-Pro**   *semi-implicit 1–step projection* with constant extrapolation of the convective direction ($u^n \cdot \nabla u^{n+1}$)

- **xl-Pro**   *semi-explicit 1–step projection* with linear extrapolation (2nd order) of the complete convective term

- **xc-Pro**   *semi-explicit 1–step projection* with constant extrapolation of the complete convective term ($u^n \cdot \nabla u^n$)

First of all, we show exemplarily some graphics containing the lift and drag values from the following tables, at least on level 6.

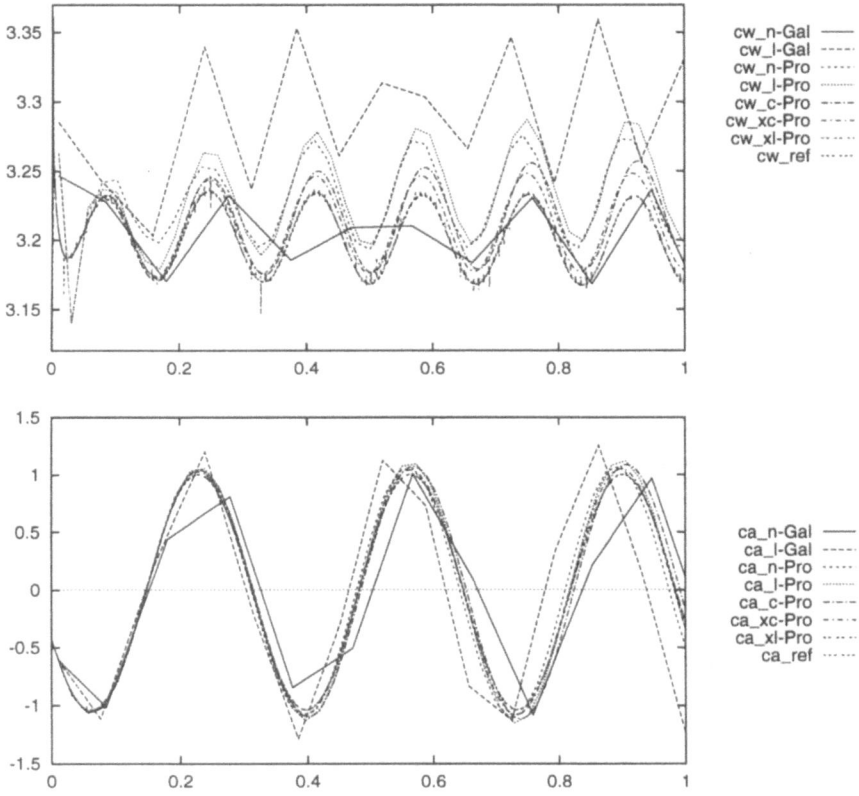

Figure 4.2: "OK" - drag (cw) and lift (ca) values on level 6

On the basis of these results and the complete set of corresponding calculations in [105], we draw the following conclusions for such kind of flow problems:

- Both semi-explicit schemes (**xl-Pro, xc-Pro**) are – at least in these calculations – much less efficient than the more implicit schemes: the needed time steps are much smaller and they further decrease for finer mesh widths. Moreover, the gain in solving only symmetric systems is almost neglegible for the resulting small time steps; larger time steps are impossible due to certain stability restrictions. Additionally, for these small time steps the amount of numerical work in solving large systems of equations is getting less dominant in comparison to the "memory expensive" matrix and defect vector assembling.

- The accuracy and the resulting total efficiency of scheme **c-Pro** which corresponds to classical projection schemes with simple (first order) lin-

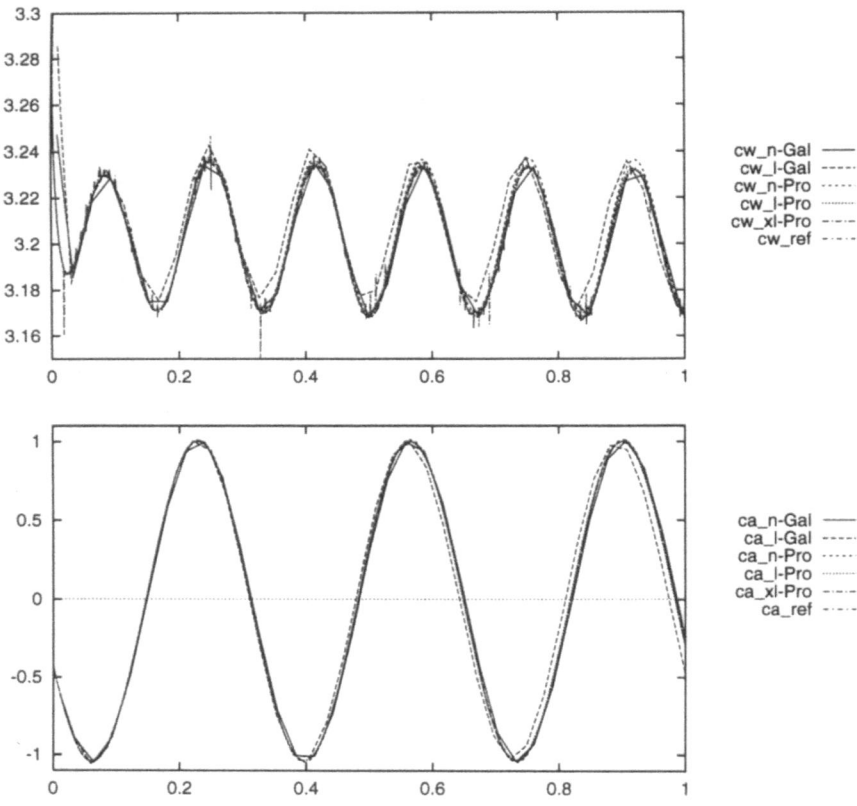

Figure 4.3: "PERF" - drag (cw) and lift (ca) values on level 6

earization of the convective term, is much less than comparable (semi-implicit) second order schemes, as **n-Pro**, **l-Pro** and **l-Gal**.

- The second order schemes **n-Pro** (classical projection scheme with fully nonlinear treatment), **l-Pro** (classical projection scheme with second order linearization), **l-Gal** (MPSC accelerated version of **l-Pro**) and **n-Gal** (full Galerkin MPSC solving a nonlinear generalized Navier-Stokes equations in each time step) are of comparable quality. The Galerkin scheme **n-Gal** requires the smallest number of time steps, but this gain is compensated by larger cost of each time step. Our favourized scheme for these tests is the "new" MPSC scheme **l-Gal** which is similar to the classical projection approach but using multigrid acceleration for the coupling of velocity and pressure. This scheme seems to have almost the same accuracy as the full Galerkin scheme **n-Gal** while it preserves

	CPU	Meth.	NT	Lift coefficient				Drag coefficient			
				l_∞	l_2	mean	peak	l_∞	l_2	mean	peak
O	760 (81/ 9%)	n-Gal	9	50%	19%	2%	1%	1%	0%	0%	0%
	342 (84/ 8%)	l-Gal	12	98%	39%	4%	8%	5%	3%	3%	4%
O	744 (56/30%)	n-Pro	33	23%	19%	3%	18%	2%	1%	1%	1%
K	315 (68/22%)	l-Pro	34	16%	12%	1%	12%	2%	1%	0%	0%
	804 (63/26%)	c-Pro	104	22%	15%	2%	11%	1%	1%	0%	0%
	938 (53/28%)	xc-Pro	210	12%	8%	1%	9%	1%	1%	0%	0%
	1,515 (53/28%)	xl-Pro	338	3%	1%	0%	1%	1%	0%	0%	0%
P	1,548 (76/12%)	n-Gal	27	7%	6%	2%	0%	1%	0%	0%	0%
E	712 (82/ 9%)	l-Gal	30	20%	9%	0%	2%	1%	0%	0%	0%
R	1,449 (47/36%)	n-Pro	97	3%	2%	1%	1%	0%	0%	0%	0%
F	740 (63/26%)	l-Pro	96	4%	1%	0%	1%	0%	0%	0%	0%
	1,515 (53/28%)	xl-Pro	338	3%	1%	0%	1%	1%	0%	0%	0%

Table 4.2: Comparative rating of the total efficiency on level 4

	CPU	Meth.	NT	Lift coefficient				Drag coefficient			
				l_∞	l_2	mean	peak	l_∞	l_2	mean	peak
O	3,031 (78/10%)	n-Gal	11	47%	22%	2%	1%	1%	0%	0%	0%
	1,522 (83/ 8%)	l-Gal	14	90%	45%	3%	23%	6%	3%	2%	2%
O	3,798 (56/30%)	n-Pro	41	13%	11%	3%	12%	2%	1%	1%	0%
K	1,543 (67/22%)	l-Pro	42	18%	11%	1%	6%	2%	1%	0%	0%
	4,562 (63/26%)	c-Pro	147	16%	11%	2%	8%	3%	1%	0%	1%
	8,655 (53/28%)	xc-Pro	372	9%	6%	0%	9%	3%	0%	0%	0%
	10,960 (53/28%)	xl-Pro	608	3%	1%	0%	1%	3%	0%	0%	0%
P	7,331 (74/12%)	n-Gal	33	6%	5%	2%	1%	1%	0%	0%	0%
E	3,397 (82/ 9%)	l-Gal	36	18%	10%	0%	1%	2%	0%	0%	1%
R	8,115 (48/36%)	n-Pro	133	1%	1%	0%	1%	3%	0%	0%	1%
F	4,282 (63/26%)	l-Pro	138	2%	1%	0%	0%	3%	0%	0%	1%
	10,960 (53/28%)	xl-Pro	608	3%	1%	0%	1%	3%	0%	0%	0%

Table 4.3: Comparative rating of the total efficiency on level 5

the excellent efficiency results from the projection schemes **n-Pro** and **l-Pro**.

- From a computational point of view, we prefer both MPSC schemes **n-Gal** and **l-Gal**. By construction, several solution steps are performed once the iteration matrices in each time step, resp., in each nonlinear step, are assembled. Consequently, 80% and more of the numerical work are spent for solution steps including multigrid, and we expect to improve these parts of the code by performing better implementations as described in the previous chapters. Hence, by approaching Peak performance for these parts of the code which consume most of the elapsed time, we expect in future even much better efficiency results due to "optimized" implementation features.

		CPU	Meth.	NT	Lift coefficient				Drag coefficient			
					l_∞	l_2	mean	peak	l_∞	l_2	mean	peak
O	K	14,876 (78/10%)	n-Gal	12	39%	23%	1%	4%	1%	0%	0%	3%
		6,910 (82/ 9%)	l-Gal	16	91%	48%	5%	23%	5%	3%	2%	0%
		16,884 (56/30%)	n-Pro	46	13%	9%	1%	11%	2%	1%	1%	2%
		6,749 (67/23%)	l-Pro	46	21%	12%	0%	6%	2%	1%	0%	2%
		22,749 (63/26%)	c-Pro	182	15%	10%	1%	7%	8%	1%	0%	0%
		38,569 (54/27%)	xc-Pro	536	6%	4%	4%	8%	0%	0%	0%	0%
		64,485 (54/27%)	xl-Pro	889	2%	1%	0%	8%	0%	0%	0%	0%
P E R F		32,538 (73/13%)	n-Gal	36	5%	4%	0%	1%	1%	0%	0%	3%
		14,358 (81/ 9%)	l-Gal	39	19%	10%	1%	1%	2%	0%	0%	2%
		42,679 (51/34%)	n-Pro	165	1%	1%	0%	0%	8%	0%	0%	0%
		21,409 (63/26%)	l-Pro	171	2%	1%	0%	0%	8%	0%	0%	0%
		64,485 (54/27%)	xl-Pro	889	2%	1%	0%	8%	0%	0%	0%	0%

Table 4.4: Comparative rating of the total efficiency on level 6

For this kind of flow (nonsteady with simple periodical behaviour, medium Reynolds number, regular spatial mesh) the second order schemes **n-Gal, n-Pro, l-Pro** and **l-Gal**, representing MPSC approaches of Galerkin type as well as SPSC-like classical projection schemes, lead to comparable results with respect to total efficiency. They all have common that they are second order schemes in all parts and that they treat the nonlinear term at least with semi-implicit approaches.

The next tests are similar to the previous example: we replace the circle by a square in the channel. Then, we are easily enabled to introduce artificial anisotropies in the mesh to simulate the numerical behaviour in configurations in which case highly refined grids are needed to resolve boundary layers. Consequently, we perform stationary as well as nonstationary tests in the following which are mainly conceptualised to examine the **robustness** behaviour of the introduced numerical schemes.

2a.) Stationary channel flow around a square:

The total length of the channel in Figure 4.5 is L=1.8, the height is H=0.75, and the length of the square inside is (again) l=0.1. We provide a parabolic inflow profile at the left inlet which is of maximum velocity 1, and we vary the viscosity parameter in the subsequent calculations.

The following pictures show the coarse meshes S1 and S2, both for level 1; all refined levels have been created via recursively connecting opposite midpoints. Mesh S1 corresponds to a moderately anisotropic grid with an

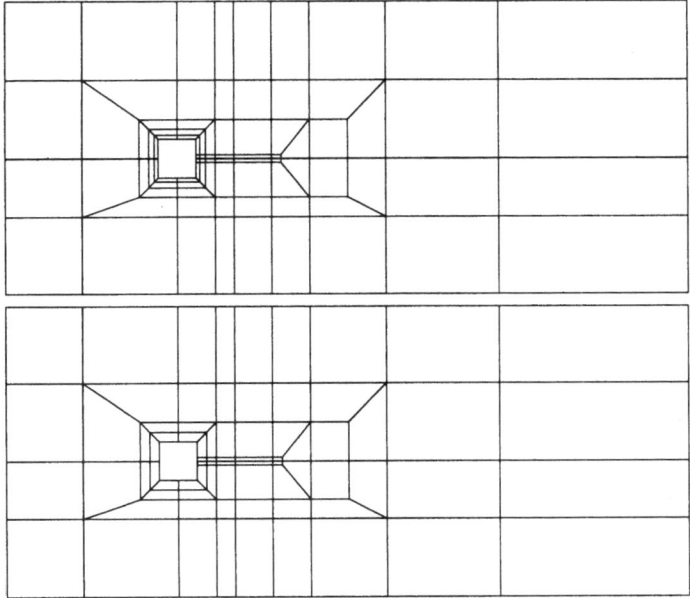

level	elements	vertices	midpoints	total unknowns
1	86	104	190	**466**
3	1,376	1,448	2,824	**7,024**
4	5,504	5,648	11,152	**27,808**
5	22,016	22,304	44,320	**110,656**
6	88,064	88,640	176,704	**441,472**

Table 4.5: Geometrical information and degrees of freedom for 'Flow around a square'

aspect ratio of about $AR \sim 10$, while the configurations S2 ($AR \sim 10^3$) and S3 ($AR \sim 10^5$) are extreme cases which are designed to simulate the behaviour if boundary layers are approximated. We omit the picture for the coarse mesh S3 since already for S2 the most inner layer of elements around the square is not "visible": S1 – S3 differ only in the **position of the most inner elements** around the square.

Test configuration:

- no slip condition at the walls and at the square, parabolic inflow with a maximum velocity of $u_{max} = 1$, natural boundary conditions at the "outflow" (right side)

- streamline–diffusion discretization of the convective terms

- weighted finite element residuals as stopping criterions

- adaptive fixed point defect correction scheme as nonlinear iteration

In the following Tables, we show the results for different schemes, if applied to stationary Navier–Stokes equations on the meshes S1 – S3. The resulting *aspect ratios AR* vary between $AR = 10$ (on S1), $AR = 10^3$ (on S2) and $AR = 10^5$ (on S3). We stop the nonlinear iteration if the maximum relative changes for the **u** and p components as well as if the residuals for the momentum and continuity equation are less then $\epsilon = 10^{-3}$. Since these values are calculated with the weighted finite element residual approach (see Section 3.3.2), these stopping criterions are related to the resulting accuracy of the discrete solution and hence they seem to be sufficient for the employed mesh widths.

For solving the resulting Oseen problems in each nonlinear step, we prescribe that the corresponding initial residual is damped by 0.1, that means we enforce to win one digit!. The following Tables present the **total number of nonlinear iterations (NNL)**, the **total number of performed linear multigrid steps (NMG)** for solving the auxiliary linear Oseen problems and the **averaged number of multigrid steps (AMG)** for one nonlinear iteration. While the nonlinear iteration is expected to be depending on the viscosity parameter ν, we hope that at least the linear multigrid rates, and hence the value for **AMG**, remains independent from the viscosity ν and the aspect ratio AR. This is the aim of our desired "discrete Black Box" solver for discretized Navier–Stokes problems.

Global MPSC/Modified Multigrid–Uzawa

In each smoothing step, we perform the diffusive preconditioner ($\sim$ diagonal pressure mass matrix) only, and we apply no presmoothing and **2** postmoothing steps in all tests.

We can draw the following conclusions which are representative for our numerical experience with **global MPSC** approaches if applied to stationary flow problems:

- As predicted by our theoretical considerations, the global MPSC scheme as smoother leads to multigrid results which are **independent of the mesh topology**: we obtain the typical rates for meshes with aspect ratio $AR = 10^1$ as well as for $AR = 10^5$.

- For low Reynolds numbers, the exclusive use of the diffusive preconditioner ("Modified Multigrid–Uzawa") is sufficient while it decreases its quality as smoother for larger Reynolds numbers. Also this fact

LEV	S1 ($AR = 10^1$) NNL / NMG / AMG	S2 ($AR = 10^3$) NNL / NMG / AMG	S3 ($AR = 10^5$) NNL / NMG / AMG
	$1/\nu = 5$		
3	4 / 6 / 1.5	4 / 6 / 1.5	4 / 6 / 1.5
4	4 / 6 / 1.5	4 / 6 / 1.5	4 / 6 / 1.5
5	5 / 7 / 1.5	4 / 7 / 1.5	5 / 8 / 1.5
	$1/\nu = 50$		
3	9 / 25 / 3.0	6 / 18 / 3.0	6 / 18 / 3.0
4	6 / 10 / 1.5	6 / 10 / 2.0	6 / 10 / 2.0
5	6 / 6 / 1.0	6 / 9 / 1.5	7 / 9 / 1.5
	$1/\nu = 500$		
3	10 / 76 / 7.5	10 / 74 / 7.5	10 / 73 / 7.5
4	11 / 51 / 4.5	11 / 49 / 4.5	11 / 54 / 5.0
5	12 / 31 / 2.5	13 / 35 / 3.0	12 / 31 / 2.5

Table 4.6: Convergence results for **global MPSC**

could be expected from the theoretical results, due to the missing convective preconditioner. The additional use of the reactive preconditioner ("Pressure–Poisson operator") leads to no improvement; it only increases the computational cost!

- Consequently, the linear multigrid rates are almost independent of the aspect ratio, while they deteriorate with increasing Reynolds number. However, for mesh width h getting smaller, the rates additionally improve since for $h \to 0$ the diffusive Stokes part in the Oseen operator gets dominant.

- The nonlinear rates of the applied *adaptive fixed point defect correction* scheme as the outer iteration seem to work independent of the grid, with respect to mesh type and mesh width; however they get worse for larger Reynolds number as could be expected.

For stationary Navier-Stokes problems with midranged Reynolds numbers, the **global MPSC** approaches may fail due to the missing convective PSC-preconditioner. In contrast, they lead to multigrid rates which are independent of the underlying spatial mesh, and hence they are optimal for Stokes problems.

Local MPSC/Standard Vanka

Next, we apply the most simple local MPSC variant, namely the classical Vanka scheme which in our context can be interpreted as elementwisely Block–Gauß–Seidel technique for the Oseen equations. In each smoothing

LEV	S1 ($AR = 10^1$) (0/8) NNL / NMG / AMG			S2 ($AR = 10^3$) (0/32) NNL / NMG / AMG			S3 ($AR = 10^5$) (0/64) NNL / NMG / AMG		
				$1/\nu = 5$					
3	4	8	2.0	4	8	2.0	4	6	1.5
4	5	18	3.5	5	16	3.0	4	14	3.5
5	5	25	5.0	5	48	9.5	5	41	8.0
				$1/\nu = 50$					
3	8	11	1.5	6	12	2.0	6	12	2.0
4	5	13	2.5	6	26	4.5	6	25	4.0
5	5	14	3.0	6	92	15.5	6	85	14.0
				$1/\nu = 500$					
3	10	19	2.0	10	20	2.0	10	20	2.0
4	11	21	2.0	11	35	3.0	11	37	3.5
5	11	24	2.0	12	138	11.5	11	141	12.5

Table 4.7: Convergence results for **Vanka** smoother in local MPSC

step, we apply no presmoothing and **n** postmoothing steps, and the value for
(**0/n**) is indicated in the following table.

While the properties of the nonlinear scheme are preserved, the linear multi-grid rates show a completely different behaviour in contrast to the previous global MPSC approach:

- Due to the elementwise Block–Gauß-Seidel/Jacobi character, the multi-grid rates depend massively on the underlying topology of the mesh: For increasing aspect ratio, we have to enlarge the number of smoothing steps, too, and still the rates are not independent of the mesh width. Additionally, decreasing the mesh width requires for anisotropic meshes the adding of more smoothing steps!

- In contrast, the dependence on the Reynolds number is weakened. As an example, we obtain even better convergence results for the Oseen equations with $1/\nu = 500$ than for the (almost) Stokes problem with $1/\nu = 5$; at least if the mesh is very regular.

While the global MPSC approach seems to have no problems with the shape of the mesh and the mesh width, this first local MPSC approach – which corresponds to the classical Vanka scheme – exhibits severe problems for locally anisotropic grids. However, in contrast to the global MPSC, it shows only weak dependence on the underlying viscosity parameter! Next, we demonstrate how to fix the mesh dependence of the local MPSC by using more appropriate patching techniques instead of working only on (pure) element level.

LEV	S1 ($AR = 10^1$)			S2 ($AR = 10^3$)			S3 ($AR = 10^5$)		
				$1/\nu = 5$					
3	4 / 8 / 2.0			4 / 7 / 2.0			4 / 7 / 2.0		
4	4 / 7 / 2.0			4 / 7 / 3.0			4 / 8 / 2.0		
5	5 / 11 / 2.0			4 / 9 / 2.0			5 / 9 / 2.0		
6	5 / 12 / 2.5			5 / 12 / 2.5			5 / 12 / 2.5		
				$1/\nu = 50$					
3	9 / 12 / 1.5			6 / 9 / 1.5			6 / 9 / 1.5		
4	6 / 11 / 2.0			6 / 10 / 1.5			6 / 12 / 2.0		
5	7 / 13 / 2.0			6 / 11 / 2.0			6 / 17 / 3.0		
6	6 / 14 / 2.5			6 / 11 / 2.0			6 / 18 / 3.0		
				$1/\nu = 500$					
3	10 / 19 / 2.0			10 / 17 / 2.0			10 / 18 / 2.0		
4	11 / 21 / 2.0			12 / 19 / 1.5			12 / 19 / 1.5		
5	11 / 21 / 2.0			13 / 19 / 1.5			11 / 23 / 2.0		
6	12 / 25 / 2.0			12 / 20 / 1.5			12 / 31 / 2.5		

Table 4.8: Convergence results for **adaptive patching** with (0/6) smoothing steps

Local MPSC/Adaptive Patching

The adaptive construction of the patches follows the algorithm of Section 2.4.2. On mesh S1 (moderate anisotropies) **no** blocking is applied while for S2 and S3 the maximum block size on level 2 is $NP = 3$ which increases up to $NP = 17$ on level 5, resp., $NP = 33$ for level 6.

Here, we see the candidate for the "discrete Black Box" tool for solving discretized Navier–Stokes problems: All linear convergence rates are – more or less – independent of the viscosity parameter ν, the mesh width h and the aspect ratio AR.

The solution of quasi-Stokes problems via global MPSC:

While the local MPSC schemes have proven to be "optimal" for stationary Oseen problems, the global MPSC approach has shown some deficiencies. Nevertheless, we show in the following examples that this technique is very well suited for stationary and particularly for nonstationary Stokes equations. For these problems, the global MPSC approach including both the diffusive as well as the reactive preconditioner (in the nonsteady case) leads to very robust <u>and</u> efficient solution schemes.

Figure 4.4 shows the coarse meshes with very different types of local anisotropies. In fact, for the "Driven Cavity" meshes (the unit squares at the left side), the *aspect ratio AR* varies from $AR = 1$ (for the equidistant cartesian grid) up to $AR = 10^6$. Nevertheless, the application of only 1 postsmoothing step, and no presmoothing at all, leads to multigrid convergence rates

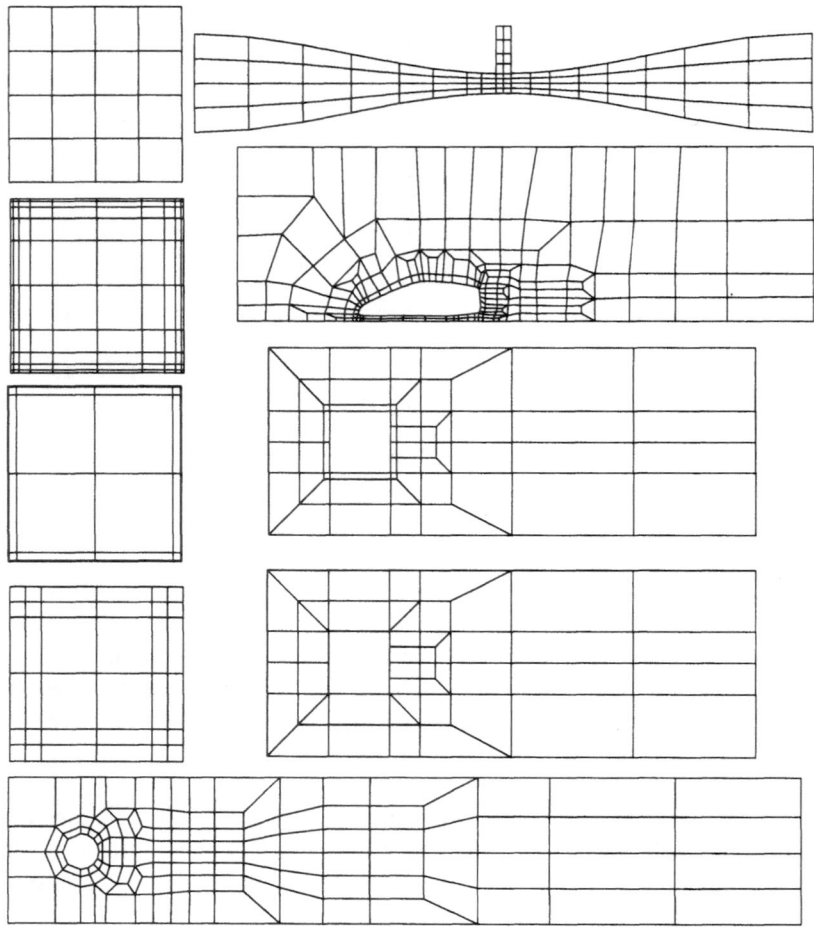

Figure 4.4: Different (coarse) meshes for the Stokes tests

$\rho_{gMPSC} \in [0.35, 0.45]$ for the Stokes problem and they further improve if we perform more smoothing steps, or if we apply the adaptive interpolation techniques (see Section 3.4).

These results show that indeed convergence rates are achievable which are independent of the domain and the used mesh. The only tool which is required for this "Black Box" tool is a **stable** Poisson solver for the velocity problems, which is available for this kind of problem (see the results in Section 3.4). Furthermore, the results show that the Stokes problem can be solved <u>without</u> Pressure–Poisson problems. In fact, the applied scheme for these calculations is nothing else than the modified Uzawa-like iteration which was described in Chapter 2.

Configuration	Level 3	Level 4	Level 5	Level 6
S1	0.386	0.402	0.367	0.334
S3	0.391	0.398	0.382	0.366
DC $(AR = 1)$	0.449	0.444	0.437	0.429
DC $(AR = 10^1)$	0.435	0.435	0.413	0.393
DC $(AR = 10^4)$	0.369	0.353	0.359	0.344
DC $(AR = 10^6)$	0.448	0.479	0.465	0.435
BENCHMARK	0.407	0.391	0.350	0.303
VENTURI PIPE	0.463	0.481	0.475	0.464
CAR	0.523	0.445	0.444	0.441

Table 4.9: Convergence rates $(0/1)$ for the Stokes problem with **global MPSC**

Level	$k = 10^2$	$k = 10^0$	$k = 10^{-2}$	$k = 10^{-5}$	$k = 10^{-8}$
3	0.394	0.447	0.207	0.001	0.001
4	0.410	0.449	0.333	0.003	0.001
5	0.377	0.392	0.381	0.013	0.001
6	0.333	0.353	0.431	0.054	0.001

Table 4.10: Convergence results (1 smoothing step) for the nonsteady Stokes problem on S1

The next test calculations concern the nonsteady Stokes equations in which case the resulting *pressure Schur complement* operator has the form $A = B^T[M+kL]^{-1}B$, with M denoting the mass matrix and L the discrete Laplacian for the velocity. In Section 2.3 we have derived the following properties if the global MPSC scheme with both preconditioners is applied:

$$\rho_{gMPSC} \quad \rightarrow \quad 0 \qquad\qquad \text{for} \quad k \rightarrow 0 \qquad\qquad (4.2)$$

$$\rho_{gMPSC} \quad \rightarrow \quad \rho_{STOKES} \qquad \text{for} \quad k \rightarrow \infty \qquad\qquad (4.3)$$

$$M + kL \quad \sim \quad O(h^2) + O(k) \qquad\qquad\qquad\qquad\qquad\qquad (4.4)$$

Consequently, the multigrid rates ρ_{gMPSC} of the global MPSC approach may depend on the *aspect ratio* and the ratio k/h^2, but they should be always bounded by the corresponding Stokes rates – $\rho_{STOKES} \leq 0.45$, see above – and they should significantly improve for smaller time steps. The corresponding numerical results are shown in Tables 4.10 and 4.11, for the meshes S1 $(AR = 10^1)$ and S3 $(AR = 10^5)$.

With the next tests for the same configuration, we aim to examine the **robustness** of the different Navier-Stokes solvers in nonsteady simulations. Here, we mainly vary the degree of anisotropy of the underlying spatial mesh.

Level	$k = 10^2$	$k = 10^0$	$k = 10^{-2}$	$k = 10^{-5}$	$k = 10^{-8}$
3	0.394	0.440	0.194	0.021	0.077
4	0.401	0.431	0.297	0.038	0.040
5	0.374	0.393	0.364	0.089	0.061
6	0.346	0.373	0.428	0.137	0.037

Table 4.11: Convergence results (1 smoothing step) for the nonsteady Stokes problem on S3

2b.) Nonstationary channel flow around a square:

The following Figure 4.5 shows the coarse mesh A1 (level 1) and the refined grid on level 3. While A1 corresponds to a rather regular grid with an *aspect ratio* of about $AR \sim 4$, we show additionally the coarse meshes for configurations A2 ($AR \sim 10$) and A3 ($AR \sim 100$). We omit the picture for the coarse mesh A4 ($AR \sim 1,000$); A2 – A4 differ only in the **position of the most inner elements** around the square. The number of elements, midpoints and hence the total number of unknowns is the same as for the stationary tests (see Table 4.5).

Test configuration:

- no slip condition at the upper and lower walls and at the square, parabolic inflow (left side) with $u_{max} = 1$, weighted by $\sin(\pi t/8)$, natural boundary conditions at the "outflow" (right side)

- viscosity parameter $\nu = 10^{-3}$, resulting maximum Reynolds number $Re = 100$

- upwind discretization of the convective terms (*adaptive upwinding*), Fractional–step–θ–scheme as time discretization

- performed for $T = [0 : 8]$, started from rest, all calculations on mesh level 5

We apply the adaptive time step control from Section 3.3.1 with corresponding functionals for simultaneously controlling velocity, pressure, drag and lift coefficients.

- $NT \sim$ number of performed macro time steps (1 step with $3k$ and 3 steps with k)

- $l_\infty \sim$ maximum absolute value, resp., l_∞-error over the calculated time interval, $l_2 \sim l_2$-value, resp., l_2-error over the calculated time interval

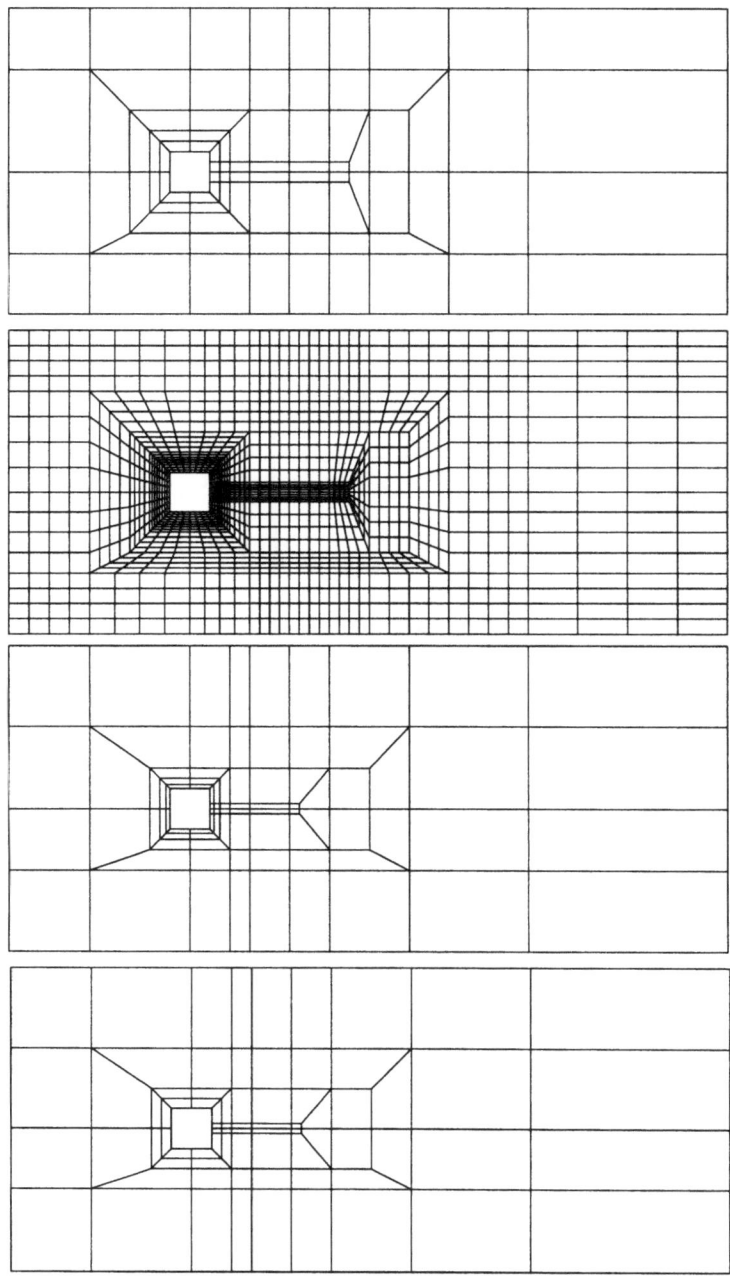

Figure 4.5: Different (coarse) meshes for the nonstationary tests

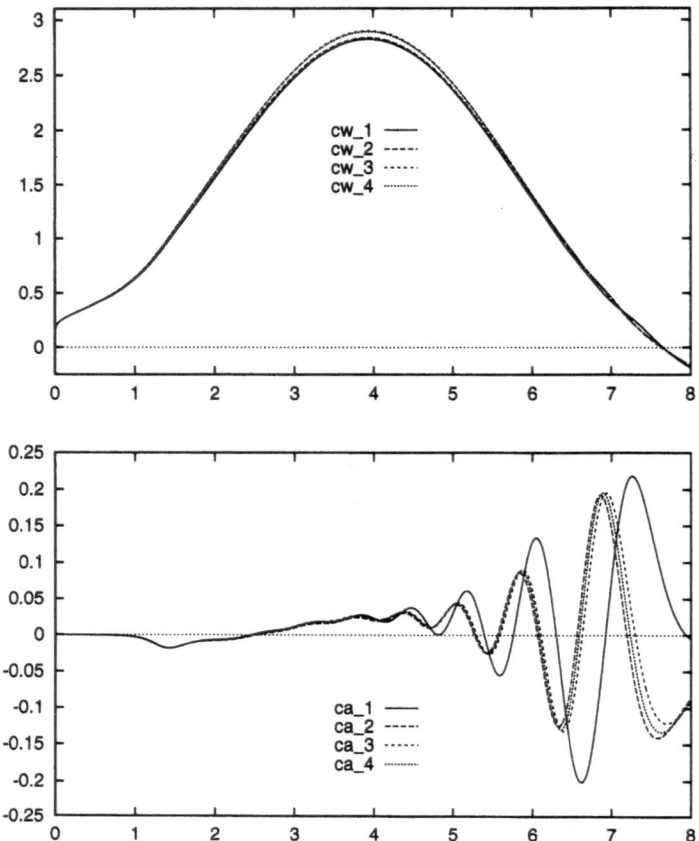

Figure 4.6: Drag (cw) and lift (ca) values for the reference calculations for $\nu = 10^{-3}$

- *mean* $\sim$ meanvalue, resp., error for the meanvalue over the calculated time interval, *peak* $\sim$ maximum value, resp., error for the maximum ·value over the time interval

For the error evaluations in integral form (l_2) we apply the summarized midpoint rule in 1D: the results have been often better than compared with the trapezoidal rule since oscillations are damped out if the comparisons between reference and actual solution are done in the midpoints of each small time interval (see the later examples). We start with showing the corresponding drag (c_w) and lift (c_a) plots for the reference solutions, with respect to each mesh A1 – A4.

Obviously, the differences for the various meshes are not large: the drag values are almost "mesh-independent", while the lift values for mesh A1 (with $AR \sim 4$ and slightly modified structures behind the square) are different from meshes A2 – A4 ($AR \sim 10 - 1,000$). For the error evaluation we restrict to compare the results for the total time interval $T = [0.1 : 8]$, hereby cancelling "strange" effects during the startup. The following calculations have been performed with $TOL = 10^{-3}$ for the time step control (see also [105] for more details). We compare the same methods as before, these are:

- **n-Gal** nonlinear *full Galerkin* approach with the global MPSC approach for Oseen subproblems and "exact" solution of the nonlinear steady equations in each time step

- **l-Gal** similar to the **n-Gal** scheme but with 2nd order linearization of the convective direction via linear extrapolation in time

- **c-Gal** similar to the **l-Gal** scheme but with first order linearization of the convective direction via constant extrapolation in time ($\mathbf{u}^n \cdot \nabla \mathbf{u}^{n+1}$)

- **n-Pro** modified *nonlinear 1–step projection* with diffusive <u>and</u> reactive preconditioner; corresponds to classical 2nd order projection-schemes ('Van Kan', 'Gresho–2')

- **l-Pro** *semi-implicit 1–step projection* with linear extrapolation (2nd order) of the convective direction, analogously to **n-Pro**

- **c-Pro** *semi-implicit 1–step projection* with constant extrapolation of the convective direction

- **xl-Pro** *semi-explicit 1–step projection* with linear extrapolation (2nd order) of the complete convective term

- **xc-Pro** *semi-explicit 1–step projection* with constant extrapolation of the complete convective term ($\mathbf{u}^n \cdot \nabla \mathbf{u}^n$)

The aim of these comparisons is not so much the resulting accuracy. Much more, we are interested in the resulting numerical behaviour with respect to the different *aspect ratios*. By physical reasons (and visible from the reference calculations), the influence should not be large, but how do the total number of time steps (NT) and the flow behaviour depend on the used mesh? Do we get solutions in all cases?

- Not all schemes are able to finish the calculations on <u>all</u> meshes. While we always obtain solutions on meshes with moderate *aspect ratios* (mesh A1 with $AR \sim 4$ and mesh A2 with $AR \sim 10$), all those schemes diverge which use the linear extrapolation backwards in time (over the last two time levels) or which are working semi-explicitly (considering the convective terms),

Meth.		NT	Lift coefficient				Drag coefficient			
			l_∞	l_2	mean	peak	l_∞	l_2	mean	peak
REF	A1	1300	0.218	0.198-0	0.148-1	0.218	2.825-0	0.496+1	0.149+1	2.825
	A2	1215	0.193	0.169-0	0.636-3	0.193	2.837-0	0.500+1	0.150+1	2.837
	A3	1243	0.196	0.165-0	0.450-2	0.196	2.901-0	0.509+1	0.153+1	2.901
	A4	1257	0.196	0.169-0	0.255-2	0.196	2.893-0	0.509+1	0.153+1	2.893
n-Gal	A1	90	26%	25%	–	5%	1%	1%	0%	0%
	A2	97	54%	55%	–	0%	1%	1%	1%	0%
	A3	98	46%	48%	–	1%	1%	1%	1%	0%
	A4	101	73%	75%	–	1%	2%	1%	0%	0%
n-Pro	A1	184	118%	110%	–	18%	2%	1%	1%	0%
	A2	159	114%	106%	–	5%	2%	1%	0%	0%
	A3	164	109%	112%	–	6%	2%	1%	0%	0%
	A4	174	140%	148%	–	7%	1%	1%	0%	0%
c-Gal	A1	115	115%	89%	–	23%	3%	2%	1%	0%
	A2	162	58%	50%	–	18%	2%	1%	1%	0%
	A3	183	41%	44%	–	20%	5%	2%	2%	5%
	A4	164	49%	41%	–	19%	3%	1%	1%	2%
c-Pro	A1	185	196%	187%	–	16%	2%	1%	1%	0%
	A2	265	109%	116%	–	47%	4%	2%	1%	0%
	A3	307	166%	161%	–	36%	4%	2%	1%	2%
	A4	287	116%	122%	–	46%	5%	2%	2%	1%
l-Gal	A1	193	144%	147%	–	55%	5%	3%	1%	0%
	A2	172	119%	117%	–	33%	4%	2%	2%	0%
l-Pro	A1	228	94%	81%	–	37%	4%	2%	1%	0%
	A2	190	86%	85%	–	6%	2%	1%	1%	0%
xl-Pro	A1	1355	229%	185%	–	54%	7%	3%	2%	1%
	A2	1999	232%	224%	–	50%	6%	2%	1%	1%
xc-Pro	A1	825	193%	180%	–	59%	6%	3%	1%	0%
	A2	1007	113%	102%	–	27%	3%	2%	1%	0%

Table 4.12: Comparative errors for the lift/drag for $\nu = 10^{-3}$ with specified time adaption

- All schemes with second order linearization of the convective terms do not seem to be able to work on highly anisotropic meshes! The examples perfectly demonstrate this fact which up to now has no theoretical explanation, but which was always present in our computations, in 2D as well as in 3D. The same is true for the semi-explicit techniques, however this is explanable with the "hidden" stability restriction which couples spatial and temporal step sizes as seen before.

Another observation which will be subject of the following examinations is the following one:

- While for the drag value almost no differences can be detected for the various schemes, large differences can be observed for approximating the lift coefficients. The MPSC schemes (n-Gal, c-Gal) reproduce nicely the shape of the curve and in particular the size of the amplitude, while the SPSC schemes (n-Pro, c-Pro) which correspond to the "classical" projection schemes, always tend to exhibit oscillations with amplitudes much too large. This experience is typical for all our numerical tests!

- The full Galerkin scheme **n-Gal** exhibits the smallest errors while requiring the smallest number of time steps at the same time. So, it is the "clear winner" of this competition.

While the schemes involving a fully nonlinear treatment or the constant extrapolation technique for linearizing the convective direction work almost independently from the underlying type of spatial mesh, the semi-explicit approaches and also the second order linearization schemes fail completely on meshes with large *aspect ratios*. The SPSC schemes tend to produce oscillations with amplitudes (much) too large.

The next tests will examine more carefully the properties of linearization strategies on anisotropic meshes. We have already demonstrated that the linear extrapolation - being second order accurate - fails for large aspect ratios. However, also the "standard" constant linearization ($\mathbf{u}^n \cdot \nabla \mathbf{u}^{n+1}$) shows certain deficiencies.

While for meshes A1 and A2 all methods show the same qualitative behaviour, some unphysical oscillations for the lift appear on meshes A3 and particularly on A4. Since the error evaluation has been performed with the described midpoint quadrature rule, the errors are surprisingly good, in contrast to the results if the trapezoidal rule was applied - which corresponds to the shown plots. It is remarkable that in spite of these high oscillations, still good approximations can be obtained for $t \geq 5$: These numerical oscillations can be weakened by smaller time steps. The same effects can be detected for the drag coefficient, but the oscillations are much smaller.

As a further conclusion from these tests we can state that <u>most</u> linearization schemes have problems on very anisotropic meshes (with respect to accuracy or stability), with or without adaptive time step control. In contrast, the fully nonlinear treatment is absolutely robust!

In the next test examples, we try to figure out how the described failures depend on the Reynolds number. Thus, the following test configuration is quite similar:

- Calculations are performed for $T = [0 : 12]$, started from rest, error measurements are done for $T = [0.5 : 12]$.

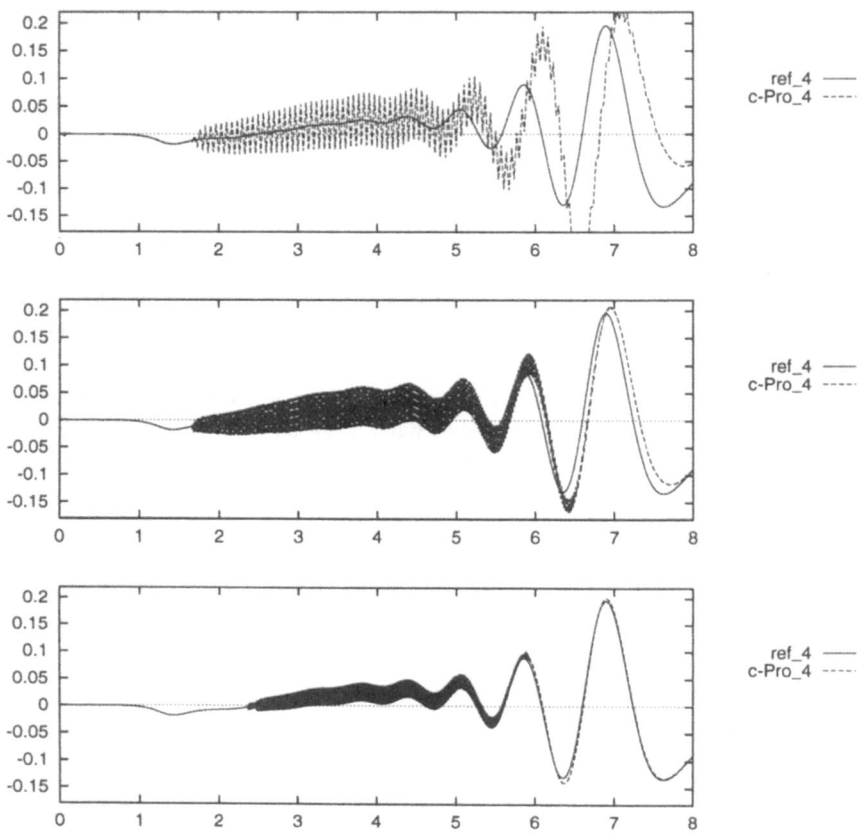

Figure 4.7: Lift values for **c-Pro** for $\nu = 10^{-3}$ with fixed $\Delta t = 1/90$ (top), $\Delta t = 1/300$ (middle) and $\Delta t = 1/900$ (bottom) on mesh A4

- Parabolic inflow (left side) is prescribed again with $u_{max} = 1$, now weighted by $\sin(\pi t/8)$ only until $t = 4$. From $t = 4$ on we prescribe the steady parabolic inflow profile with $u_{max} = 1$, and due to the large viscosity parameter $\nu = 1$, the flow gets stationary.

Again, we start with showing the corresponding drag (c_w) and lift (c_a) plots for the reference solutions, with respect to each mesh A1 – A4.

Since the Reynolds number is very small, we do not expect much differences with respect to the different kinds of linearization: we omit the semi-explicit methods from the following tests. Instead, we are interested in testing the various techniques for treating the incompressibility constraint. Therefore, we add the following scheme:

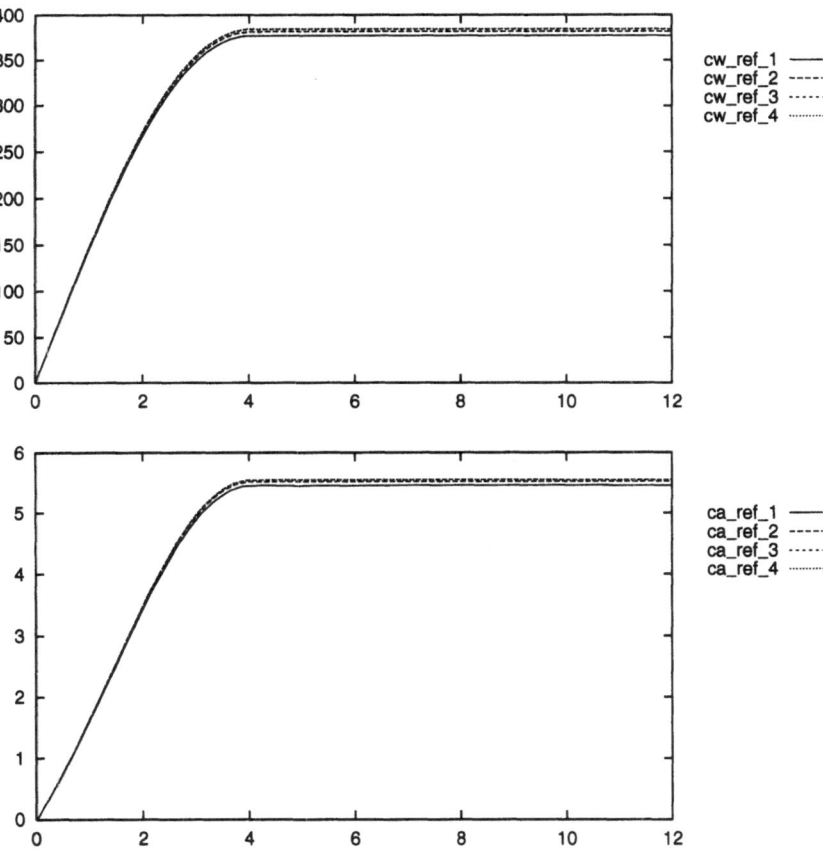

Figure 4.8: Drag (cw) and lift (ca) values for the reference calculations for $\nu = 1$

- **old-Pro** <u>classical</u> *(nonlinear) 1–step projection* with reactive pre-conditioner only which corresponds to the "Pressure Poisson problem" (without pressure mass matrix)

There are essentially two different approaches as Navier-Stokes solvers for this problem:

1. **SPSC-schemes** (as "classical" pressure correction methods: **old-Pro, n-Pro, c-Pro, l-Pro**) versus **MPSC-schemes** (**n-Gal, c-Gal, l-Gal**) which spend more work in solving the Stokes-like systems in each time step.

2. Schemes applying the **"reactive"** preconditioner only ("classical" pressure correction methods involving the Pressure–Poisson operator only:

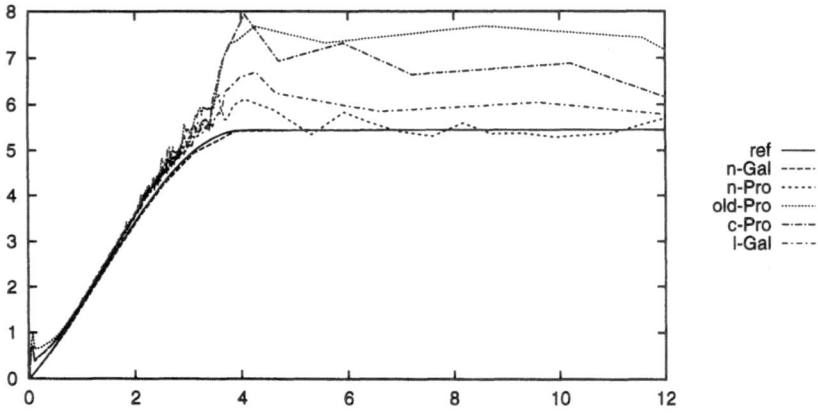

Figure 4.9: Lift values on mesh A1 for $\nu = 1$

old-Pro) versus schemes involving both **"reactive"** and **"diffusive"**
preconditioners (**n-Gal, n-Pro**).

The results in [105] and the subsequent plots of the drag and particularly the
lift coefficients allow the following conclusions:

1. As soon as MPSC schemes (**n-Gal, c-Gal, l-Gal**) are applied which
   aim to "solve" the coupled Stokes, resp., Oseen-like systems in each
   time step, the resulting quality is better than compared with the SPSC
   schemes (**old-Pro, n-Pro, c-Pro, l-Pro**). The treatment of the
   (small) nonlinearity can be almost neglected, and full Galerkin as well
   as linearization techniques lead to almost the same results.

2. The influence of the type of mesh is small. In fact, on the strongly
   anisotropic meshes A3 and A4 the results are sometimes even better
   than on the regular grids A1 and A2.

3. While the exclusive use of the "classical" reactive preconditioner
   ("Pressure–Poisson operator") obviously is not sufficient for this small
   Reynolds number (see the plots for scheme **old-Pro**), the additional
   "diffusive" operator, which is usually used in Uzawa-like methods, guarantees a much better numerical behaviour. Since the additional numerical cost are neglegible, we strongly recommend to use this second preconditioner in projection-like, resp., in pressure correction-like schemes.

We finish with Table 4.13 showing the final values at $t = 12$ which should
have come near to the steady state value. These examples clearly demonstrate
the disadvantageous behaviour of the exclusive use of the "classical" reactive
preconditioner for small Reynolds numbers (scheme **old-Pro**) and of the
SPSC-schemes (**Pro**) compared with the MPSC-schemes (**Gal**).

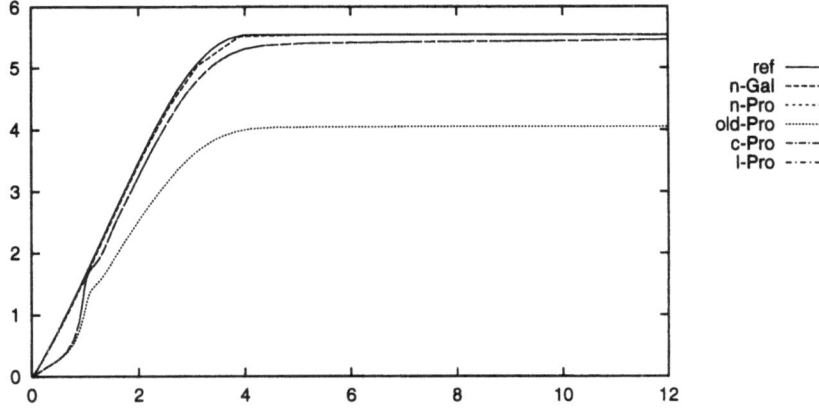

Figure 4.10: Lift values on mesh A4 for $\nu = 1$

For the next tests we further increase (!!!) the viscosity parameter to $\nu = 10^3$ which corresponds to a **very small Reynolds number**. This flow may be viewed as an example for simulating a very viscous fluid as oil, but nonstationary through nonsteady boundary conditions.

Consequently, the nonlinearity can be neglected such that corresponding Stokes calculations give the same results. We omit the plots of the corresponding drag (c_w) and lift (c_a) curves for the reference solutions which look (up to some scaling) identical as for $\nu = 1$. In the following, we restrict to the methods **n-Gal, n-Pro, old-Pro** which represent:

- fully coupled MPSC methods, solving the Stokes-like systems in each time step (**n-Gal**).

- "classical" projection SPSC schemes with the "reactive" preconditioner only (**old-Pro**).

- modified projection SPSC techniques with both preconditioners (**n-Pro**).

From the previous Table (and the Figures in [105]), the following conclusions can be drawn:

1. The Galerkin scheme **n-Gal** which **solves "exactly"** the Stokes-like systems in each time step, works best for this kind of flow. The numerical behaviour is independent of the underlying mesh, and the results are more or less the same as for $\nu = 1$.

2. The classical projection scheme **old-Pro** with the "reactive" preconditioner only, that means performing only once the Pressure Poisson

Meth.		NT	Lift coefficient		Drag coefficient	
			value	error	value	error
REF	A1	70	5.448	–	3.771+2	–
	A2	69	5.518	–	3.814+2	–
	A3	71	5.545	–	3.840+2	–
	A4	69	5.543	–	3.844+2	–
n-Gal	A1	13	5.450	0%	3.771+2	0%
	A4	13	5.546	0%	3.844+2	0%
n-Pro	A1	329	5.702	5%	3.775+2	0%
	A2	336	5.533	0%	3.819+2	0%
	A3	54	5.552	0%	3.839+2	0%
	A4	43	5.467	1%	3.824+2	1%
old-Pro	A1	313	7.172	31%	4.183+2	11%
	A2	338	6.072	10%	3.750+2	2%
	A3	58	5.639	2%	3.113+2	19%
	A4	41	4.051	27%	2.750+2	28%
l-Gal	A1	13	5.447	0%	3.771+2	0%
	A4	13	5.546	0%	3.844+2	0%
l-Pro	A1	318	5.785	6%	3.812+2	1%
	A4	43	5.467	1%	3.824+2	1%
c-Gal	A1	13	5.446	0%	3.771+2	0%
	A4	13	5.547	0%	3.844+2	0%
c-Pro	A1	303	6.155	13%	3.934+2	4%
	A4	43	5.467	1%	3.823+2	1%
xc-Gal	A1	15	5.449	0%	3.772+2	0%
	A4	15	5.547	0%	3.844+2	0%
xc-Pro	A1	297	6.083	12%	3.843+2	2%
	A4	43	5.467	1%	3.824+2	1%

Table 4.13: Comparative errors for the lift and drag for $\nu = 1$ at $t = 12$

operator as preconditioner in each time step, fails completely and cannot even finish the calculation. Even very small (fixed) time step sizes still produce huge errors.

3. The adding of the "diffusive" preconditioner of Uzawa type in the scheme n-Pro leads to slightly better results, but the resulting SPSC scheme is not comparable with the MPSC approach. The dependence on the underlying mesh is not clear.

4. The failures for the SPSC schemes are not only with respect to "local" drag and lift values. Instead of it, the pressure is even globally wrong (see the results in [105]).

5. For this kind of flow (very small Reynolds number), only the MPSC approach leads to satisfying results while the "classical" SPSC techniques

Type		NT	Lift coefficient		Drag coefficient	
			value	error	value	error
n-Gal	A1	20	3.513+3	0%	3.769+5	0%
	A2	20	3.563+3	0%	3.813+5	0%
	A3	20	3.583+3	0%	3.839+5	0%
	A4	20	3.583+3	0%	3.843+5	0%
n-Pro	A1	53	2.077+4	277%	1.036+6	172%
	A2	53	2.356+4	470%	6.880+5	81%
	A3	53	4.840+3	35%	3.722+5	3%
	A4	52	1.918+3	65%	3.223+5	16%
n-Pro	A1	1200	3.592+3	4%	3.769+5	0%
(fixed time steps)	A2	1200	3.663+3	3%	3.812+5	0%
	A3	1200	3.594+3	0%	3.839+5	0%
	A4	1200	3.589+3	0%	3.843+5	0%
old-Pro	A1	1200	1.311+4	217%	8.442+5	124%
(fixed time steps)	A2	1200	7.364+3	106%	6.293+5	65%
	A3	1200	3.852+3	8%	2.839+5	26%
	A4	1200	3.051+3	15%	2.543+5	34%

Table 4.14: Comparative errors for the lift and drag for $\nu = 1,000$ at $t = 12$

fail! The applied time step control for the projection-like schemes **n-Pro, old-Pro** is <u>useless</u> for this kind of flow!

While for large Reynolds numbers the linear extrapolation techniques get in trouble on anisotropic meshes, the SPSC schemes, without and also with both preconditioners, fail for very low Reynolds number. Only the **MPSC approaches** in combination with **fully nonlinear techniques** seem to work efficient and robust, with respect to complex meshes as well as different viscosity parameters.

Next, we examine the dependence of the numerical solution behaviour for the 2nd order linearization techniques on the prescribed viscosity parameter. We have already shown that for large Reynolds numbers ($\nu = 10^{-3}$) the linear extrapolation backwards in time - for evaluating the convective direction - fails on very anisotropic meshes (A3, A4) while no problems are visible for small Reynolds numbers ($\nu \geq 1$).

In the following, we restrict on the mesh A4 (with aspect ratio 1,000), and perform the same flow simulations as before for different viscosity parameters to shed more light into the indicated strange numerical behaviour. Furthermore, we only test the following schemes which all are potentially 2nd order accurate in time:

- **n-Gal**   nonlinear *full Galerkin* approach with the global MPSC approach for Oseen subproblems and "exact" solution of the nonlinear steady equations in each time step

- **l-Gal**   similar to the **n-Gal** scheme but with 2nd order linearization of the convective direction via linear extrapolation in time

- **n-Pro**   modified *nonlinear 1–step projection* (SPSC) with diffusive <u>and</u> reactive preconditioner; corresponds to classical 2nd order projection-schemes ('Van Kan')

- **l-Pro**   *semi-implicit 1–step projection* (SPSC) with linear extrapolation (2nd order) of the convective direction, analogously to **n-Pro**

- **l-Pro1**   incomplete *semi-implicit 1–step projection* (MPSC) with linear extrapolation (2nd order) of the convective direction, analogously to **l-Pro**: exactly 1 (!) MPSC multigrid sweep instead of full iteration as in **l-Gal**

Figure 4.11 demonstrates that for the viscosity parameter $\nu = 1/100$ no problems occur with large aspect ratios. All schemes show about the same convergence behaviour and approximate with sufficient accuracy the nonstationary phase of the flow as well as the steady state solution. Next, we decrease the viscosity parameter to $\nu = 1/250$: While the classical projection scheme **l-Pro** diverges shortly before reaching the maximum Reynolds number at $t = 4$, method **l-Pro1** improves slightly but still fails. The "full" MPSC approach **l-Gal** seems to be able to continue. Nevertheless, at about $t = 10$, this schemes begins to oscillate, too, and finally it diverges! In contrast, both **fully nonlinear** approaches **n-Gal, n-Pro** show no problems at all. Choosing a smaller time stepping criterion can fix the problems, particularly for scheme **l-Gal** if the steady limit is reached.

Further, no differences could be observed related to the application of the Crank-Nicolson (CN) scheme instead of the Fractional-step-θ (FS) method. However, a fully implicit treatment of the convective terms as applied in the Implicit Euler (IE) scheme (that means, the convective term at the actual time level is weighted with $\theta = 1$ instead of $\theta = 0.5$ in the CN method) helps in stabilizing the schemes, particularly in reaching the steady state. But the accuracy of this "mixed" approach (CN for the reactive and diffusive parts, IE for the convective part) is not clear for fully nonsteady problems.

The final test is performed for $\nu = 1/500$ which – in contrast to both previous configurations – exhibits a nonsteady flow behaviour with the typical periodical vortex shedding behind the square. We restrict to the "hardest" mesh A4 only, with an aspect ratio of about $AR = 1,000$. Figure 4.12 shows the corresponding lift values for the reference solution on this mesh, and the "catastrophic" results for the linearized schemes **l-Pro, l-Pro1, l-Gal**.

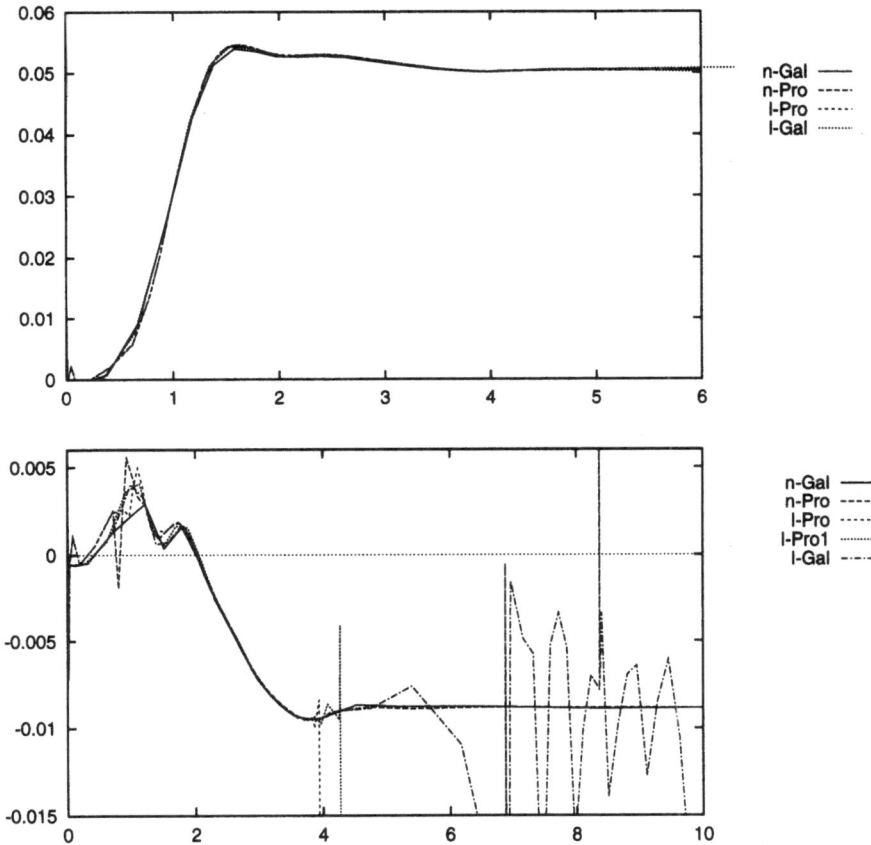

Figure 4.11: Lift values on mesh A4 for $1/\nu = 100$ (top) and $1/\nu = 250$ (bottom)

As can be seen, all these methods diverge between $t = 2$ and $t = 3$, and do not even reach the moment when the Reynolds number receives its maximum value (at $t = 4$). Figure 4.13 shows the corresponding results if we perform the same simulations with the Implicit Euler method for the complete momentum equation (IE) and for the convective part only (CN/IE). The range of stability can be increased, and both linearized variants of the MPSC scheme l-Gal do even finish the calculation up to $t = 12$. However, the corresponding plots for the lift indicate a loss of accuracy and indicate first order schemes only.

Next, we compare the schemes n-Gal ("full Galerkin") and n-Pro ("modified projection"), both with fully nonlinear treatment. Additionally, we look at scheme c-Gal which is a MPSC scheme with (first order) constant linearization of the convective direction (see Figure 4.14).

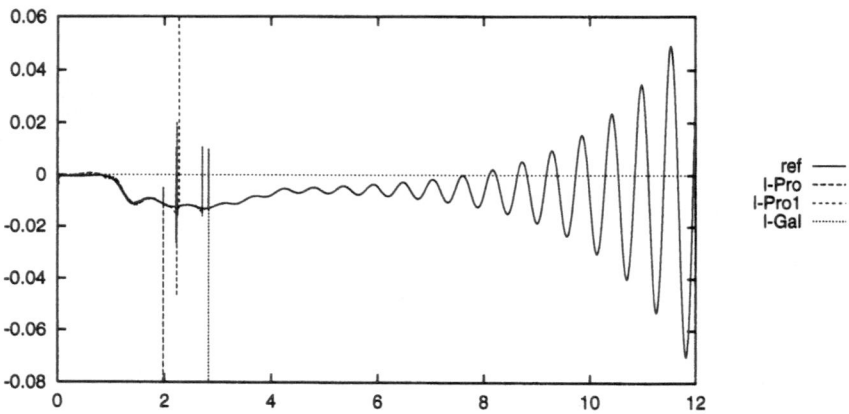

Figure 4.12: Lift values for linearized schemes and reference solution on A4 for $1/\nu = 500$

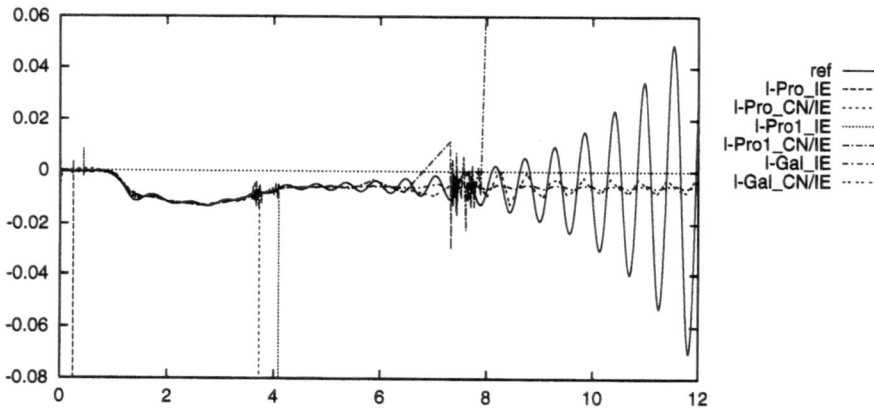

Figure 4.13: Lift values for modified schemes and reference solution on A4 for $1/\nu = 500$

The "projection-like" schemes **n-Pro** and **c-Gal** tend to unphysically large oscillations while the full Galerkin scheme produces curves with amplitudes too small. If we increase the time stopping criterion to $\varepsilon = 10^{-4}$, we can observe that the amplitudes of the "projection-like" schemes decrease while the corresponding values of **n-Gal** increase. This effect that the "projection-like" schemes tend to unphysically large oscillations, is examined more carefully in the 'Virtual Album' (see the included CDROM) which demonstrates how such effects can be "used" for spectacular CFD simulations while their quantitative behaviour is completely wrong!

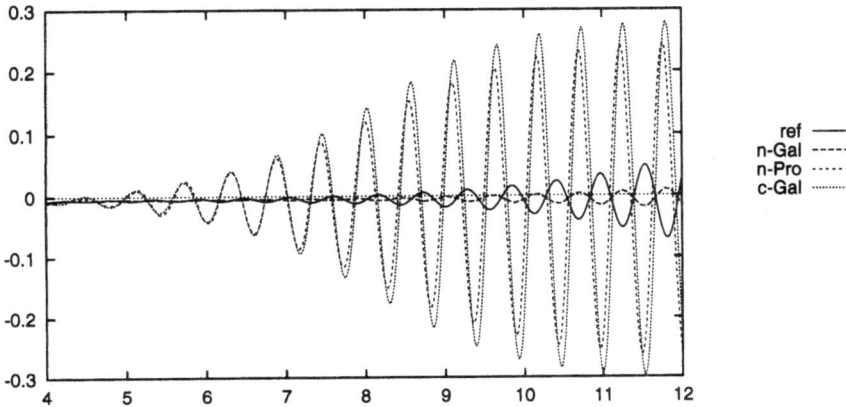

Figure 4.14: Lift values for time stopping criterion $\varepsilon = 10^{-3}$ for $1/\nu = 500$

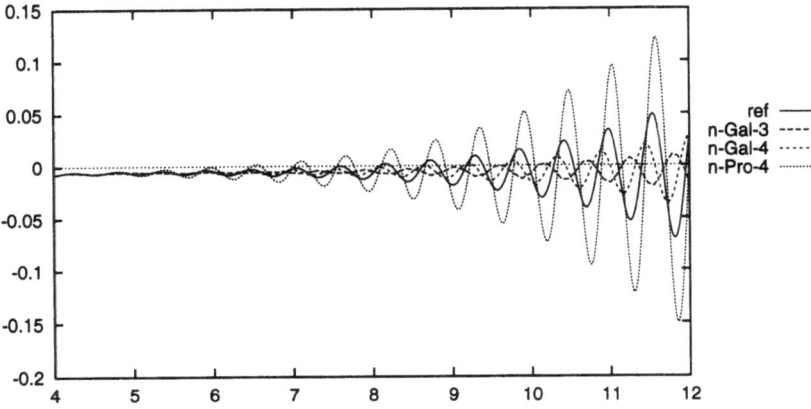

Figure 4.15: Lift values for time stopping criterions $\varepsilon = 10^{-3}$ and $\varepsilon = 10^{-4}$ for $1/\nu = 500$

A further alternative approach to the MPSC Galerkin scheme **n-Gal** is the following variant **n-Gal-con** which determines the corresponding velocity approximation with the main emphasis on the continuity equation instead of solving the momentum equation with the given pressure as in **n-Gal** (see Section 2.3): This modified approach always provides discretely divergence-free approximations, but no advantages are visible! See the following computations which provide corresponding results.

With this set of Navier-Stokes solvers, we finish our robustness tests with showing the selected time step sizes which result from the applied adaptive time step control.

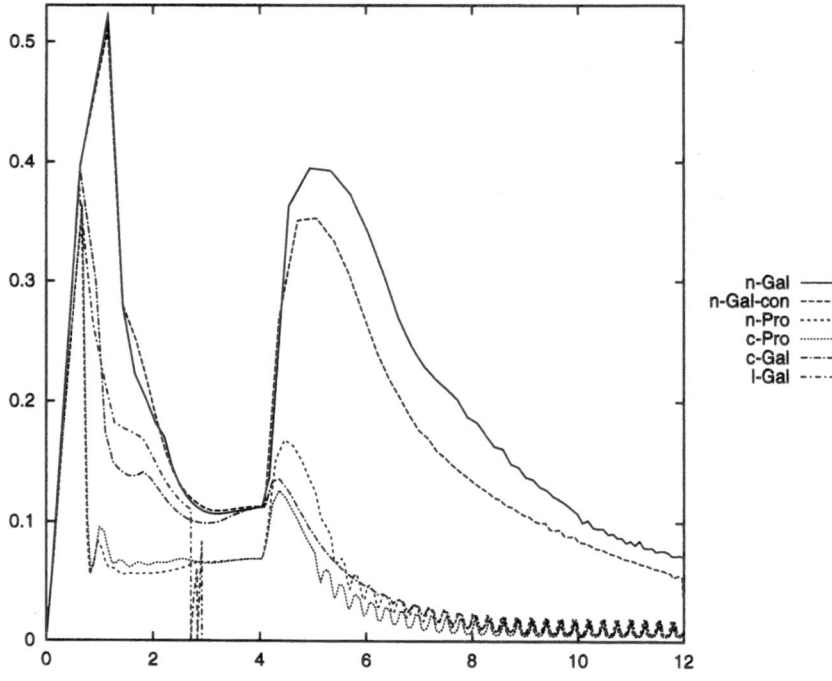

Figure 4.16: Actually applied macro time step sizes for $\varepsilon = 10^{-3}$ for $1/\nu = 500$

Meth.	$T = 4$	$T = 6$	$T = 12$
n-Gal	28	6 (**34**)	48 (**82**)
n-Gal-con	28	7 (**35**)	60 (**95**)
n-Pro	60	23 (**83**)	559 (**642**)
c-Pro	56	35 (**91**)	754 (**845**)
c-Gal	34	25 (**59**)	508 (**567**)

Table 4.15: Performed number of macro time steps until $t = 4$, from $t = 4$ to $t = 6$, and from $t = 6$ until $t = 12$ (in brackets the total number of time steps used up to that time)

The nonlinearity is not dominating during the start phase until $t = 4$, in contrast to the incompressibility, such that the MPSC approaches **n-Gal**, **n-Gal-con** and **c-Gal** are clearly superior. Subsequently, the Reynolds number reaches its maximum value such that the fully nonlinear scheme **n-Pro** improves. However, the full Galerkin MPSC schemes **n-Gal** demonstrate again the optimal behaviour for both separate problems, even independent of the underlying mesh. This result is typical for many other numerical tests and indicates why this approach is thought to be part of our desired "discrete Black Box" solver.

The final test calculations aim to examine the resulting accuracy and efficiency of the different methods if they are applied to **nonsteady flows with complex behaviour**. In contrast to the previous simulations which "only" showed periodical vortex shedding behind an obstacle, we present now results for a flow which does not seem to exhibit such an easy flow pattern in space and time (see also the examples in the 'Virtual Album').

3.) Nonstationary flow through a Venturi pipe:

The test configuration is almost identical with the framework which has been the basis for the numerical comparisons in the papers [102]; see Section 3.2 for the resulting flow patterns. An interesting flow quantity in this simulation is the flux through the upper small channel. In "real life", this Venturi pipe simulates a small device, i.e. in sailing boats, to suck the water from the boat. If the speed is high enough, then – due to the Bernoulli principle – the narrowing section enforces a low pressure which creates a flux through this small device, out of the boat. Therefore, we are mainly interested in controlling the flux through this device, as function in time and its mean value.

The following picture shows the coarse mesh (level 1) and the refined mesh on level 3; the refined levels have been created via recursively connecting opposite midpoints. All calculations have been carried through on the level 6.

Test configuration:

- no slip condition at the upper and lower walls and at the sides of the small upper channel, parabolic inflow (left side) with a maximum velocity of $u_{max} = 1$, natural boundary conditions at the "outflow" (right side) and at the small upper "in/outflow"

- viscosity parameter $\nu = 10^{-3}$, resulting Reynolds number $Re \approx 5,000$

- (*adaptive*) upwind discretization of the convective terms, Fractional-step-θ-scheme as time discretization, performed for $T = [0, 20]$, started with a fully developed solution

Applied time step control:

We apply the adaptive time step control from Section 3.2.3 with the following functionals for simultaneously controlling the velocity field and flux through

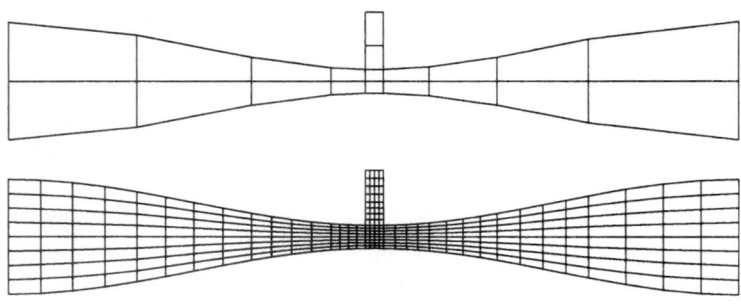

level	vertices	elements	midpoints	total unknowns
1	34	20	53	**126**
3	373	320	692	**1,704**
6	20,897	20,480	41,376	**103,232**

Table 4.16: Geometrical information for 'Flow through a Venturi pipe'

the small upper device. The term 10^{-n} in the subsequent tables has the following meaning:

- relative error (l_2) for the velocity smaller than 10^{-n+1}

- relative error for the flux through the upper small device smaller than 10^{-n}

Error evaluation and reference solutions:

For the error evaluation, we first perform reference calculations on the same spatial mesh and store the resulting flux (*flux*) and the value of the pressure in an interior point (*pp*) as function in time. For the error evaluation we restrict to compare the results for the total time interval $T = [0.5 : 20]$, neglecting effects during the startup.

- $NT \sim$ number of performed macro time steps (1 step with $3k$ and 3 steps with k)

- $l_\infty \sim$ maximum absolute value, resp., l_∞-error over the calculated time interval, $l_2 \sim l_2$-value, resp., l_2-error over the calculated time interval

- *mean* $\sim$ meanvalue, resp., error for the meanvalue over the calculated time interval, *peak* $\sim$ maximum value, resp., error for the maximum value over the time interval

In the following Figure, we show the corresponding flux (*flux*) and pressure (*pp*) plots for the reference solution, with respect to grid level 6. As indicated, the flow behaviour seems to be quite complex. We compare the following methods which have been introduced before, namely:

- **n-Gal** nonlinear *full Galerkin* approach with the global MPSC approach for Oseen subproblems and "exact" solution of the nonlinear steady equations in each time step

- **l-Gal** similar to the **n-Gal** scheme but with 2nd order linearization of the convective direction via linear extrapolation in time

- **c-Gal** similar to the **l-Gal** scheme but with first order linearization of the convective direction via constant extrapolation in time $(\mathbf{u}^n \cdot \nabla \mathbf{u}^{n+1})$

- **n-Pro** modified *nonlinear 1–step projection* with diffusive <u>and</u> reactive preconditioner; corresponds to classical 2nd order projection-schemes ('Van Kan', 'Gresho–2')

- **l-Pro** *semi-implicit 1–step projection* with linear extrapolation (2nd order) of the convective direction, analogously to **n-Pro**, resp., **l-Gal**

- **c-Pro** *semi-implicit 1–step projection* with constant extrapolation of the convective direction

- **xl-Pro** *semi-explicit 1–step projection* with linear extrapolation (2nd order) of the complete convective term

- **xc-Pro** *semi-explicit 1–step projection* with constant extrapolation of the complete convective term $(\mathbf{u}^n \cdot \nabla \mathbf{u}^n)$

We can start with the following conclusions:

1. The semi-explicit approaches **xl-Pro, xc-Pro** may be of second order accuracy, but due to inherent stability restrictions the chosen time steps are not allowed to pass a certain limit. This is obvious from the presented results; the applied time step control which is based on accuracy reasons only fails since a "hidden" restriction due to an implicitly given CFL condition is dominant. In fact, the applied time step control is even contradictory since – by accuracy reasons – larger time steps may be chosen which lead to worse flow behaviour due to loosing stability. Thus, only the choice of $\epsilon = 10^{-5}$ leads to satisfying results for **xl-Pro** and **xc-Pro**, but with time steps much smaller than for the other tested schemes.

2. The 2nd order linearization variants **l-Pro, l-Gal** exhibit some stability problems for large time steps $(\epsilon = 10^{-3})$. Due to the second order character, the time step control attempts to select large time steps, but the performed linearization strategy does not seem to be sufficiently stable.

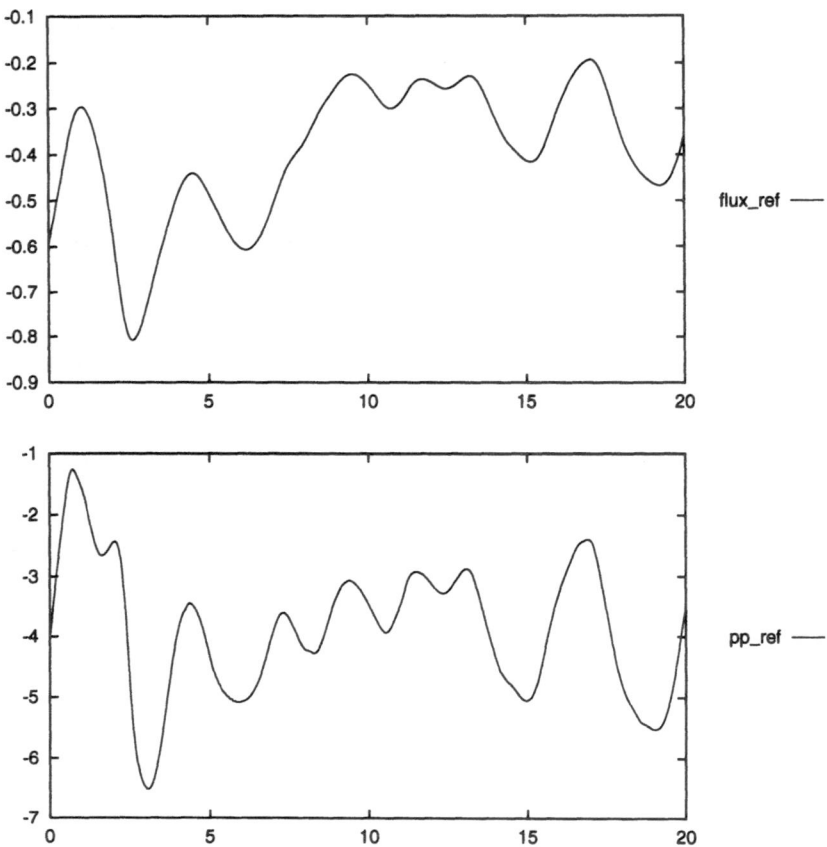

Figure 4.17: Flux (flux) and pressure point (pp) values for the reference calculation

3. Both 2nd order MPSC schemes, **n-Gal** and **l-Gal**, provide the best results with respect to accuracy. The full Galerkin scheme **n-Gal** with an "exact" treatment of the nonlinearity is the best with respect to all threshold parameters in the time step control, while the linearized version **l-Gal** gets in trouble for time steps being too large.

4. Both "classical" projection schemes **n-Pro** (with fully nonlinear treatment) and **l-Pro** (via linear extrapolation) lead to comparable results, particularly for small time steps, but only scheme **n-Pro** is stable for all threshold parameters.

5. The simple MPSC scheme **c-Gal** which uses the last velocity value as linearization for the convective term (first order!), seems to be surprisingly good in comparison to the fully nonlinear approach **n-Pro**.

Meth.	ϵ	NT	Flux value				Pressure value			
			l_∞	l_2	mean	peak	l_∞	l_2	mean	peak
REF		6500	0.81-0	1.85-0	-0.3911	-0.1934	6.50-0	1.79+1	-3.905	-1.257
n-Gal	10^{-3}	38	20%	15%	1%	3%	29%	19%	1%	1%
	10^{-4}	117	14%	5%	1%	0%	23%	7%	1%	0%
	10^{-5}	312	2%	1%	0%	0%	4%	2%	0%	0%
n-Pro	10^{-3}	138	57%	44%	24%	15%	74%	42%	19%	8%
	10^{-4}	263	20%	11%	2%	1%	28%	14%	3%	2%
	10^{-5}	630	6%	3%	0%	0%	10%	4%	0%	0%
l-Gal	10^{-3}	341	80%	87%	64%	9%	–	176%	68%	–
	10^{-4}	222	16%	11%	2%	1%	23%	13%	0%	3%
	10^{-5}	508	2%	1%	0%	0%	3%	1%	0%	0%
l-Pro	10^{-3}	306	72%	57%	45%	0%	–	239%	57%	–
	10^{-4}	253	39%	27%	2%	0%	56%	33%	0%	11%
	10^{-5}	649	6%	4%	0%	0%	10%	5%	0%	0%
c-Gal	10^{-3}	67	51%	40%	4%	12%	77%	48%	7%	33%
	10^{-4}	198	21%	18%	1%	0%	31%	23%	1%	1%
	10^{-5}	768	11%	6%	1%	0%	13%	8%	1%	0%
c-Pro	10^{-3}	90	60%	53%	34%	29%	81%	58%	28%	34%
	10^{-4}	250	44%	29%	7%	1%	63%	35%	10%	2%
	10^{-5}	769	10%	8%	2%	0%	12%	10%	2%	0%
xl-Pro	10^{-3}	–	–	–	–	–	–	–	–	–
	10^{-4}	–	–	–	–	–	–	–	–	–
	10^{-5}	6941	16%	11%	0%	2%	27%	14%	0%	22%
xc-Pro	10^{-3}	–	–	–	–	–	–	–	–	–
	10^{-4}	2932	9%	5%	0%	1%	292%	30%	0%	243%
	10^{-5}	2730	11%	7%	1%	1%	20%	8%	0%	4%

Table 4.17: Comparative errors for the flux and a selected pressure value, for $T = [0.5 : 20]$

Beside the "error values" in the previous table, we also have to examine the flow behaviour in the "picture norm". As pointed out in the previous chapters, the rating of the quality of nonstationary behaviour is difficult on the basis of error measurements only. Looking at some of the following simulations, they reproduce almost the correct curve plots in time, only slightly shifted but still capturing the main properties of the flow. These differences have huge influence on the measured errors, but the flow behaves very well in a qualitative way which might be sufficient for this kind of complex dynamics. So, both measures have to be taken into account.

While for the stopping criterion $\epsilon = 10^{-3}$ the Galerkin scheme n-Gal seems to be able to capture the main flow features with respect to flux and pressure values, the other schemes fail for this "rough" criterion; at least if we consider the pressure approximations (see also [105]). Decreasing the stopping criterion to $\epsilon = 10^{-4}$ allows not only to use the Galerkin scheme n-Gal, but additionally the variants n-Pro and l-Gal give reasonable approximations. The semi-explicit approach works fine for calculating the flux values, but fails for approximating the pressure. Only even harder stopping criterions

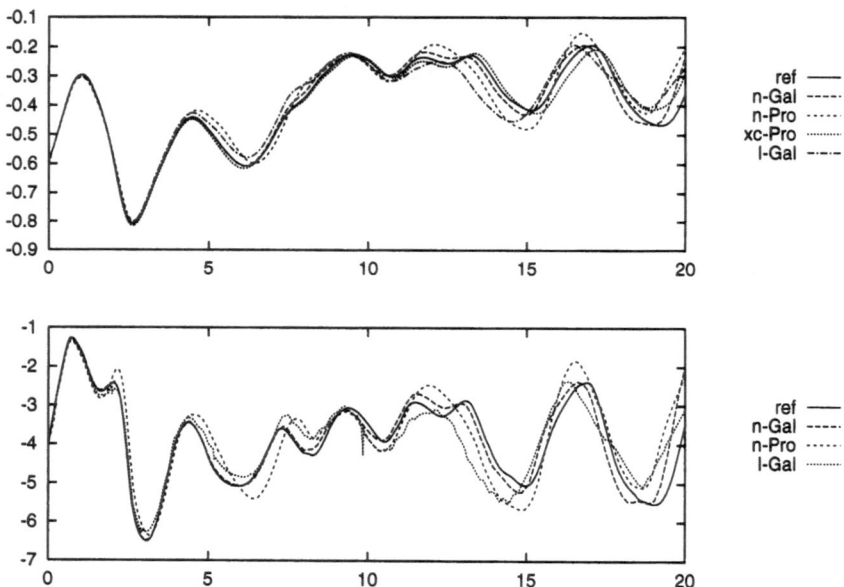

Figure 4.18: Flux (top) and pressure point (bottom) values for $\epsilon = 10^{-4}$

			Flux value				Pressure value			
	Meth.	NT	l_∞	l_2	mean	peak	l_∞	l_2	mean	peak
	n-Gal	38	20%	15%	1%	3%	29%	19%	1%	1%
	l-Gal	222	16%	11%	2%	1%	23%	13%	0%	3%
O	n-Pro	263	20%	11%	2%	1%	28%	14%	3%	2%
K	l-Pro	649	6%	4%	0%	0%	10%	5%	0%	0%
	c-Gal	768	11%	6%	1%	0%	13%	8%	1%	0%
	c-Pro	769	10%	8%	2%	0%	12%	10%	2%	0%
	xc-Pro	2730	11%	7%	1%	1%	20%	8%	0%	4%
	xl-Pro	6941	16%	11%	0%	2%	27%	14%	0%	22%
P	n-Gal	117	14%	5%	1%	0%	23%	7%	1%	0%
E	l-Gal	508	2%	1%	0%	0%	3%	1%	0%	0%
R	n-Pro	630	6%	3%	0%	0%	10%	4%	0%	0%
F	l-Pro	649	6%	4%	0%	0%	10%	5%	0%	0%

Table 4.18: Comparative rating of the schemes for $T = [0.5 : 20]$

($\epsilon = 10^{-5}$), which enforce smaller time steps, allow all schemes to produce satisfying approximations.

Finally, we rearrange again the tables in order to reach a better comparison of the resulting **"total" efficiency**. Based on the previous results from the tables and shown figures, we provide our subjective ratings of the examined schemes. "OK" stands again for an acceptable error, while "PERF" means that the approximate and the reference solution are almost identical.

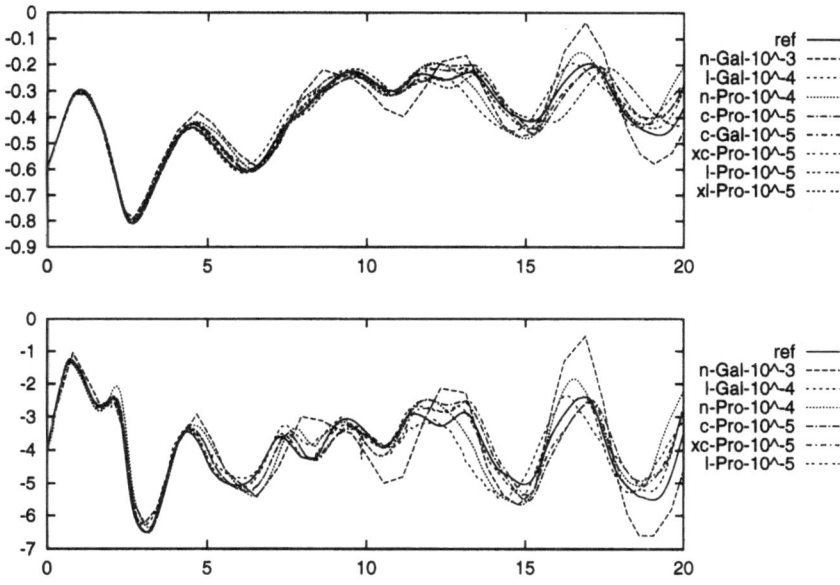

Figure 4.19: "OK" - flux (top) and pressure point (bottom) values

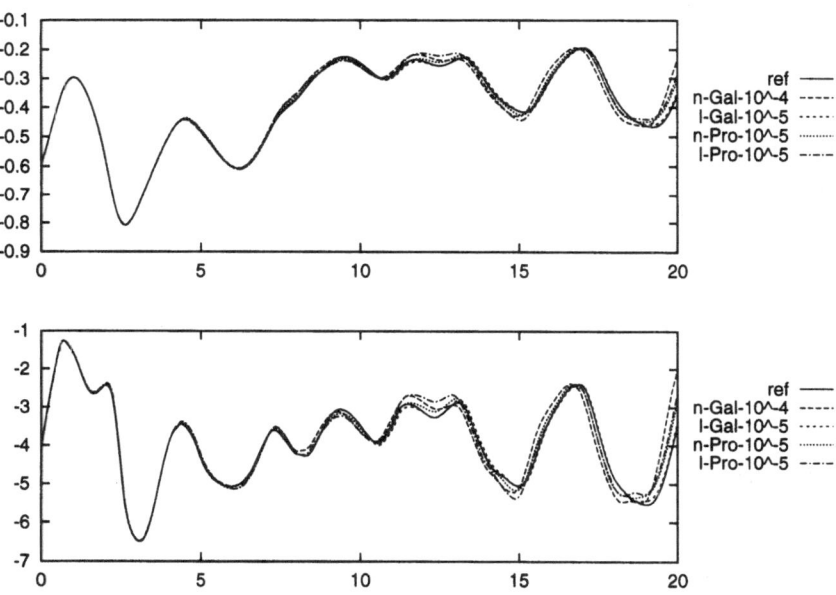

Figure 4.20: "PERF" - flux and pressure point values

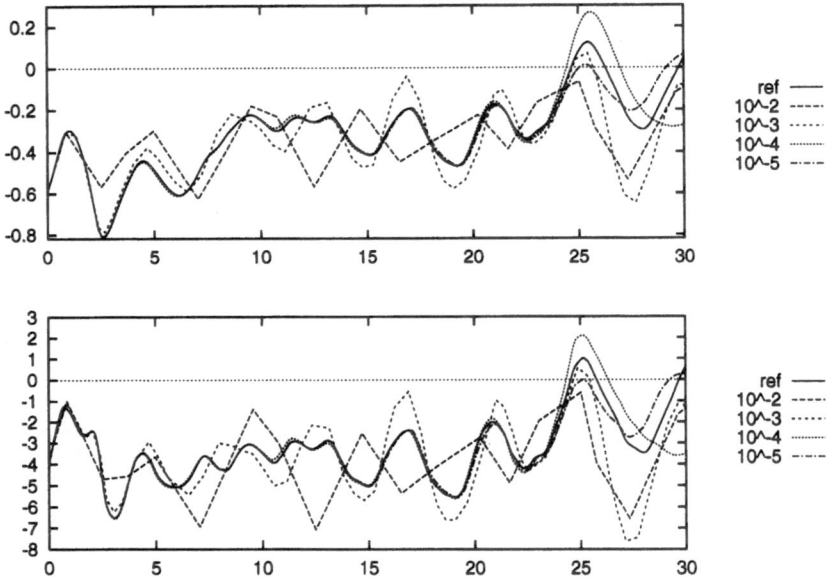

Figure 4.21: Flux and pressure point values for $\epsilon = 10^{-n}, n = 2, \ldots, 5$ with scheme **n-Gal**

Obviously, the "winner" in this competition is the full Galerkin scheme **n-Gal**. By far, the necessary number of time steps is the smallest to obtain reasonable approximations. Even if we cannot present the resulting CPU seconds for all schemes (see later), the reader may believe that similar results as given in Tables 4.2 – 4.4 are valid. Moreover, from a computational point of view, we expect in future even better efficiency results for the approach **n-Gal**, due to improved implementation techniques (see also the *FEAST* project [1]).

We finish with some plots of the resulting flux and pressure point approximations which show the resulting performance of the Galerkin scheme **n-Gal** for different threshold parameters in directing the time step control. Even for $\epsilon = 10^{-2}$, which results in 22 macro time steps only for calculating up to $t = 30$ even, the general flow patterns are still visible. While for harder criterions in the time step control – with resulting smaller time steps – the global MPSC approach for solving the Oseen subproblems in each nonlinear step is our favourite, for the simulation with very large time steps the local MPSC approach leads to better efficiency rates. So, both MPSC techniques are absolutely necessary.

With respect to accuracy, stability and (computational) efficiency our favourized scheme for simulating nonstationary flow is the Galerkin approach **n-Gal**. This method involves a fully nonlinear treatment as well as an "exact" solution of the resulting Oseen problems via MPSC techniques in each time step. All tests show that this approach is very accurate in time, even for complex "non-periodical" flow, and the numerical behaviour is independent of the underlying domain and mesh (large aspect ratios!) and the viscosity parameter.

4.2 Conclusions from the numerical simulations

Based on the complete "numerical proofs" in [105] and the previous exemplary results, we provide in short form the final ratings. We begin with some characteristics and corresponding pros and cons of the introduced schemes for the five "extreme" flow regimes. In the steady cases we set $k = 1$ in the velocity matrix (4.5); in the nonstationary settings we use $\alpha = 1$:

$$Su + kBp = \mathbf{f}, \quad B^T\mathbf{u} = g \quad \text{with} \quad S := \alpha M + \nu\theta_1 kL + k\theta_2 K \qquad (4.5)$$

We did not always provide explicit comparisons of the computational complexity since our computations have been performed on different platforms. We were unable to restrict to only one special computer type since our complete tests consumed much more than 100.000.000 (!!!) CPU seconds to obtain a rigid basis for our comparative ratings. Though, the reader may believe our explicit comments concerning the resulting computational results. Further information about the corresponding software package FEATFLOW and many "colorful" flow examples ("Virtual Album of Fluid Motion") can be taken from the included CDROM or from:

http://www.iwr.uni-heidelberg.de/~featflow

1.) Steady flow for low Reynolds numbers, resp., Stokes flow:

- **global MPSC:**   optimal !!!

- — "Optimal" in the sense that the convergence rates for Stokes problems are independent of ν and the underlying domain and mesh. That means, the rates are insensitive against large aspect ratios and other spatial anisotropies.

- — The multigrid rates can be bounded for all parameters α. In fact, the "pure" Stokes case $\alpha = 0$ is the limit case which results in rates smaller than $\rho_{gMPSC} \leq 0.5$ if at least one smoothing step is performed.

- — If α is growing, the convergence rates significantly improve, with $\rho_{gMPSC} = O(\frac{1}{\alpha})$.

- — These results are only obtained if we apply both preconditioners, for the reactive and the diffusive part, in an additive way. This is a very simple, but significant modification of classical projection approaches.

- **local MPSC:** $\boxed{\textbf{sufficient up to optimal !!!}}$

  - — "Optimal" in the sense that the convergence rates for the Stokes problems are independent of ν, α and the underlying domain and mesh.

  - — The rates are independent of large aspect ratios and other spatial anisotropies, but can be only achieved if the *adaptive patchwise* preconditioners as modifications of the *Vanka* smoother are applied.

  - — If α is growing, no improvement can be stated.

- **Conclusion:**

  Both schemes have the potential to provide "discrete Black Box" tools which satisfy all requirements concerning efficiency and robustness for arbitrary settings.

2.) Steady flow for Reynolds numbers of medium size:

- **Nonlinearity:** $\boxed{\textbf{satisfying !!!}}$

  - — The *adaptive fixed point defect correction* method leads to satisfying results. The nonlinear convergence behaviour is independent of the aspect ratios and other geometrical details.

- The nonlinear rates deteriorate with decreasing ν and improve with increasing α. Both effects could be expected.

- **global MPSC:** $\boxed{\textbf{(not always) sufficient !!!}}$

- Sufficient in the sense that the convergence rates for linear Oseen problems are independent of the underlying domain and mesh. They are insensitive against large aspect ratios and other spatial anisotropies.

- The (linear) multigrid rates depend on ν and cannot be bounded at all. This is due to the fact that we still miss the convective preconditioner.

- If α is growing, the convergence rates significantly improve. The same happens with decreasing h since the diffusive part is getting dominant.

- **local MPSC:** $\boxed{\textbf{sufficient up to optimal !!!}}$

- "Optimal" in the sense that the convergence rates for linear Oseen problems are independent of ν, α (this might be a disadvantage, too) and the underlying domain and mesh.

- The rates are independent of large aspect ratios and other spatial anisotropies if the *adaptive patchwise* preconditioners are applied.

- In fact, the rates may even improve for increasing Reynolds numbers in comparison to the pure Stokes case.

- **Conclusion:**

  The *global MPSC* approach which is based on modified projection-like techniques via constructing global preconditioners for the *pressure Schur complement* is not stable without the corresponding convective preconditioner. This is a significant failure which excludes this approach from "discrete Black Box" tools for this flow regime. Instead, the *local MPSC* approach (including the *Vanka* scheme) has the potential to provide "discrete Black Box" tools which together with the applied nonlinear iterative techniques satisfy many requirements concerning numerical efficiency. If we demand for absolute robustness, we have to perform the *adaptive patchwise* preconditioners.

3.) Nonsteady Stokes flow (via nonsteady boundary conditions):

- **global MPSC:** | **optimal !!!**

    - "Optimal" in the sense that the convergence rates and the time step behaviour are independent of ν and the underlying domain and mesh. If k is getting smaller, the convergence rates significantly improve, with $\rho_{gMPSC} \approx O(k)$.

    - <u>Both</u> preconditioners, for the reactive and the diffusive part, should be used in the proposed additive way.

- **local MPSC:** | **satisfying, but not optimal !!!**

    - Satisfying in the sense that the convergence rates for Stokes problems are independent of ν, k and the underlying domain and mesh.

    - If k is decreasing, no improvement can be stated. Consequently, the disadvantage of this approach is enforced by computational reasons since the resulting run-time behaviour for small time steps is not adequately improved.

- **1-step Projection:** | **ranging from not sufficient to optimal !!!**

    - The time stepping behaviour for small ν is almost identical with the coupled schemes involving the "exact" solution of the resulting Stokes problems in each time step.

    - For large ν, the additional diffusive preconditioner is absolutely necessary, otherwise the projection schemes fail as predicted. Altogether, the coupled schemes (*global MPSC*) are clearly preferable for large viscosity parameters.

    - The acceleration of the 1-step projection via multigrid can lead to significant improvements, not only for the numerical but also with respect to the computational behaviour.

- **Conclusion:**

    While the coupled approaches have no problems at all, the 1-step projection schemes may degenerate for large viscosity parameters. A very surprising result is that the numerical solution of nonsteady flow problems due to <u>very large</u> viscosity parameters can be an extremely hard problem, if additionally significant grid anisotropies occur. While the pure "classical" projection schemes fail, with or without time step control, the *global* and *local MPSC* schemes (with *adaptive patchwise* preconditioning) are candidates for "discrete Black Box" solvers. In contrast, for small parameters ν, the 1-step projection schemes may show an even superior run-time behaviour.

4.) Nonsteady flow for midrange Reynolds numbers (via nonsteady b.c.'s):

- **Nonlinearity:** **sufficient, sometimes optimal !!!**

  - The *adaptive fixed point defect correction* method leads to satisfying results. The nonlinear convergence behaviour is independent of aspect ratios and other geometrical details. They deteriorate with decreasing ν, but they dramatically improve with decreasing k as expected.

  - Linearization techniques via linear or constant extrapolation backwards in time may work, but they show no advantageous behaviour.

- **global MPSC:** **sometimes not satisfying !!!**

  - Satisfying in the sense that the convergence rates for the Oseen problems are independent of the underlying domain and mesh.

  - The multigrid rates depend on ν due to the fact that we miss the convective preconditioner. Hence, the numerical cost are large if one tries to solve "exactly" the resulting Oseen equations in each nonlinear step. A remedy is to weaken this solution process by performing a few multigrid steps only or by prescribing weaker stopping criterions. However, then the number of nonlinear iteration steps has to be increased, or the time step has to be diminished. All propositions lead to non-optimal time stepping schemes.

- **local MPSC:** **almost optimal !!!**

- "Optimal" in the sense that the convergence rates for linear Oseen problems are independent of ν, k (this might be a disadvantage, too) and the underlying domain and mesh.

- The rates are independent of large aspect ratios and other spatial anisotropies only if the *adaptive patchwise* preconditioners are applied.

• **1-step Projection:** | **sometimes not sufficient !!!** |

- The 1-step projection with fully nonlinear treatment and the application of both reactive and diffusive preconditioners leads to quite efficient time stepping schemes.

- Based on small time steps k, the time stepping schemes work accurate and efficient. The resulting total numerical amount is not clear since the cost per time step may be small due to the necessary small time step but the overall number of time steps is correspondingly large.

- The acceleration of the 1-step projection schemes via multigrid leads to significant improvements but the numerical cost may increase. In fact, the calibration of the multigrid parameters may be a delicate task for this flow regime.

• **Conclusion:**

While the projection-like techniques show robustness problems which result in smaller time steps or in increasing numerical cost for solving the Oseen problems, the *local MPSC* approach has the potential to provide "discrete Black Box" tools which together with the applied nonlinear iterative techniques satisfy all requirements concerning numerical efficiency. If we demand for absolute robustness, we have to perform the *adaptive patchwise* preconditioners. Linearization may work fine in combination with the (classical) *1-step projection* schemes due to the automatically chosen smaller time steps.

5.) Fully nonstationary high Reynolds number flow:

• **Nonlinearity:** | **optimal !!!** |

- The *adaptive fixed point defect correction* method leads to very good results. The nonlinear convergence behaviour is independent of the aspect ratios and other geometrical details, and it is related to k.

– Linearization via linear extrapolation is optimal with respect to accuray and efficiency, but the mesh is not allowed to contain large anisotropies.

– Linearization via constant extrapolation is much too inaccurate.

- **global MPSC:** | **almost optimal !!!** |

– "Optimal" in the sense that the convergence rates for the Oseen problems are independent of ν (since k is small) and the underlying domain and mesh.

– The multigrid convergence rates significantly improve for small time steps.

- **local MPSC:** | **not satisfying !!!** |

– The convergence rates for the linear Oseen problems are independent of ν, k and the underlying domain and mesh.

– Since the convergence rates for the linear problems are independent of k, the numerical complexity for fully nonsteady flow is too large.

- **1-step Projection:** | **optimal !!!** |

– The 1-step projection schemes exhibit optimal numerical complexity.

– The only problem may be a rigorous control of the time steps.

- **Conclusion:**

"Classical" projection-like methods are optimal for highly nonstationary flow. The multigrid acceleration is not absolutely necessary since the 1-step schemes have about the same accuracy but the numerical complexity is slightly smaller. We prefer the fully nonlinear treatment. In that case only that the structure of the mesh is known to be "nice" (no large aspect ratios), the linear extrapolation in time should be performed which leads to the most efficient time stepping scheme.

6.) Nonstationary configurations for varying Reynolds numbers:

- **Nonlinearity:**   satisfying !!!

  - The *adaptive fixed point defect correction* method leads to satisfying results with nonlinear convergence rates independent of the aspect ratios and other geometrical details. Additionally, they are related to time step size k.
  - Linearization via linear extrapolation should be applied only on meshes without large anisotropies, otherwise the schemes may fail.
  - Linearization via constant extrapolation is too inaccurate for large Reynolds numbers.

- **local/global MPSC:**   almost optimal !!!

  - For very low or high Reynolds numbers the *global MPSC* perform optimally, while for midrange viscosity parameters the *local MPSC* schemes perform best.
  - The resulting linear convergence rates are independent of the underlying domain and mesh and improve significantly for small time steps. There is an upper bound with respect to the complete range of Reynolds numbers.
  - If we switch adaptively between the *global* and *local MPSC* schemes during the calculation, the resulting method is optimal.

- **1-step Projection:**   not always sufficient !!!

  - Insufficient for large viscosities which may even lead to divergence of the schemes.
  - There may arise problems with the time step control for large viscosity parameters.

- **Conclusion:**

  1. Classical projection-like methods with the proposed modifications (diffusive preconditioner, discrete projection, MPSC acceleration) are optimal for highly nonstationary flow only.

2. The *coupled Galerkin* approach is able to "simulate" the solution schemes which have proven to be the favourized methods for the examined "extreme cases" of flow behaviour. Consequently, the *fixed point defect correction* method in combination with the described adaptive switching between the *global* and *local MPSC* schemes leads to optimal results, with respect to efficiency, robustness and accuracy, and with corresponding optimal computational complexity.

3. The *coupled Galerkin scheme* is an essential component of our aimed "discrete Black Box" solver. Additionally, there is the future potential to apply rigorous error control mechanisms.

4. An adaptive time step control is absolutely necessary. However, the simple technique which we performed so far is an error indicator only which may fail and which cannot guarantee a certain accuracy. Nevertheless, this approach is up to now an essential component of our "discrete Black Box" solver.

Chapter 5

Conclusions and outlook

The aim of this book is the discussion of mathematical, algorithmic and computational aspects for the efficient numerical simulation of the laminar incompressible Navier–Stokes equations. As an important part of modern *Computational Fluid Dynamics* (CFD), these equations have to be solved as "standalones" or as part of even more complex problems. However, as we have demonstrated in Chapter 1, 'Motivation for current research', the efficient solution process is even today still a grand-challenging problem, particularly in the case of "real life" applications.

Typical aspects which have to be considered for the design of "high–performance" solution tools are the following ones. All of them have proven to be necessary components which have to fit perfectly together. These mathematical and computational topics are:

1. stable and accurate **discretizations** in space and time, if necessary allowing stabilization techniques for the convective parts and the incompressibility constraint.

2. flexible treatment of **boundary conditions**.

3. adaptive **error control** mechanisms.

4. efficient **nonlinear solvers** and/or linearization techniques in time.

5. efficient **solvers** for linear Stokes and/or **Oseen–like equations**.

6. efficient **linear solvers** for scalar subproblems with respect to velocity and pressure.

7. optimized **implementation** techniques with regard to modern processor technologies.

The detailed description of the underlying numerical and computational problems and of possible solution strategies was part of the previous chapters. Summarizing our mathematical studies, the following "techniques/strategies" have shown to be candidates for successful CFD tools:

Summary of the necessary CFD components:

1) Discretization techniques: see Sections 3.1 and 3.2

We prefer the $\tilde{Q}1/Q0$ Stokes element (nonconforming rotated multilinear velocities, constant pressure in cells) since second order accuracy is provided, hereby satisfying the *Babuška–Brezzi condition* independent of the shape and size of the used triangulation. In addition, robust upwind or streamline-diffusion mechanisms can be applied for higher Reynolds numbers. In combination with the MPSC techniques, it can be shown that this element pair is almost "optimal" for fully nonstationary flows with respect to the computational cost. Further, this element pair is the finite element analogue to some widely used staggered grid discretization which has shown to be very successful. However, this FEM Stokes element can be easily applied on general meshes, too, and allows a rigorous error analysis based on variational arguments.

For time discretization, we suggest to use the Fractional–step–θ scheme which in comparison to the standard Crank–Nicolson–scheme is of second order accuracy <u>and</u> strongly *A–stable*, and this with about the same numerical effort. Both diffusive and convective parts of the momentum equation should be treated in this fully implicit way.

2) Boundary conditions: see Section 3.5

We recommend to use the theoretically analyzed and numerically tested natural boundary conditions (*do nothing condition*) whenever the "right" Dirichlet values for the velocity are not given explicitly. They allow the formulation of "pressure drop problems" as well as "flux problems", and require the typical finite element setting only.

We apply the standard techniques for implementing boundary conditions in combination with the described *fictitious boundary* techniques if iterative solvers are used. Both together allow the straightforward simulation on very complex domains, including moving boundary parts and small scale structures.

3) Error control mechanisms: see Sections 3.1 and 3.2

While possible error control strategies in space have been introduced only briefly (in fact: this is one of our main points of future interest), we have shown how "simple" (that means without large implementation effort) error indicators for the time step size can be incorporated. They can be applied for arbitrary user-defined functionals (l_2-errors, drag/lift coefficients, etc.) and are designed for the use in fully **implicit (!)** approaches.

4) Treatment of the nonlinearity: see Section 3.3

For the iterative treatment due to the nonlinear convective part we recommend the *adaptive fixed point defect correction method* as a variant of quasi–Newton schemes. This method has proven to be very robust, allowing the efficient numerical solution for a wide range of discretization techniques, with linear multigrid as "Black Box" preconditioner.

On meshes containing moderate aspect ratios ($AR \leq 10$) the linear extrapolation backwards in time as linearization technique is a quite good alternative – it also provides second order accuracy in time, a robust temporal behaviour and the smallest amount of numerical cost – but on highly anisotropic meshes this semi-implicit approach fails and the fully nonlinear treatment is a "must"!

5) Solvers for linear Stokes and/or Oseen equations: see Chapters 2 and 4

We have introduced the *Multilevel Pressure Schur Complement* (MPSC) schemes which allow the generalization and improvement of many existing Navier–Stokes solvers. They include, as *global MPSC*, well-known methods as projection, pressure correction or fractional step schemes and also Uzawa-like techniques, while the *local MPSC* approach contains methods as the *Vanka* smoother. Embedding them rigorously into the standard multigrid context, they provide robust and efficient solution schemes for the complete range of flow configurations, including aspects as varying time step sizes and viscosity parameters, size and shape of the spatial mesh and complexity of the domain. In fact, these MPSC techniques are beside the other described items mainly responsible for the high efficiency of our CFD tools as it has been demonstrated in the extensive numerical tests.

6) Solvers for linear scalar subproblems: see Section 3.4

The derived multigrid solvers are able to solve **all (!)** resulting scalar Pressure–Poisson problems as well as reaction–convection–diffusion problems for the velocity, being symmetric as well as nonsymmetric. The combination of the special adaptive grid transfer and coarse grid operator techniques together with appropriate smoothing operators and renumbering strategies provide "Black Box" tools which are necessary in the complete framework of solving the incompressible Navier–Stokes equations.

7) Implementation techniques: see Chapters 1, 2 and 4

As indicated in the previous chapters, our "new" implementation concept is based on the encouraged use of **hierarchical** data structures, **hierarchical** solvers as well as **hierarchical** matrix (!) structures. Consequently applied, they should guarantee the typical multigrid rates with respect to robustness and efficiency behaviour, however maintaining – at the same time – more of the high Peak performance on modern processors than compared to most standard approaches. This software project is the mentioned FEAST package which is just under development (for the current status of all our software see the Web–site below!).

Summarizing all these issues, we are recently able to demonstrate (numerically and with computational results) how the "optimal" solution process may look like for the discretized incompressible Navier–Stokes equations, for almost arbitrarily given spatial meshes. This is the first step in deriving "optimal" CFD tools and allows us, at the moment, at least the "guaranteed" efficient solution of the discretized Navier–Stokes equations, on more or less general meshes. Hence, we are prepared for doing the next step: the adding of some error control concepts of the described type.

Additionally, all benchmarks and other numerical comparisons demonstrate that we also have to work in improved implementation techniques: the modern processors provide several hundreds of MFLOP Peak performance, but how to realize them at least approximately in finite element CFD codes? Our new software package FEAST is intended to realize many of the given numerical and computational ideas, and in fact some successful test implementations have been already carried through. The most prominent key words in the context of the FEAST project are *ScaRC* ("Scalable Recursive Clustering") as generalized *Domain Decomposition/Multigrid* solvers, including directly *parallelism* and *adaptivity* concepts, which are supported by special *hierarchical data, solver and matrix structures.*

Besides, we also plan to improve the current *FEATFLOW* version, and recently we have added components for nonnewtonian flows, Boussinesq approximations and moving boundary components. So, the interested reader is advised to look at our FEATFLOW Homepage

http://www.iwr.uni-heidelberg.de/~featflow

which always describes most of our current research activities (papers, software, demos, etc.), including our '**Virtual Album of Fluid Motion**'. Or, continue with the following description of the enclosed CDROM.

Chapter 6

The enclosed CDROM

In August 1998, we announced the availability of the new Version 1.1 of our incompressible flow solver package **FEATFLOW 1.1** and of the **Virtual Album of Fluid Motion (v1.1)**.

As in Version 1.0, the **FEATFLOW** package contains the complete sources for our fully coupled (CC2D/CC3D) and projection-like (PP2D/PP3D) FEM solvers for the stationary and nonstationary incompressible Navier-Stokes equations. The package also includes the full documentation (POST-SCRIPT and HTML) and tools for grid generation/modification (new: an english manual for OMEGA2D) and postprocessing (for AVS 5 and GMV which is for "free"!).

The complete package and the **Virtual Album** can be found at the enclosed CDROM or can be downloaded from the FEATFLOW Homepage:

> http://www.iwr.uni-heidelberg.de/~featflow

Beside the flow solver components, the basic FEM packages FEAT2D and FEAT3D are included, too. All software can be automatically installed in an interactive way, on most UNIX workstations (SUN SPARC and ULTRA, IBM RS/6000 and POWERPC, SGI, HP, DEC) as well as on LINUX platforms (PC's with PENTIUM or ALPHA CHIP). Instructions for the Installation process can be found at the given Homepage and the enclosed CDROM.

In addition to Version 1.0, we also added (test) versions of:

- the Boussinesq solver BOUSS as generalization of PP2D

- the nonnewtonian (with Power Law) version BOUSS-POWERLAW

- the nonnewtonian (with Power Law) version CC2D-POWERLAW

- the variant CC2D-MOVBC with "fictitious boundary components"

- the test version of CP2D which is the basis for most calculations in this book

Moreover, the first official version of our **Virtual Album of Fluid Motion (v1.1)** is finished with MANY examples for incompressible flow configurations which all are stored as MPEG movies! In combination with the **FEATFLOW 1.1** software, all "Pages" ($\approx$ one "flow configuration") in the **Virtual Album** provide the reader with computational data which allow the execution of such flow simulations by each user!

All these software components and a part of the **Virtual Album** (from 4 GB completely!!!) are available on the enclosed CDROM (tested for UNIX, LINUX, WINDOWS NT and APPLE platforms). Those movies which are not included in the CDROM are automatically downloaded (if desired!) from the given FEATFLOW Homepage.

So, insert the CDROM and load the file `index.html` or `start.html` into your WWW browser (NETSCAPE or similar). Then, you only need time and a working MPEG player ...

If problems occur, or if the software package has to be updated, or if the latest movies from our "Movie Gallery" are wanted, please, contact us at:

http://www.iwr.uni-heidelberg.de/~featflow

Bibliography

[1] Altieri, M., Becker, Chr., Kilian, S., Oswald, H., Turek, S., Wallis, J.: *Some Basic Concepts of* **FEAST**, to appear in: Proc. 14th GAMM Seminar 'Concepts of Numerical Software', Kiel, January 1998, *Notes on Numerical Fluid Mechanics*, Vieweg

[2] Altieri, M., Becker, Chr., Turek, S.: *On the realistic performance of components in iterative solvers*, submitted to: Proc. FORTWIHR Conference, Munich, March 1998, LNCSE, Springer-Verlag

[3] Axelsson, O., Barker, V.A.: *Finite Element Solution of Boundary Value Problems*, Academic Press, 1984

[4] Babuška, I., Rheinboldt, W.C.: *Error estimates for adaptive finite element computations*, SIAM J. Num. Anal., 15, 736–754 (1978)

[5] Bank, R.E., Weiser, A.: *Some a posteriori error estimators for elliptic partial differential equations*, Math. Comput., 44, 283–301 (1985)

[6] Bank, R.E., Welfert, B., Yserentant, H.: *A class of iterative methods for solving saddle point problems*, Numer. Math., 56, 645–666 (1990)

[7] Becker, R., Rannacher, R.: *Finite element discretization of the Stokes and Navier–Stokes equations on anisotropic grids*, Proc. 10th GAMM-Seminar, Kiel, January 14–16, 1994 (G. Wittum, W. Hackbusch, eds.), Vieweg

[8] Becker, R., Rannacher, R.: *Weighted a posteriori error control in FE methods*, Proc. Enumath–95, Paris, 18–22 Sept., 1995

[9] Becker, R., Rannacher, R.: *A Feed-Back Approach to Error Control in Finite Element Methods: Basic Analysis and Examples*, Preprint 96–52, University of Heidelberg, SFB 359, 1996

[10] Becker, R.: *An adaptive finite element method for the incompressible Navier–Stokes equations on time–dependent domains*, Ph.D. Thesis, University of Heidelberg, 1995

[11] Becker, R.: *Weighted error estimators for Finite Element approximations for the incompressible Navier–Stokes equations*, to appear in: Comput. Methods Appl. Mech. Engrg., 1998

[12] Blum, H., Harig, J., Müller, S., Turek, S.: **FEAT2D** . *Finite element analysis tools. User Manual. Release 1.3*, 1992

[13] Braess, D.: *Finite Elemente*, Springer Verlag, Berlin–Heidelberg, 1992

[14] Braess, D., Sarazin, R.: *An efficient smoother for the Stokes problem*, Applied Numerical Mathematics, 23, 3–19 (1997)

[15] Bramble, J.H.: *Multigrid Methods*, Pitman research notes in mathematical sciences, 294, Longman, London, 1993

[16] Brenner, S.C., Scott, R.: *The Mathematical Theory of Finite Element Methods*, Springer Verlag, Berlin, 1994.

[17] Brezzi, F., Fortin, M.: *Mixed and Hybrid Finite Element Methods*, Springer Verlag, Berlin–Heidelberg–New York, 1991

[18] Briggs, W.: *A multigrid tutorial*, SIAM, 1987

[19] Bristeau, M.O., Glowinski, R., Periaux, J.: *Numerical methods for the Navier–Stokes equations: Applications to the simulation of compressible and incompressible viscous flows*, Report UH/MD–4, University of Houston, 1987, in: Computer Physics Report 1987

[20] Cahouet, J., Chabard, J.P.: *Some fast 3D solvers for the generalized Stokes problem*, Int. J. Numer. Meth. Fluids, 8, 269–295 (1988)

[21] Chan, T., Mathew, T.: *Domain Decomposition methods*, Acta Numerica, 1992

[22] Chorin, A.J.: *Numerical solution of the Navier–Stokes equations*, Math. Comp., 22, 745–762 (1968)

[23] Ciarlet, Ph.G.: *The Finite Element Method for Elliptic Problems*, North–Holland, Amsterdam, 1976

[24] Crouzeix, M., Raviart, P.A.: *Conforming and non–conforming finite element methods for solving the stationary Stokes equations*, R.A.I.R.O. **R–3**, 77–104 (1973)

[25] Cuvelier, C., Segal, A., Steenhoven, A.: *Finite element methods and Navier–Stokes equations*, D. Reidel Publishing Company, Dordrecht, 1986

[26] Douglas, C.C.: *Caching in with multigrid algorithms: Problems in two dimensions*, IBM Research Report RC 20091, IBM, Yorktown Heights, NY, 1995 (can be obtained via MGNET)

[27] Elman, H.C.: *Multigrid and Krylov subspace methods for discrete Stokes equations*, Int. J. Numer. Meth. Fluids, 22, 755–770 (1996)

[28] Elman, H.C., Golub, G.H.: *Inexact and preconditioned Uzawa algorithms for saddle point problems*, SIAM J. Numer. Anal., 31, 1645–1661 (1994)

[29] Engelman, M.S., Haroutunian, V., Hasbani, I.: *Segregated finite element algorithms for the numerical solution of large–scale incompressible flow problems*, Int. J. Numer. Meth. Fluids, 17, 323–348 (1993)

[30] Eriksson, K., Johnson, C. : *Adaptive finite element methods for parabolic problems:I. A linear model problem*, SIAM J. Num. Anal. 28, 43-77 (1991)

[31] Eriksson, K., Estep, D., Hansbo, P., Johnson, C.: *Computational Differential Equations*, Cambridge University Press, 1996

[32] Feistauer, M.: *Mathematical methods in fluid dynamics*, Pitman Monographs and Surveys in Pure and Applied Mathematics 67, Longman Scientific & Technical, 1993

[33] Ferziger, J.H., Peric, M.: *Computational Methods for Fluid Dynamics*, Springer Verlag, Berlin–Heidelberg, 1996

[34] Führer, C.: *A posteriori error control for the numerical approximation of hyperbolic problems*, Ph.D. Thesis, University of Heidelberg, 1997

[35] Galdi, G.: *An Introduction to the Mathematical Theory of the Navier-Stokes equations, Vol. I: Linearized steady problems*, Springer Tracts in Natural Philosophy, Volume 38, Springer 1994

[36] Girault, V., Raviart, P.A.: *Finite Element Methods for Navier–Stokes Equations*, Springer Verlag, Berlin–Heidelberg, 1986

[37] Gjesdal, T., Lossius, M.E.H.: *Comparison of pressure correction smoothers for multigrid solution of incompressible flow*, can be obtained via MGNET

[38] Glowinski, R.: *Numerical Methods for Nonlinear Variational Problems*, Springer, New York–Berlin–Heidelberg–Tokyo, 1984

[39] Glowinski, R.: *Finite element methods for the numerical simulation of incompressible viscous flow. Introduction to the control of the Navier-Stokes equations*, Lectures in Applied Mathematics, Volume 28 (1991)

[40] Glowinski, R., Periaux, J.: *Numerical methods for nonlinear problems in fluid dynamics*, Proc. Intern. Seminar on Scientific Supercomputers, Paris, Feb. 2–6, 1987, North–Holland

[41] Gresho, P.M.: *On the theory of semi–implicit projection methods for viscous incompressible flow and its implementation via a finite element method that also introduces a nearly consistent mass matrix, Part 1: Theory*, Int. J. Numer. Meth. Fluids, 11, 587–620 (1990). *Part 2: Implementation*, Int. J. Numer. Meth. Fluids, 11, 621–659 (1990)

[42] Gresho, P.M., Lee, L.L., Sani, R.L.: *On the time-dependent solution of the incompressible Navier-Stokes equations in two and three dimensions*, in "Numerical Methods in Fluids", Pineridge Press Ltd., Swansea, U.K., Vol 1, 27–80 (1980)

[43] Gresho, P.M.: *Some current CFD issues relevant to the incompressible Navier-Stokes equations*, Comp. Meth. Appl. Mech. Eng., 87, 201–252 (1991)

[44] Gresho, P.M., Sani, R.L.: *Résumé and remarks on the open boundary condition minisymposium*, Int. J. Numer. Meth. Fluids, 18, 983–1008 (1994)

[45] Gresho, P.M., Sani, R.L.: *Incompressible Flow and the Finite Element Method*, Wiley, Chichester, 1998

[46] Griebel, M., Dornseifer, T., Neunhoeffer, T.: *Numerische Simulation in der Strömungsmechanik*, Vieweg, Braunschweig/Wiesbaden, 1995

[47] Griffiths, D.: *An approximately divergence–free 9–node velocity element (with variations) for incompressible flows*, Int. J. for Num. Meth. in Fluids, 1, 323–346 (1981)

[48] Grimmer, A.: *A comparative study of transport–oriented discretizations of convection–dominated problems with application to the incompressible Navier–Stokes equations*, Ph.D. Thesis, University of Heidelberg, 1997

[49] Hackbusch, W.: *Multi–Grid Methods and Applications*, Springer Verlag, Berlin–Heidelberg, 1985

[50] Hackbusch, W.: *On the feedback vertex set problem for a planar graph*, Computing, Vol. 58 Nr. 2, 129–155 (1997)

[51] Harig, J., Schreiber, P., Turek, S.: **FEAT3D** . *Finite element analysis tools in 3 dimensions. User Manual. Release 1.2*, 1994

[52] Heywood, J.G.: *On uniqueness questions in the theory of viscous flow*, Acta Math., 136, 61-102 (1976)

[53] Heywood, J.G.: *The Navier–Stokes equations: On the existence, regularity and decay of solutions*, Indiana Univ. Math. J., 29, 639-681 (1980)

[54] Heywood, J.G.: *Auxiliary flux and pressure conditions for Navier-Stokes problems*, Proc. IUTAM Symp., Paderborn 1979 (R. Rautmann, ed.), pp. 223–234, Lecture Notes in Mathematics, 771, Springer 1980

[55] Heywood, J.G., Rannacher, R.: *Finite element approximation of the nonstationary Navier-Stokes problem. Part 1: regularity of solutions and second-order error estimates for spatial discretization*, SIAM J. Numer. Anal., 19, 275–311 (1982), *Part 2: Stability of solutions and error estimates uniform in time*, ibidem, 23, 750–777 (1986), *Part 3: Smoothing property and higher order error estimates for spatial discretization*, ibidem, 25, 489–512 (1988), *Part 4: error analysis for second-order time discretization*, ibidem, 27, 353–384 (1990)

[56] Heywood, J., Rannacher, R., Turek, S.: *Artificial boundaries and flux and pressure conditions for the incompressible Navier–Stokes equations*, Int. J. Numer. Meth. Fluids, 22, 325–352 (1996)

[57] Hughes, T.J.R.: *The Finite Element Method*, Prentice–Hall, Englewood Cliffs, New York, 1987

[58] Hughes, T.J.R., Franca, L.P., Balestra, M.: *A new finite element formulation for computational fluid mechanics: V. Circumventing the Babuska–Brezzi condition: A stable Petrov–Galerkin formulation of the Stokes problem accomodating equal order interpolation*, Comp. Meth. Appl. Mech. Eng., 59, 85–99 (1986)

[59] Johnson, C.: *Numerical Solution of Partial Differential Equations by the Finite Element Method*, Cambridge University Press, 1987

[60] Johnson, C.: *Error estimates and adaptive time-step control for a class of one-step methods for stiff ordinary differential equations*, SIAM J. Numer. Anal., 908-926 (1988)

[61] Johnson, C.: *The streamline diffusion finite element method for compressible and incompressible fluid flow*, in "Finite Element Method in Fluids VII", Huntsville, 1989

[62] Johnson, C., Rannacher, R., Boman, M.: *Numerics and hydrodynamic stability: Towards error contol in CFD*, SIAM J. Numer. Anal., 32 (1995)

[63] Kilian, S., Turek, S.: *An example for parallel ScaRC and its application to the incompressible Navier-Stokes equations*, to appear in: Proc. ENUMATH-97, Heidelberg, October 1997

[64] Kloucek, P., Rys, F.S.: *On the stability of the fractional step-θ-scheme for the Navier–Stokes equations*, SIAM J. Numer. Anal., 31, 1312–1335 (1994)

[65] Lötzbeyer, H., Rüde, U.: *Patch-adaptive multilevel iteration*, to appear in BIT

[66] Lube, G.: *Stabilized Galerkin finite element methods for convection dominated and incompressible flow problems*, Numerical Analysis and Mathematical Modelling, Banach Center Publications, Volume 29, Institute of Mathematics, Polish Academy of Sciences, Warszawa 1994

[67] Lube, G., Tobiska, L.: *A nonconforming finite element method of streamline-diffusion type for the incompressible Navier–Stokes equation*, J. Comp. Math., 8, 147–158 (1990)

[68] Malek, J., Turek, S.: *Non–newtonian flow prediction by divergence free finite elements*, SAACM - Vol. 3, n. 3, 165–180 (1993)

[69] McCormick, S.F.: *Multigrid Methods*, SIAM Frontiers In Applied Mathematics, 1987

[70] Minev, P.D., van de Vosse, F.N., Timmermans, L.J.P.: *An approximate projection scheme for incompressible flow using spectral elements*, to appear in: Int. J. Numer. Meth. Fluids

[71] Minev, P.D., van de Vosse, F.N., Timmermans, L.J.P.: *A spectral element projection scheme for incompressible flow with application to shear–layer stability studies*, in: Proceedings of ICOSAHOM'95, Houston Journal of Mathematics (1996)

[72] Müller–Urbaniak, S.,: *Eine Analyse des Zwischenschritt-θ-Verfahrens zir Lösung der instationären Navier–Stokes Gleichungen*, Ph. D. Thesis, University of Heidelberg, 1993

[73] Müller, S., Prohl, A., Rannacher, R., Turek, S.: *Implicit time-discretization of the nonstationary incompressible Navier–Stokes equations*, Proc. 10th GAMM-Seminar, Kiel, January 14–16, 1994 (G. Wittum, W. Hackbusch, eds.), Vieweg

[74] Ohmori, K., Ushijima, T.: *A technique of upstream type applied to a linear nonconforming finite element approximation of convective diffusion equations*, R.A.I.R.O. Numer. Anal., 18, 309–332 (1984)

[75] Oswald, H., Turek, S.: *A parallel multigrid algorithm for solving the incompressible Navier–Stokes equations with nonconforming finite elements in three dimensions*, to appear in: Proc. Parallel CFD'96, Capri/Italy, 1996

[76] Patankar, S.V.: *Numerical Heat Transfer and Fluid Flow*, Hemisphere, Washington D.C., 1980

[77] Pironneau, O.: *Finite Element methods for Fluids*, Wiley, Masson, Paris, 1989

[78] Pironneau, O.: *On the transport–diffusion algorithm and its applications to the Navier–Stokes equations*, Numer. Math., 38, 309–332 (1982)

[79] Prohl, A.: *Projection and Quasi–Compressibility Methods for Solving the Incompressible Navier–Stokes equations*, B.G.Teubner, Stuttgart, 1997

[80] Prohl, A.: *Über Projektionsverfahren erster und zweiter Ordnung zur Lösung der instationären Navier–Stokes–Gleichungen*, Ph.D. Thesis, University of Heidelberg, 1995

[81] Rannacher, R.: *Numerical analysis of the Navier–Stokes equations*, Appl. Math., 38, 361–380 (1993)

[82] Rannacher, R.: *On Chorin's projection method for the incompressible Navier–Stokes equations*, in "Navier–Stokes Equations: Theory and Numerical Methods" (R. Rautmann, et al., eds.), Proc. Oberwolfach Conf., August 19–23, 1991, Springer, 1992

[83] Rannacher, R., Schreiber, P., Turek, S.: *Numerische Modellierung von Gasbrennern: Berechnung schwachkompressibler Gasströmungen*, in: Mathematik – Schlüsseltechnologie für die Zukunft (K.-H. Hoffmann, W. Jäger, T. Lohmann, H. Schunck, eds.), Springer, 1996

[84] Rannacher, R., Turek, S.: *A simple nonconforming quadrilateral Stokes element*, Numer. Meth. Part. Diff. Equ., 8, 97–111 (1992)

[85] Roos, H.G., Stynes, M., Tobiska, L.: *Numerical Methods for Singularly Perturbed Differential Equations, Convection–Diffusion and Flow Problems*, Springer Verlag, 1996

[86] Saad, Y., Schultz, M.H.: *GMRES: A generalized minimal residual method for solving nonsymmetric linear systems*, SIAM J. Sci. Statist. Comput., 7, 856–869 (1986)

[87] Schäfer, M., Turek, S. (with support by F. Durst, E. Krause, R. Rannacher): *Benchmark computations of laminar flow around cylinder*, in E.H. Hirschel (editor), *Flow Simulation with High-Performance Computers II*, Volume 52 of *Notes on Numerical Fluid Mechanics*, 547–566, Vieweg, 1996.

[88] Schäfer, M., Turek, S. (with Rannacher, R., Breuer, M., Durst, F., Rodi, W.): *Definition of Benchmark Problems: Incompressible turbulent flow*, distributed by the authors

[89] Schieweck, F.: *A parallel multigrid algorithm for solving the Navier-Stokes equation*, Impact Comp. Sci. Engnrg., 5, 345–378 (1993)

[90] Schieweck, F.: *Parallele Lösung der stationären inkompressiblen Navier-Stokes Gleichungen*, Habilitation Thesis, University of Magdeburg, 1996

[91] Schieweck, F., Tobiska, L.: *A nonconforming finite element method of upstream type applied to the stationary Navier-Stokes equation*, MMAN, 23, 627–647 (1989)

[92] Schieweck, F., Tobiska, L.: *An optimal order error estimate for an upwind discretization of the Navier-Stokes equations*, to appear

[93] Schreiber, P.: *A new finite element solver for the nonstationary incompressible Navier-Stokes equations in three dimensions*, Ph.D. Thesis, University of Heidelberg, 1996

[94] Schreiber, P., Turek, S.: *An efficient finite element solver for the nonstationary incompressible Navier-Stokes equations in two and three dimensions*, Proc. Workshop "Numerical Methods for the Navier-Stokes Equations", Heidelberg, Oct. 25–28, 1993, Vieweg

[95] Schreiber, P., Turek, S.: *Multigrid results for the nonconforming Morley element*, Technical report SFB 359, 67, 1993

[96] Temam, R.: *Theory and Numerical Analysis of the Navier-Stokes Equations*, North-Holland, 1977

[97] Thomasset, F.: *Implementation of Finite Element Methods for Navier-Stokes Equations*, Springer, New York, 1981

[98] Tobiska, L.: *Full and weighted upwind finite element methods*, In: Splines in Numerical Analysis, ed. by J.W.Schmidt and H. Späth, Internationales Seminar ISAM 89 in Weissig, Akademie Verlag Berlin, 1989

[99] Tobiska, L.: *Numerical methods in singularly perturbed problems*, In: International Seminar in Applied Mathematics, ed. by H.G. Roos, A. Felgenhauer, L. Angermann, Internationales Seminar ISAM 91 in Dresden, 1991

[100] Tobiska, L., John, V., Maubach, J.M., : *Nonconforming streamline-diffusion-finite-element methods for convection-diffusion problems*, Preprint Nr. 10, 1996, University of Magdeburg, Fakultät für Mathematik

[101] Turek, S.: *Tools for simulating nonstationary incompressible flow via discretely divergence-free finite element models*, Int. J. Numer. Meth. Fluids, 18, 71–105 (1994)

[102] Turek, S.: *A comparative study of time stepping techniques for the incompressible Navier–Stokes equations: From fully implicit nonlinear schemes to semi–implicit projection methods*, Int. J. Numer. Meth. Fluids, 22, 987–1011 (1996)

[103] Turek, S.: *On discrete projection methods for the incompressible Navier–Stokes equations: An algorithmical approach*, Comput. Methods Appl. Mech. Engrg., 143, 271–288 (1997)

[104] Turek, S.: *Ein robustes und effizientes Mehrgitterverfahren zur Lösung der instationären, inkompressiblen 2–D Navier–Stokes–Gleichungen mit diskret divergenzfreien finiten Elementen*, Ph.D. Thesis, University of Heidelberg, 1991

[105] Turek, S.: *Multilevel Pressure Schur Complement techniques for the numerical solution of the incompressible Navier–Stokes equations*, Habilitation Thesis, University of Heidelberg, 1997

[106] Turek, S.: *On ordering strategies in a multigrid algorithm*, Proc. 8th GAMM–Seminar, Kiel, January 24–26, 1992, Notes on Numerical Fluid Mechanics, Volume 41, Vieweg, Braunschweig, 1993

[107] Turek, S.: *Numerik in der Strömungsmechanik: Numerische Effizienz versus Gigaflops*, Phys. Bl., 52, Nr. 11, 1137–1139, 1996

[108] Turek, S.: *Recent Benchmark Computations of Laminar Flow Around a Cylinder*, in Proc. 3rd World Conference in Applied Computational Fluid Mechanics, Freiburg, 1996

[109] Turek, S.: *Multigrid techniques for a divergence–freefinite element discretization*, East-West J. Numer. Math., Vol. 2, No. 3, 229–255 (1994)

[110] Turek, S.: *Multigrid techniques for simple discretely divergence–free finite element spaces*, in: Hemker/Wesseling: Multigrid Methods IV, 321–332, 1994

[111] Turek, S.: **FEATFLOW** *Finite element software for the incompressible Navier–Stokes equations: User Manual, Release 1.1*, University of Heidelberg, 1998

[112] Turek, S.: *Konsequenzen eines numerischen 'Elch Tests' für Computersimulationen*, Discussion paper, 1998

[113] Van der Vorst, H.: *BI–CGSTAB: A fast and smoothly converging variant of BI–CG for the solution of nonsymmetric linear systems*, SIAM J. Sci. Stat. Comput., 13, 631–644 (1992)

[114] Vanka, S.P.: *Implicit multigrid solutions of Navier–Stokes equations in primitive variables*, J. Comp. Phys., 65, 138–158 (1985)

[115] Van Kan, J.: *A second-order accurate pressure-correction scheme for viscous incompressible flow*, SIAM J. Sci. Stat. Comp., 7, 870–891 (1986)

[116] Varga, R.S.: *Matrix Iterative Analysis*, Prentice–Hall, Englewood, New Jersey, 1961

[117] Verfürth, R.: *A Review of A Posteriori Estimation and Adaptive Mesh-Refinement Techniques*, Advances in Numerical Mathematics, Wiley/Teubner, New York-Stuttgart, 1996

[118] Wittum, G.: *On the robustness of ILU–smoothing*, SIAM J. Sci. Stat. Comp., 10, 699–717 (1989)

[119] Wittum, G.: *The use of fast solvers in computational fluid dynamics*, in: "Numerical Methods in Fluid Mechanic" (P. Wesseling, eds.), Vieweg, Braunschweig, 1990

[120] Xu, J.: *Iterative methods by space decomposition and subspace correction*, SIAM Review, 34, 581–613 (1992)

[121] Young, D.M.: *Iterative solution of large linear systems*, Academic Press, 1971

[122] Yserentant, H.: *Old and new convergence proofs for multigrid methods*, Acta Numerica, 285–326, 1993

[123] Zhou, G.: *How accurate is the streamline–diffusion method?*, Preprint 95–22, University of Heidelberg, SFB 359, 1995

[124] Zienkiewicz, O.C., Zhu, J.Z.: *A simple error estimator and adaptive procedure forpractical engineering analysis*, Int. J. Numer. Methods Engrg., 24 (1987)

General Remarks

Lecture Notes are printed by photo-offset from the master-copy delivered in camera-ready form by the authors. For this purpose Springer-Verlag provides technical instructions for the preparation of manuscripts. See also *Editorial Policy*.

Careful preparation of manuscripts will help keep production time short and ensure a satisfactory appearance of the finished book. The actual production of a Lecture Notes volume normally takes approximately 12 weeks.

Authors receive 50 free copies of their book. No royalty is paid on Lecture Notes volumes.

For conference proceedings, editors receive a total of 50 free copies of their volume for distribution to the contributing authors.

Authors are entitled to purchase further copies of their book and other Springer mathematics books for their personal use, at a discount of 33,3 % directly from Springer-Verlag.

Commitment to publish is made by letter of intent rather than by signing a formal contract. Springer-Verlag secures the copyright for each volume.

Addresses:

Professor M. Griebel
Institut für Angewandte Mathematik
der Universität Bonn
Wegelerstr. 6
D-53115 Bonn, Germany
e-mail: griebel@iam.uni-bonn.de

Professor D. E. Keyes
Computer Science Department
Old Dominion University
Norfolk, VA 23529–0162, USA
e-mail: keyes@cs.odu.edu

Professor R. M. Nieminen
Laboratory of Physics
Helsinki University of Technology
02150 Espoo, Finland
e-mail: rniemine@csc.fi

Professor D. Roose
Department of Computer Science
Katholieke Universiteit Leuven
Celestijnenlaan 200A
3001 Leuven-Heverlee, Belgium
e-mail: dirk.roose@cs.kuleuven.ac.be

Professor T. Schlick
Department of Chemistry and
Courant Institute of Mathematical
Sciences
New York University
and Howard Hughes Medical Institute
251 Mercer Street, Rm 509
New York, NY 10012-1548, USA
e-mail: schlick@nyu.edu

Springer-Verlag, Mathematics Editorial
Tiergartenstrasse 17
D-69121 Heidelberg, Germany
Tel.: *49 (6221) 487-185
e-mail: peters@springer.de
http://www.springer.de/math/
peters.html

Editorial Policy

§1. Submissions are invited in the following categories:

i)   Research monographs
ii) Lecture and seminar notes
iii) Reports of meetings

Those considering a project which might be suitable for the series are strongly advised to contact the publisher or the series editors at an early stage.

§2. Categories i) and ii). These categories will be emphasized by Lecture Notes in Computational Science and Engineering. **Submissions by interdisciplinary teams of authors are encouraged.** The goal is to report new developments – quickly, informally, and in a way that will make them accessible to non-specialists. In the evaluation of submissions timeliness of the work is an important criterion. Texts should be well-rounded and reasonably self-contained. In most cases the work will contain results of others as well as those of the authors. In each case the author(s) should provide sufficient motivation, examples, and applications. In this respect, articles intended for a journal and Ph.D. theses will usually be deemed unsuitable for the Lecture Notes series. Proposals for volumes in this category should be submitted either to one of the series editors or to Springer-Verlag, Heidelberg, and will be refereed. A pro-visional judgment on the acceptability of a project can be based on partial information about the work: a detailed outline describing the contents of each chapter, the estimated length, a bibliography, and one or two sample chapters – or a first draft. A final decision whether to accept will rest on an evaluation of the completed work which should include

– at least 100 pages of text;
– a table of contents;
– an informative introduction perhaps with some historical remarks which should be accessible to readers unfamiliar with the topic treated;
– a subject index.

§3. Category iii). Reports of meetings will be considered for publication provided that they are both of exceptional interest and devoted to a single topic. In exceptional cases some other multi-authored volumes may be considered in this category. One (or more) expert participants will act as the scientific editor(s) of the volume. They select the papers which are suitable for inclusion and have them individually refereed as for a journal. Papers not closely related to the central topic are to be excluded. Organizers should contact Lecture Notes in Computational Science and Engineering at the planning stage.

§4. Format. Only works in English are considered. They should be submitted in camera-ready form according to Springer-Verlag's specifications. Electronic material can be included if appropriate. Please contact the publisher. Technical instructions and/or TEX macros are available on http://www.springer.de/author/tex/help-tex.html; the name of the macro package is "LNCSE – LaTEX2e class for Lecture Notes in Computational Science and Engineering". The macros can also be sent on request.

Lecture Notes
in Computational Science and Engineering

For further information on these books please have a look at our mathematics catalogue at the following URL: http://www.springer.de/math/index.html

Author's Remark

In August 1998, we announced the availability of the new Version 1.1 of our incompressible flow solver package FEATFLOW 1.1 and of the Virtual Album of Fluid Motion (v1.1) which can be found on the enclosed CD-ROM.

So, insert the CD-ROM and load the file index.html or start.html into your WWW browser. Then, you only need time and a working MPEG player...

If problems occur, or if the software package has to be updated, or if the latest movies from our "Movie Gallery" are wanted, please contact us at:

http://www.iwr.uni-heidelberg.de/~featflow

Stefan Turek